Global Climate Change -
The Technology Challenge

ADVANCES IN GLOBAL CHANGE RESEARCH

VOLUME 38

For other titles published in this series, go to
www.springer.com/series/5588

Frank T. Princiotta

Editor

Global Climate Change - The Technology Challenge

 Springer

Editor
Frank T. Princiotta
Air Pollution Prevention and Control Division,
National Risk Management Research Laboratory
U.S. Environmental Protection Agency
109 TW Alexander Drive
Research Triangle Park
NC, USA
princiotta.frank@epa.gov

ISBN 978-90-481-3152-5   e-ISBN 978-90-481-3153-2
DOI 10.1007/978-90-481-3153-2
Springer Dordrecht Heidelberg London New York

Library of Congress Control Number: 2011929040

Printed on acid-free paper

Springer is part of Springer Science+Business Media (www.springer.com)

Preface

The industrial *revolution* in the eighteenth and nineteenth century has had a profound impact on every aspect of human activity. This technology revolution catalyzed a transition from a manual and animal labor based economy towards an energy driven manufacturing economy. This revolution started with the mechanization of the textile industries, the development of iron-making techniques and the increased use of coal for steam engines and furnaces. It has evolved to include the generation of large quantities of electric power, derived primarily from the combustion of coal and other fossil fuels, construction and operation of energy intensive buildings and the construction and utilization of energy-intensive transportation systems on the land, in the sea and in the air. A key impact of this revolution has been dramatically enhanced per capita income, with an associated sixfold growth in population to the current 6.8 billion. Unfortunately, a by-product of this revolution has been the massive generation of greenhouse gases, most importantly, Carbon Dioxide (CO_2). There has been a consistent increase of anthropogenic CO_2 emissions since the beginning of the industrial revolution. Over the period of 2000–2008, there has been an acceleration of CO_2 emissions associated with strong economic growth in China and other Asian countries yielding increased demand for coal-based electricity and petroleum based cars and trucks. In 2008, humanity emitted almost 30 billion tons of CO_2. Emissions of such a magnitude are unsustainable, and if not dramatically reduced, can yield potentially catastrophic climate change. The goal of this book is to consider the challenges for another technology-based *revolution*, this one based on the development and wide scale utilization of low carbon technology.

The scope of the book evolved from papers and presentations by Frank Princiotta, the book editor, who attempted to cover the subject of the climate change mitigation challenge and the availability of key technologies for all the key sectors. It became apparent to him that the subject was too broad and complex, to be adequately covered in a single publication by a single author. When Springer Publications suggested that a book might be the appropriate venue for this important and complex subject, and his agency, the Environmental Protection Agency[1]

---

[1]Please note: The views expressed in this book are those of the authors and do not necessarily reflect the views or policies of the U.S. Environmental Protection Agency. Mention of trade names or commercial products does not constitute Agency endorsement or recommendation for use.

agreed, the book project was initiated. This effort would not have been possible without the enthusiastic & diligent participation by the talented chapter authors from the private, academic & public sectors, willing to devote their valuable time to this project. Note that Chapter 5, Renewable Energy: Status and Prospects, was graciously contributed by the International Energy Agency, from Energy Technology Perspectives (2006). Also, special recognition should be given to Gloria Fuller, USEPA, whose administrative contributions were invaluable during all stages of the book's development.

Chapter 1 examines the greenhouse gas mitigation challenge and summarizes the status of key technologies in all the key energy sectors. It quantifies the reductions in emissions that will be necessary to avoid unacceptable climate change, and the technologies that have the potential to play an important role in mitigating CO_2 emissions. It concludes, that in order to avoid the potentially catastrophic impacts of global warming, the recent 3% CO_2 annual global emission growth rate must be transformed to a 2–3% declining annual rate, as soon as possible. This will require a rapid and radical *revolutionary* transformation of the world's energy production and end use systems. It concludes that the current generation of energy technologies, are not capable of achieving the level of mitigation required. Next generations of renewable, low carbon generation and end use technologies will be needed. Their status and prospects are summarized for each key sector, e.g., power generation.

Subsequent chapters dig in more deeply in describing technological challenges for each of the key energy sectors. They consider the status of key technologies needed to protect the planet from serious climate change impacts. Current and emerging technologies are characterized for their mitigation potential, status of development and potential environmental impacts. The status of technologies relevant to *Power generation, mobile sources, industrial and building sectors are evaluated in detail.* The importance and unique challenges for rapidly developing countries, such as China, India and Mexico are discussed in a separate chapter. Current global research and development efforts for key technologies are discussed. It is concluded that it will be necessary to substantially upgrade and accelerate the current worldwide RDD&D effort on both emerging energy technologies and those enabling technologies needed to improve mitigation effectiveness and economics. A chapter examining the potential environmental characteristics of evolving energy technologies, concludes that It will also be necessary to carefully evaluate the potential environmental characteristics of next generation technologies to avoid unacceptable health and ecological impacts.

Finally, given the monumental technological challenge associated with transforming the world's energy system, geoengineering options, i.e., intentional anthropogenic modifications of the earth's thermal balance, are evaluated, since *if* deemed feasible and successfully deployed, they have the potential to allow more time for the necessary energy system transformation.

Contents

Contributors

Anthony Baratta
US Nuclear regulatory Commission, Atomic Safety//. and Licensing Board Panel, MS T3-F23, Washington DC, 20555-0001, USA
ab2@psu.edu

Ananth Chikkatur
Belfer Center for Science and International Affairs, John F. Kennedy School of Government, Harvard University, Cambridge MA, 02138, USA
ICF International, Fairfax VA, 22031, USA
ap_chikkatur@yahoo.com

Cynthia L. Gage
Air Pollution Prevention and Control Division, National Risk Management Research Laboratory, Office of Research and Development, US Environmental Protection Agency, Research Triangle Park NC, USA

Gayle S. W. Hagler
Air Pollution Prevention and Control Division, National Risk Management Research Laboratory, Office of Research and Development, US Environmental Protection Agency, Research Triangle Park NC, USA

Brooke L. Hemming
Air Pollution Prevention and Control Division, National Risk Management Research Laboratory, Office of Research and Development, US Environmental Protection Agency, Research Triangle Park NC, USA
and
Global Change Research Program, National Center for Environmental Assessment, Office of Research and Development, US Environmental Protection Agency, Washington DC, USA
hemming.brooke@epa.gov

Jim Jetter
Air Pollution Prevention and Control Division, National Risk Management Research Laboratory, Office of Research and Development, US Environmental Protection Agency, Cincinnati OH, USA

James R. Katzer
Chemical and Biological Engineering, Iowa State University,
Ames IA, 50012, USA
jrksail@comcast.net

David Marr
Air Pollution Prevention and Control Division, National Risk Management
Research Laboratory, Office of Research and Development,
US Environmental Protection Agency, Cincinnati OH, USA

C. Andrew Miller
Air Pollution Prevention and Control Division, National Risk Management
Research Laboratory, Office of Research and Development,
US Environmental Protection Agency, Research Triangle Park NC, USA
miller.andy@epa.gov

Clyde Owens
Air Pollution Prevention and Control Division, National Risk Management
Research Laboratory, Office of Research and Development,
US Environmental Protection Agency, Cincinnati OH, USA

Frank Princiotta
Air Pollution Prevention and Control Division, National Risk Management
Research Laboratory, Office of Research and Development, US Environmental
Protection Agency, Durham NC, USA
and
Air Pollution Prevention and Control Division, National Risk Management
Research Laboratory, US Environmental Protection Agency,
T.W. Alexander Drive 109 27709, Research Triangle Park NC, USA
frank@epamail.epa.gov

Bruce Rising
Siemens Energy, Orlando FL, USA
bruce.rising@siemens.com

Ravi K. Srivastava
Air Pollution Prevention and Control Division, National Risk Management
Research Laboratory, Office of Research and Development, US Environmental
Protection Agency, Cincinnati OH, 45268, USA

Bob Thompson
Air Pollution Prevention and Control Division, National Risk Management
Research Laboratory, Office of Research and Development, US Environmental
Protection Agency, Cincinnati OH, USA

Elineth Torres
Sector Policy and Programs Division, Office of Air Quality Planning and
Standards, US Environmental Protection Agency, Research Triangle Park
NC, 27711, USA

Samudra Vijay
Air Pollution Prevention and Control Division, National Risk Management
Research Laboratory, Office or Research and Development, US Environmental
Protection Agency, Research Triangle ParkNC, 27711, USA
and
Sam Analytic Solutions, LLC, 614 Willingham RdMorrisville NC, 27560,
USAAir Pollution Prevention and Control Division, National Risk Management
Research Laboratory, Office of Research and Development, US Environmental
Protection Agency, Cincinnati OH, 45268, USA
and
Sam Analytic Solutions, LLC, 614 Willingham Road Morrisville NC, 27560, USA
sam@samanalyticsolutions.com

Michael P. Walsh
International Council on Clean Transportation, Washington DC, USA
mpwalsh@igc.org

Chapter 1
Global Climate Change and the Mitigation Challenge*

Frank T. Princiotta[†]

Abstract *This chapter aims to provide a succinct integration of the projected warming the earth is likely to experience in the decades ahead, the emission reductions that may be needed to constrain this warming, and the technologies needed to help achieve these emission reductions.* Transparent modeling tools and the most recent literature are used, to quantify the challenge posed by climate change and potential technological remedies. The chapter examines forces driving CO_2 emissions, how different emission trajectories could affect warming this century, a sector-by-sector summary of mitigation options, and R&D priorities. It is concluded that it is too late too avoid substantial warming; the best result that appears achievable, would be to constrain warming to about 2°C (range of 1.3–2.7°C) above pre-industrial levels by 2100. In order to constrain warming to such a level, the current annual 3% CO_2 emission growth rate needs to transform rapidly to an annual decrease rate of from 2% to 3% for decades. Further, the current generation of energy generation and end use technologies are capable of achieving less than half of the emission reduction needed for such a major mitigation program. New technologies will have to be developed and deployed at a rapid rate, especially for the key power generation and transportation sectors. Current energy technology research, development, demonstration and deployment programs fall far short of what is required.

---

*The findings included in this chapter do not necessarily reflect the view or policies of the Environmental Protection Agency. Mention of trade names or commercial products does not constitute Agency endorsement or recommendation for use.

[†]© US Government 2011

F.T. Princiotta (✉)
Air Pollution Prevention and Control Division, U.S. Environmental Protection Agency,
Office of Research and Development, National Risk Management Research Laboratory,
Research Triangle Park, NC, USA
e-mail: princiotta.frank@epa.gov

F.T. Princiotta (ed.), *Global Climate Change - The Technology Challenge*,
Advances in Global Change Research 38, DOI 10.1007/978-90-481-3153-2_1,
© Springer Science+Business Media B.V. 2011

1.1   Introduction

In February, 2007, the Intergovernmental Panel on Climate Change (IPCC) [1] concluded:

- Warming of the climate system is unequivocal, as is now evident from observations of increases in global average air and ocean temperatures, widespread melting of snow and ice, and rising global average sea level.
- Most of the observed increase in globally averaged temperatures since the mid-twentieth century is very likely due to the observed increase in anthropogenic greenhouse gas concentrations.
- The combined radiative forcing due to increases in carbon dioxide, methane and nitrous oxides…is very likely to have been unprecedented in more than 10,000 years.
- The total temperature increase from 1850–1899 to 2001–2005 is 0.76°C.
- Depending on the assumed greenhouse gas emission, warming in 2095, relative to pre-industrial levels, is projected to be 1.6–6.4°C.

Given these findings, this chapter will examine the critical global energy sector with the aim of evaluating the ability of technologies to moderate projected warming. Factors that lead to increasing emissions of CO_2 the critical greenhouse gas will be analyzed, and the anticipated importance of key countries will be discussed. Then, CO_2 emissions will be projected into the future for key sectors and warming will be projected with consideration of model uncertainties. The chapter will summarize the state of the art of key technologies and R&D priorities for each of four key sectors that can contribute to mitigating such emissions (Note that in this chapter, all CO_2 concentrations will be in ppmv, abbreviated as ppm, and all warming will be realized or transient warming, unless specifically identified, as opposed to equilibrium warming. (Equilibrium warming is the ultimate global warming after an infinite time interval, resulting from an elevated concentration of greenhouse gases.)

1.2   Climate Change and Other Global Environmental Impacts: The Sustainability Challenge

In addition to the emission of greenhouse gases, there are a number of other anthropogenic activities that can yield unacceptable long term global impacts. They include deforestation, ecosystem deterioration, resource depletion, and ocean contamination. Individually and certainly collectively, such impacts may be inconsistent with the long-term viability of humanity. In other words, they may be incompatible with long-term sustainability. Long-term environmental sustainability can be defined as the ability of humanity to indefinitely live compatibly with the Earth. Sustainable development refers to a systematic approach to achieving human development in a way that sustains planetary resources, based on the recognition

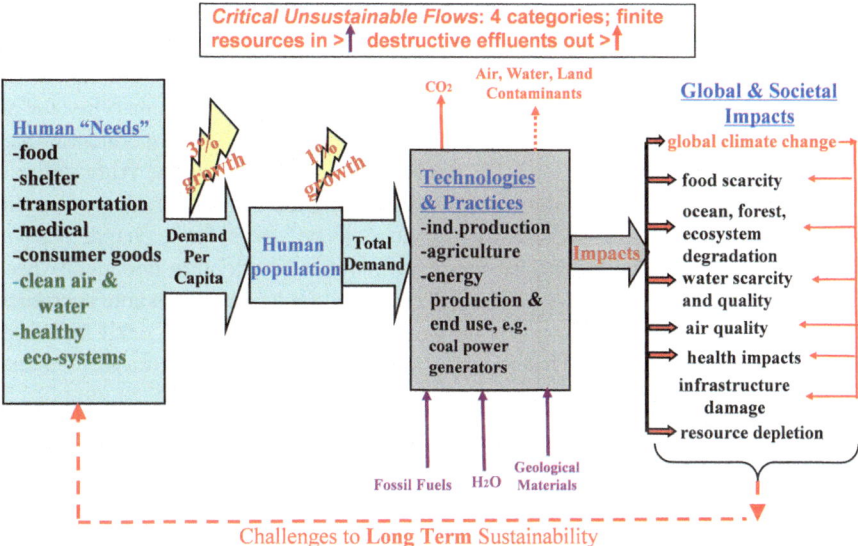

Fig. 1.1   Global Climate Change: A key challenge to long-term sustainability

that human consumption is currently occurring at a rate that is beyond Earth's capacity to support it. Population growth and the developmental pressures spawned by an increasing demand for resource intensive goods, foods and services are altering the planet in ways that threaten the long-term well being of humans and other species.

Figure 1.1 illustrates the role that climate change plays in challenging long-term sustainability. It also indicates the factors that are responsible for potentially unsustainable global impacts, including climate change. Such impacts have the potential to modify the home planet so that it is inhospitable to the needs of the growing population, expected to pass ten billion later this century [1].

The following discusses the implications of the figure from left to right. The root cause of potential deleterious impacts is the technological challenge of meeting human "needs," which are growing dramatically, especially in developing nations. Over time, the developed nations have expanded their list of "needs" to include: personal transportation, large residences with energy intensive heating, cooling and lighting requirements, a diet heavily oriented toward meat production and a growing array of consumer goods. Developing countries such as China and India, with large populations, are moving in the same direction. Although it is difficult to quantify the growth rate of such per capita "needs," It is reasonable to roughly relate it to per capita annual economic growth, which has been over 3% in recent years. The problem is further magnified by the fact that the global population is growing at roughly 1% per year. At these growth rates, the overall demand for such needs will *double every 20 years*.

The middle of the figure indicates that these human needs are met by means of a large array of industrial, agricultural, and energy technologies and practices. Although there are a multitude of inputs and outputs associated with humankind's "Technologies and Practices," the major threats to long term sustainability for an advanced level of civilization are shown on the figure: fossil fuel, mineral and water depletion and the emissions of CO_2 and other greenhouse gases. Although air, water, and waste contamination are serious deleterious products of our current consumer oriented infrastructure, there appears to be a reasonable chance that we can modify our industrial infrastructure to maintain a tolerable impact of these contaminants over the long term. The U.S., EU, and Japan have been able keep such impacts at close to tolerable levels despite population (in the U.S.) and industrial growth in recent decades. That is why these impacts are shown in a *dashed line* format.

On the right hand side of the figure is a listing of key impacts to the Earth associated with the technology and practices currently used to meet human needs. As indicated by the red return arrows, climate change has the potential to exacerbate global impacts associated with non-energy related technologies and practices. Ocean and forest degradation are examples of such amplification. Climate change can also yield unique impacts, e.g., infrastructure damage, due to seawater rise and storm damage. As indicated by the return flow at the bottom of the graphic, in a business as usual scenario, these impacts will challenge the ability of humanity to meet its needs over the long term, *challenging long-term sustainability*.

So what are the potential remedies? The figure suggests the following mitigative possibilities. First, humanity could downscale its "needs" to reduce the demand on technology with its associated environmental impacts. Examples would include a transition to smaller houses, more mass transit and fewer cars, and a modification of our resource intensive diets. To the extent we scale back our needs; we cut back on resource depletion, reduce greenhouse gas emissions, and also reduce the impact associated with other environmental impacts, such as air and water pollution and ocean and forest degradation. Second, humanity could consider a move toward population stabilization. Third, we could fundamentally change the technology we use to meet our needs. This can be achieved by a major transition to: low carbon energy production, more efficient end use technologies, pollution prevention (green chemistry), and renewables and reuse/recycling. The net effect is to dramatically minimize environmental impact per unit of production.

A holistic view of long-term sustainability cannot ignore our ever-growing demands on fossil fuels, water, and other finite geological resources. To the extent we back off on "need" requirements or slow population growth, we slow depletion of these critical resources. To the extent we rely on technology to help achieve long term sustainability, we must account for the resource depletion characteristics of such technologies, if we are to have a chance of achieving long term sustainability.

The balance of this chapter will focus on the *third* option, applied more narrowly to energy production and use, with the aim of mitigating climate change and its potential to yield planetary impacts inconsistent with long-term sustainability.

The primary focus of this and subsequent chapters will be on the mitigation of CO_2, since it is estimated to be responsible for about 80% of the radiative forcing responsible for global warming.

1.3 Anthropogenic Drivers of CO_2 Emissions

The World Resources Institute, WRI [2], has examined the factors that have driven CO_2 emissions for key countries in the 1992–2002 time period. The factors considered are: *Gross Domestic Product* (GDP) *per capita, population, carbon intensity* (i.e., carbon emissions per unit of energy), and *energy intensity* (i.e., energy usage per unit of GDP). The relationship is as follows: Carbon emissions = GDP per capita × population × carbon intensity × energy intensity. The sum of the *rates* of these factors approximates the annual Carbon (and CO_2) emission *growth rate*. WRI data [2] has been used to generate Fig. 1.2, which shows how these factors have influenced the annual growth rate of CO_2 for selected countries during this 10-year period. As can be seen for the *world*, despite *decreases* in the energy use per unit of GDP, the CO_2 growth rate has been about 1.4% per year. The rate for the *U.S.* also has been about 1.4%, but the growth rate for *China* and *India* has been about 4% per year driven by economic growth, and for India, population growth as well.

However, a more recent analysis by Raupach [3] concluded that in the period 2000–2004, CO_2 worldwide emissions have increased more rapidly than in previous years, at an annual growth rate of *3.2%*. This is more than twice the growth rate of the 1992–2002 period. Rapidly developing economies in China and other Asian

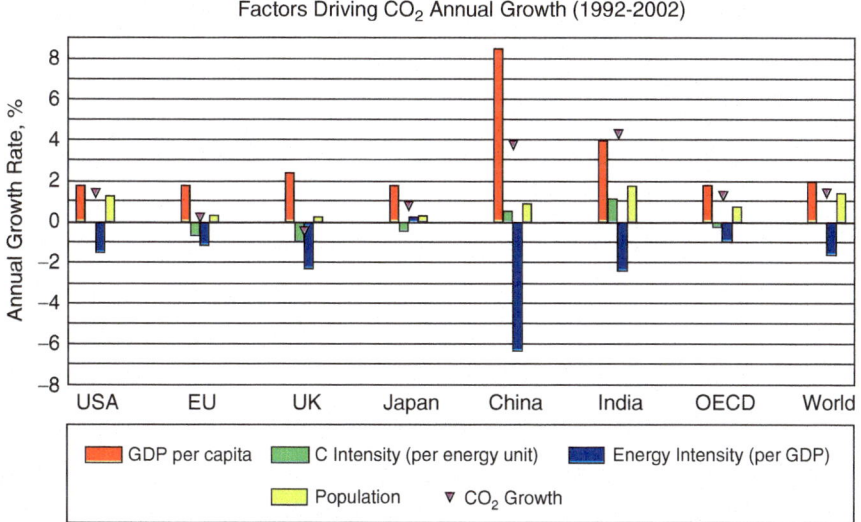

Fig. 1.2 Factors driving CO_2 growth rate for selected countries for 1992–2002 period

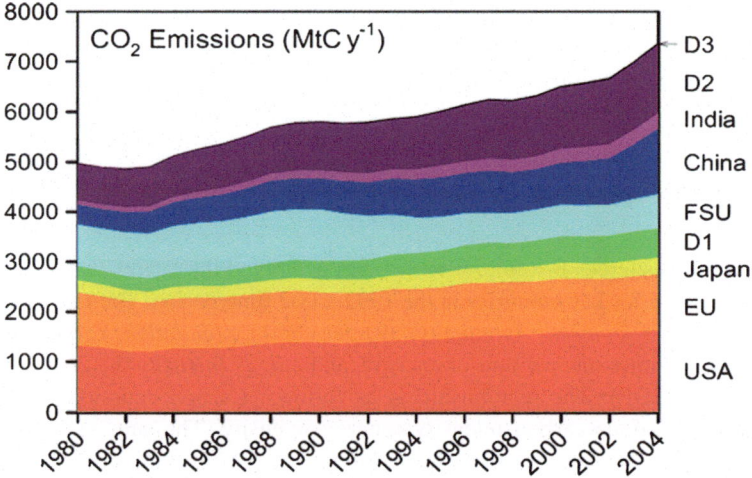

Fig. 1.3 1980–2004 CO$_2$ emission data by country. Note: *FSU* republics of the former Soviet Union, *D1* 15 other developed nations, including Australia, Canada, S. Korea and Taiwan, *D2* 102 actively developing countries, from Albania to Zimbabwe and *D3* 52 least developed countries, from Afghanistan to Zambia

countries are particularly significant. China is currently constructing the equivalent of two, 500-MW, coal-fired power plants per week and a capacity comparable to the entire United Kingdom power grid, each year [4]. Developing economies, together forming 80% of the world's population, accounted for 73% of the global growth in CO$_2$ emissions in 2004. However, these economies accounted for only 41% of emissions themselves and only 23% of emissions since the start of the Industrial Revolution around 1,800. Figure 1.3, Raupach [3], summarize these global emission trends for 1980–2004. Figure 1.4 was derived using country level data from this reference. This indicates the importance of China's industrial growth, as the major factor driving this increased growth rate in recent years. In October of 2009, analyzing the most recent data, Le Quere [5] concluded that global emissions have grown at 3.25% annually for the 2000–*2008* period. Therefore, this high growth rate has continued for the last 8 years that data is available. It is expected that there will be a temporary respite from these large growth rates given the world-wide recession that started in early 2008.

1.4 What Levels of Warming Are Projected: What Are the Uncertainties?

A credible base case, or business as usual (BAU), scenario must be established if we are to estimate warming with any confidence between now and the year 2100. IPCC [1], IEA [6, 7], and Hawksworth [8] have all postulated such scenarios that

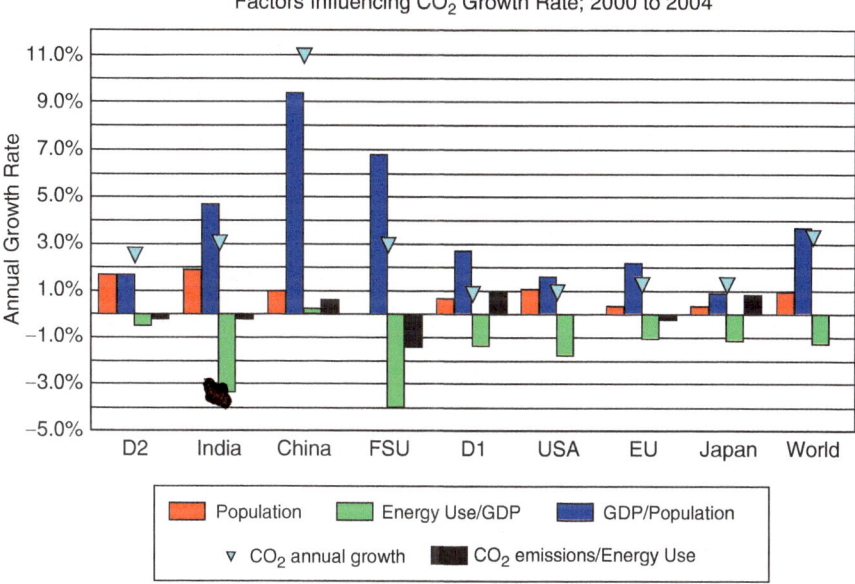

Fig. 1.4 Factors influencing CO_2 emission growth rate for selected countries for 2000–2004

allow such estimates. The IEA base scenario was selected as the basis for this analysis, because it does not assume major technology changes over time. Since it was limited to 2050, the projection was extended to 2100 by assuming reduced emission growth rates between 2050 and 2100. This scenario assumes the following CO_2 growth rates in the specified time intervals: 2000–2030, 1.6%; 2030–2050, 2.2% (from IEA); 2050–2075, 1.2%; and 2075–2100, 0.7%. Note that the reduced 2050–2100 growth rate assumption was based on projected declines in global population growth rates, but relatively stable GDP, carbon intensity and energy intensity growth rates.

Figures 1.5 and 1.6 present model-generated graphics of both CO_2 concentrations and warming from pre-industrial times projected to 2100, assuming this emission scenario. The Model for the Assessment of Greenhouse-Induced Climate Change, MAGICC (version 5.3) [9], was used to generate these projections. An earlier version of this model was used by the IPCC in its Third Assessment Report to evaluate impact of various emission scenarios. MAGICC is a set of coupled gas-cycle, climate, and ice-melt models that allows the determination of the estimated global-mean temperature resulting from user-specified emissions scenarios, which the author generated. Note that in both figures, which were generated directly by the model, the uncertainty range is included, as calculated by the model. As can be seen, warming uncertainties are much higher than for concentration projections. The main uncertainty factor for warming projections is the extent to which the earth is sensitive to permanent increase of CO_2 concentration (i.e., how much does the global equilibrium temperature change as a function of elevated CO_2 concentrations).

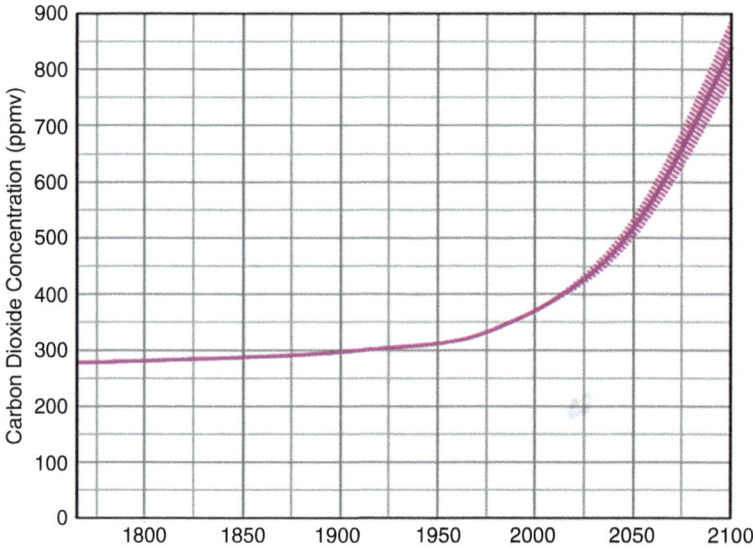

Fig. 1.5 Projected CO_2 concentrations for base case

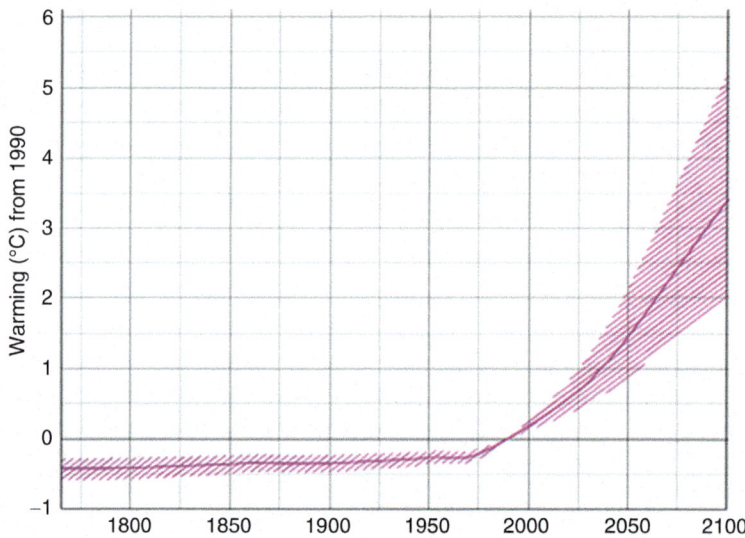

Fig. 1.6 Projected warming for base case

For a doubling of CO_2 levels from pre-industrial levels, also known as *equilibrium climate sensitivity*, IPCC [1], Wigley [9], and others state that this is quite uncertain, and their estimates range from 1.5°C to 6.0°C. This is the default range assumed by MAGICC when calculating warming ranges. The model assumes a default value of 3.0°C for the *most likely* atmospheric sensitivity.

Warming is projected to continue after 2100. When one accounts for continued warming projected into the next century, the equilibrium, or eventual warming, is projected to range from 2.3°C to 10.1°C with the best guess at 4.8°C above 1990 levels; this assumes an ultimate steady state 850 ppm CO_2 concentration.

As mentioned earlier, new data indicates that the recent annual global CO_2 emission growth rate is 3.25% in the 2000–2008 time frame. However, model calculations for Figs. 1.5 and 1.6, assumed a 2000–2030 growth rate of 1.6%, consistent with mainstream projections. Figure 1.7 illustrates the impact of assuming a 3.0% growth rate in this critical period. As can be seen, it would substantially increase the atmospheric CO_2 concentrations and global warming. Equilibrium warming, which would occur during the next century, would be from 3.0°C to 12.8°C, with the best guess 6.2°C above 1990 levels.

1.5   Achievable Mitigation Levels

Figure 1.8 presents the recent IPCC [10] analysis relating projected warming from 1990 to 2100 to the following global impacts: fresh water availability, ecosystem damage, food supplies, seawater rise, extreme weather events, and human health impacts. The author has added projected warming ranges for a credible business-as-usual case and an aggressive global mitigation case. *Note that for both ranges, it was projected that global annual emissions would grow at a 1.6% rate until 2030 or until mitigation starts, not the most recent (2000–2008) 3.25% growth rate.* Figure 1.9 is a modified version of Fig. 1.8, and shows the potential impact of a 3% growth rate in emissions until mitigation. The mitigation option in this case also assumed 1% annual reductions that would start in 2025. Delayed mitigation amplifies the effect of the high growth rate, because it allows greater quantities of CO_2 to be emitted before mitigation, over a longer time period.

For both base cases (Figs. 1.8 and 1.9) temperature increases in these range would result in potentially severe impacts, especially if the temperature increase is in the middle to upper end of the uncertainty range. Note that for the 3% growth case both the base and mitigation ranges are substantially greater with potentially more severe impacts. Also note, the upper end of the base case is off the IPCC chart, indicating the potential seriousness of impacts if warming is on the high end of the uncertainty range.

Using the MAGICC/SCENGEN model [9], Figs 1.10 and 1.11 projects 2100 warming for the 3% growth base and mitigation cases; and Figs. 1.12 and 1.13 project annual precipitation changes for the same two cases. The SCENGEN model generated these geographically explicit climate change projections using the MAGICC results, together with climate change information utilizing the four General Circulation Models listed on the figures. Projections are based on the default conditions inherent in the model.

As Figs. 1.10 and 1.11 indicate, warming is projected to be more severe over land and at highest and lowest latitudes. For example, Fairbanks, Alaska is projected to

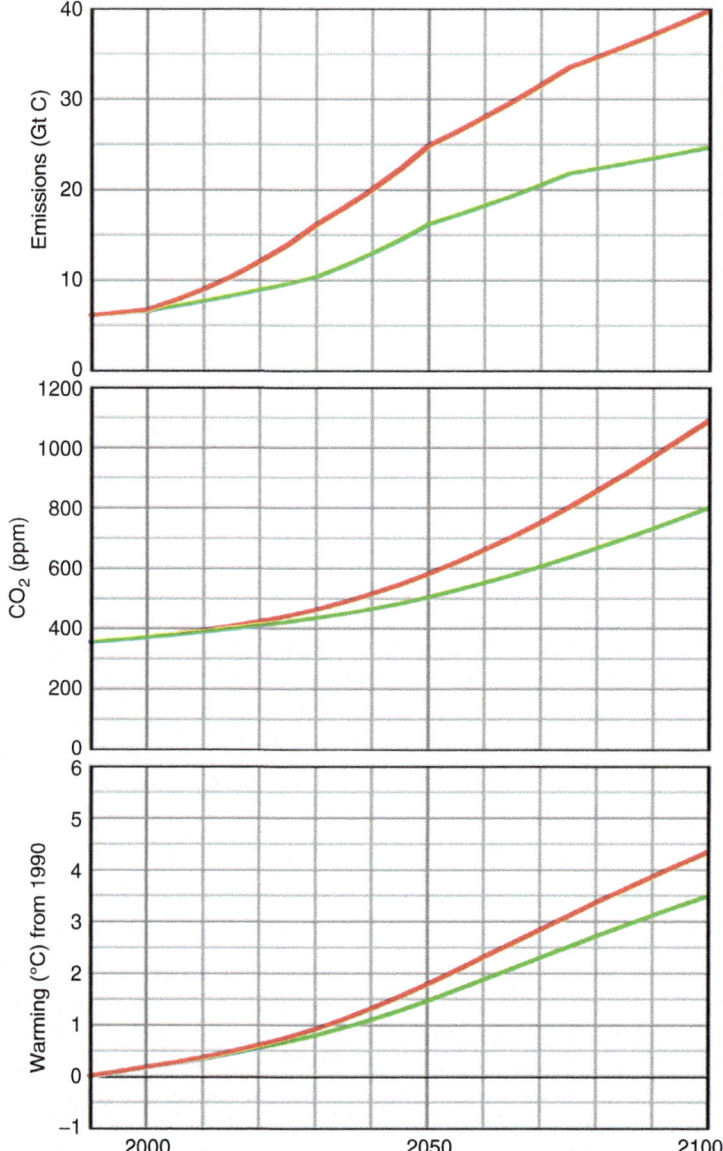

Fig. 1.7 Two global emission scenarios: original (*green*): IEA base case assumed 1.6% annual CO$_2$ emission growth rate from 2000 to 2030; Revised (*red*): annual growth rate of 3.0% from 2000 to 2030

see warming of 8.5°C for the base case and 5.1°C for the mitigation case. It is also apparent that despite mitigation, substantial warming is still projected. Figures 1.12 and 1.13 project major changes in precipitation patterns for base and mitigation cases. The models project that in, general, currently dry regions will get dryer, and

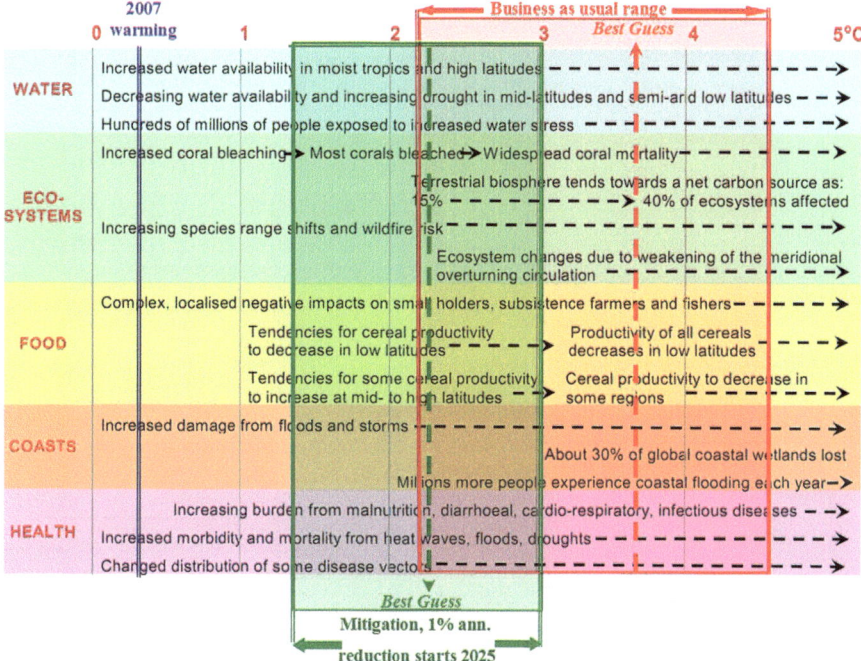

Fig. 1.8 Projected impacts as a function of 2100 warming (0–5°C) from 1990; 1.6% early emission growth rate. Note: 1.6% growth rate to 2030. Entries are placed so the *left* hand side of text indicates approximate onset of impact, *black lines* link impacts, and *dotted arrows* indicate impacts increase with increasing warming

relatively wet regions will see enhanced precipitation. For example, Los Angeles is projected to receive 48% less precipitation in 2100 for the base case, and 30% less for the mitigation case, whereas Chicago is projected to see substantial precipitation increases. Note that evaporation for all regions will increase due to warmer temperatures, exacerbating the potential for serious drought conditions for those areas with reduced precipitation.

Impacts associated with climate change [1] could include the following: water could become scarce for millions of people, wide-scale ecosystem extinctions, lower food production in many areas, loss of wetlands, infrastructure damage, storms and floods, and increased health impacts from infectious diseases. Although not included in Figs. 1.8 and 1.9, IPCC [1] also projects the potential for declining air quality in cities, due to warmer/more frequent hot days and nights over most land areas.

As noted, despite substantial CO_2 emission mitigation, substantial warming is projected, especially if the high emission growth rate continues and serious mitigation is not initiated until 2025. Therefore, limiting warming to about 2.0°C (range of 1.3–2.7°C) from 1990 values is likely the best result achievable even with a major CO_2 global mitigation program. Limiting warming to 2.5 ± 0.7°C is probably

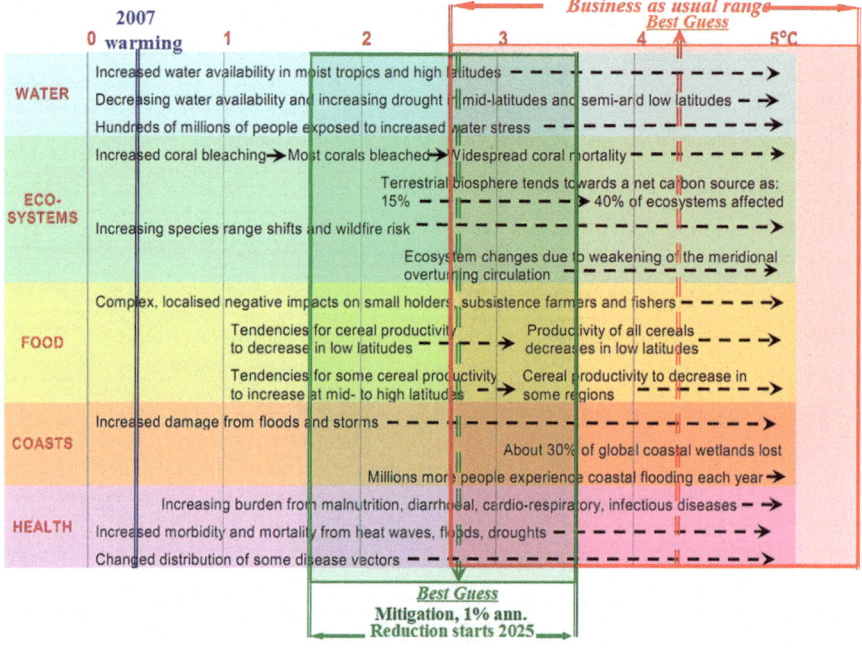

Fig. 1.9 Projected impacts as a function of 2100 warming (0–5°C,) from 1990; 3% early emission growth rate (Note: 3.0% growth rate to 2030. Entries are placed so the *left* hand side of text indicates approximate onset of impact, *black lines* link impacts, and *dotted arrow* indicate impacts increase with increasing warming)

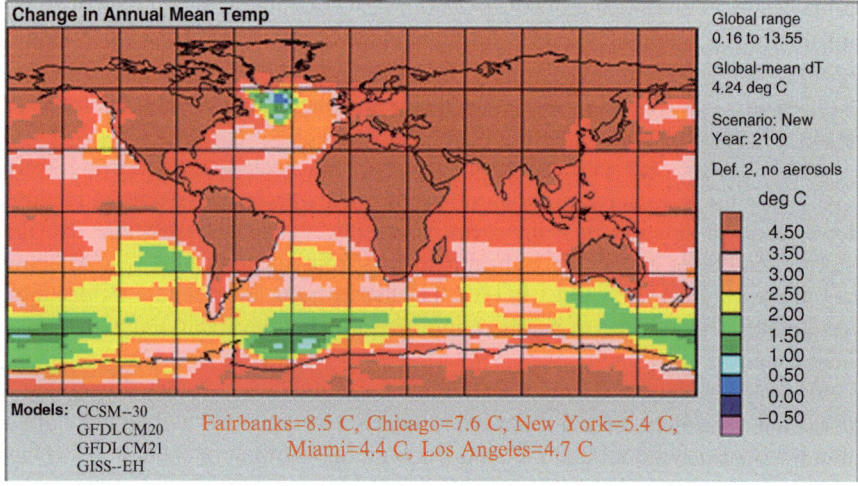

Fig. 1.10 Projected 1990–2100 warming for Base Case; Projected, 3% early emission growth rate

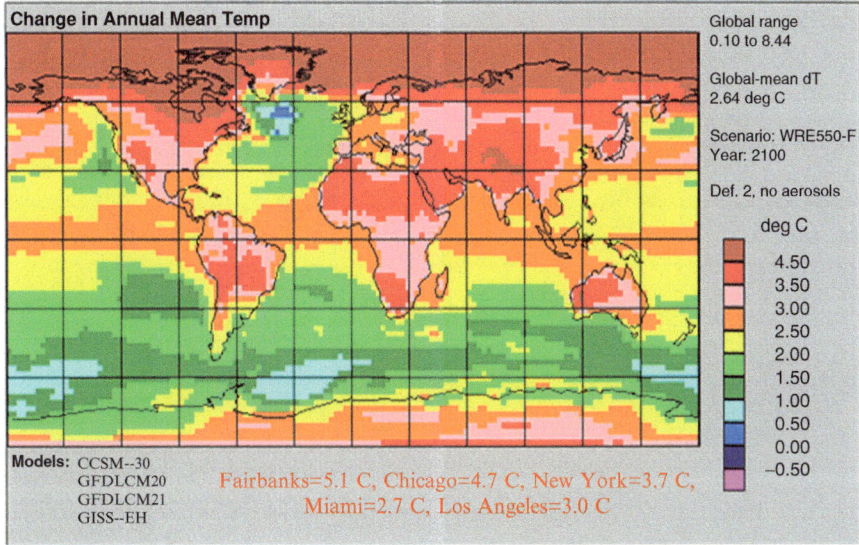

Fig. 1.11 Projected 1990–2100 warming for Mitigation Case (1% annual emission decrease for 75 years, starting 2025)

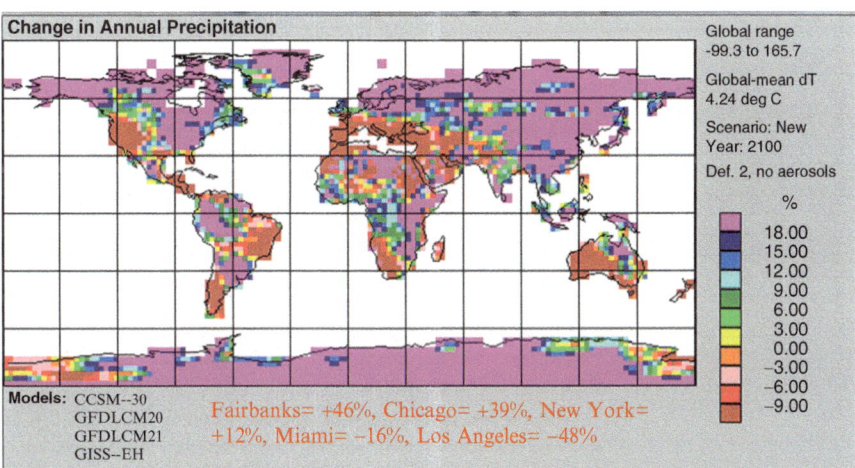

Fig. 1.12 Projected 1990–2100 annual precipitation change base case, 3% early emission growth

a more realistic goal given recent increases in emissions and the unavailability of key technologies.

To more carefully explore the factors influencing the ability to constrain warming, emission scenarios were evaluated to see what reduction levels, starting in what year, would limit warming to the 2–3°C (±0.7°C) range from the pre-industrial period. Figures 1.14–1.16 were generated utilizing a large number of MAGICC runs.

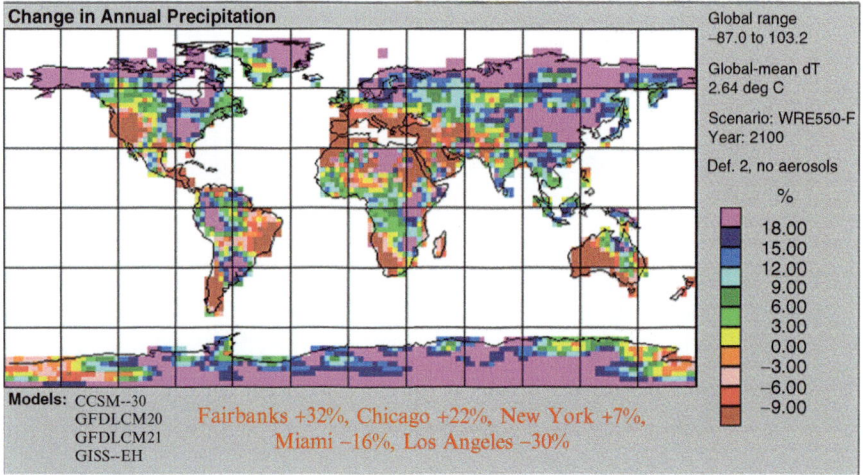

Fig. 1.13 Projected 1990–2100 annual precipitation change for mitigation case (1% annual emission decrease for 75 years, starting 2025)

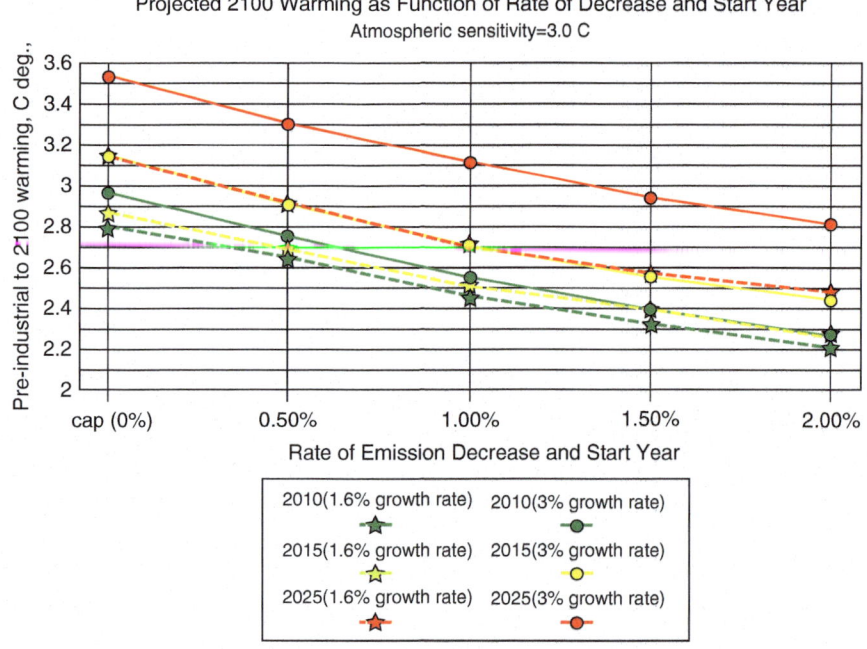

Fig. 1.14 2100 global warming as a function of near term CO_2 emission growth rates until mitigation year (1.6% or 3%), and year emission reduction starts (2010, 2015, or 2025)

They allow selection of combinations of emission growth reductions and start years needed to limit warming in 2100 to a given level. Figure 1.14 illustrates the impact of the faster 3% business as usual (BAU) growth rate, which yields additional

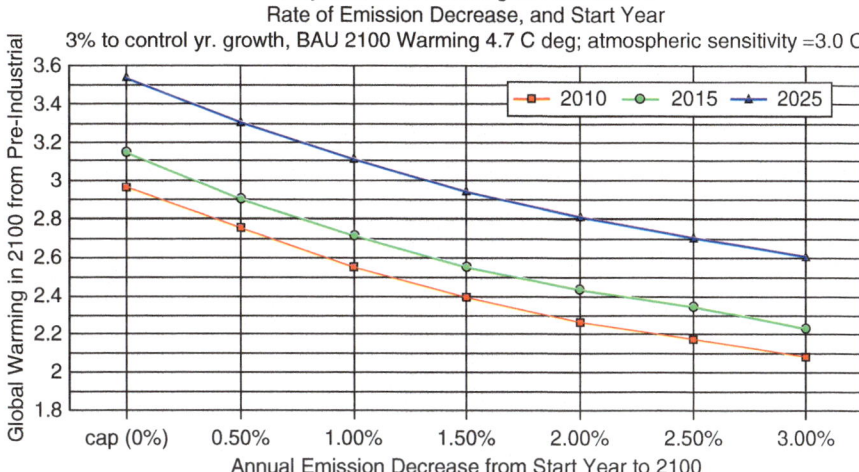

Fig. 1.15 2100 warming (°C) as function of annual emission decrease rate and year mitigation starts (assumes 3% growth rate until mitigation starts)

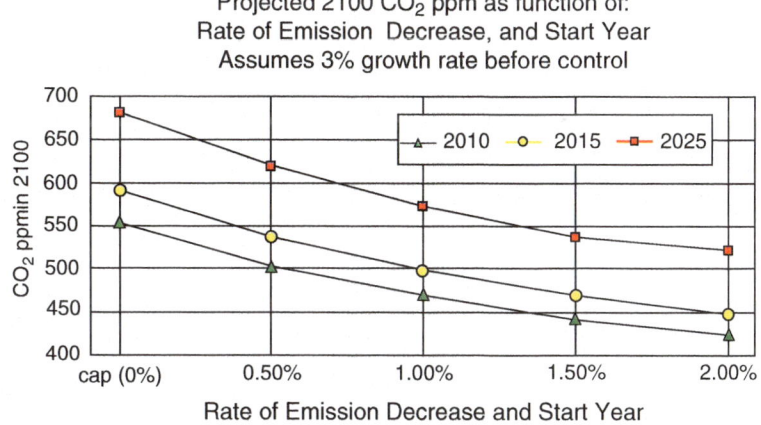

Fig. 1.16 CO_2 atmospheric concentrations in ppm in 2100 as a function of annual emissions reduction rate and the year reductions start

warming, relative to the 1.6% BAU case. As can be seen, additional warming increases as the start year for mitigation is delayed. Figures 1.15 and 1.16 focus just on the 3% base scenario and project 2100 warming and CO_2 concentrations, respectively. Note that an annual decrease of 0.00% means emissions are held constant, at the start year until 2100. Also note that, in order to simplify the analysis, it is assumed that there is an *immediate* change in growth rate from the base case, to a decreasing emission growth rate at the control "start year". In reality, there would be a transition period between the positive and negative growth rates. Therefore, from this perspective, Figs. 1.15 and 1.16 should be considered somewhat optimistic, since emissions would not be avoided at the ultimate rate, during this transition period.

As can be seen, major *annual decreases* in emissions will be necessary if a warming target below 2.5±0.7°C is to be achieved. Note that the earlier this reduction starts, the less the annual reduction rate has to be to meet a given warming target.

For example, if such a program would have started in 2010, reductions would need to be about 1% annually for 90 years to limit warming to about 2.5±0.7°C; whereas if such a program were to start in 2025, annual reductions would need to be in the order of 3% per year for 75 years. Again, it must be noted that there is a large range of uncertainty in the resulting temperature for a given maximum CO_2 concentration. Figure 1.17 illustrates this, by displaying the range of projected warming, from 1990, for a particular emission scenario (i.e., an annual <u>decrease</u> of 1%, starting in 2010, with a BAU growth of 1.6%, projected to constrain concentrations to the 440–480 ppm range). Note that Fig. 1.17 projects warming from 1990; about 0.4°C must be added to estimate warming from the pre-industrial era to be consistent with Figs. 1.14 and 1.15 Also, note an aggressive methane mitigation program could yield additional warming reduction of about 0.3°C in this time frame. Figure 1.18 quantifies the major challenge such reductions represent, relative to the IEA base case (1.6% growth to 2030) emission trends

It should be again noted, that if the world community continues to increase CO_2 emissions at the rate of 3% per year over the next two decades, warming mitigation will be more difficult. Figure 1.19 illustrates the consequences of a higher emission

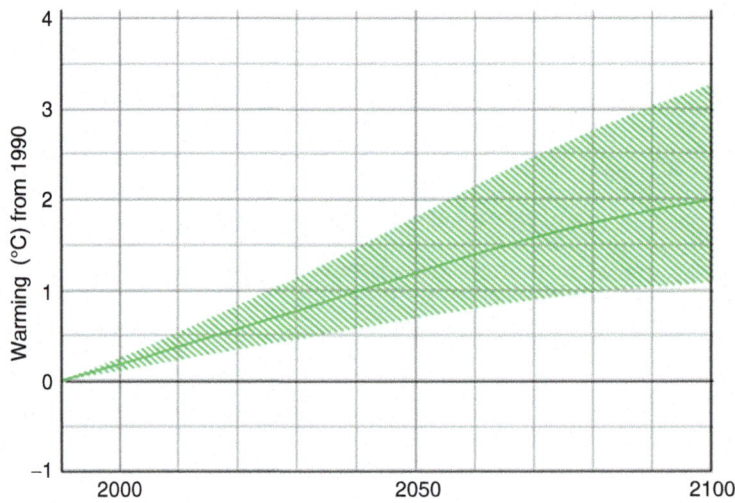

Fig. 1.17 Projected warming range for a 1% annual decrease in CO_2 emissions started in 2010

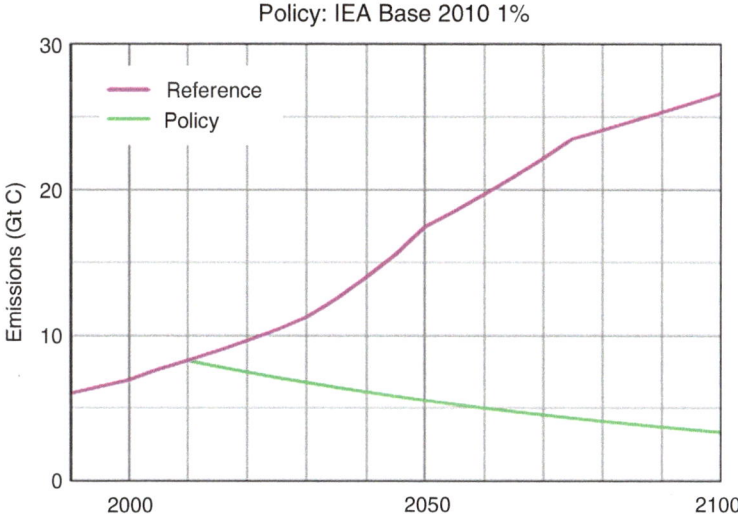

Fig. 1.18 Base case (*red*) and early mitigation case (*green*); Gt Carbon (3.67 Gt CO_2 per Gt C). Note: the area between the *curves* represents the amount of carbon avoidance needed to achieve the target temperature versus the base case: over 1 trillion tons of carbon or over 3.7 trillion tons of CO_2 over the 90-year period

growth rate prior to a mitigation program started in 2025. Mitigation is less successful in moderating warming when the program is initiated after 25 years of a 3% growth rate, compared to the 1.6% growth rate of the IEA base case. As Fig. 1.13 indicated, this penalty becomes less severe the earlier the mitigation program is initiated.

1.6   The Mitigation Challenge: Which Sectors and Gases Are Most Important?

In order to identify the most productive mitigation strategies, it is necessary to understand the current and projected sources of CO_2 and the other greenhouse gases. The author has derived the information in Fig. 1.20 from IEA [6]. This graphic projects world CO_2 emissions by sector. The emission growth rates are consistent with the business as usual base case, discussed previously: 1.6% from 2000 to 2030, and 2.2% from 2030 to 2050. It suggests that power generation and transportation sources are the fastest growing sectors and controlling these sources will be the key to any successful mitigation strategy. There is historical evidence that, as a country develops economically, it uses greater quantities of electrical power and experiences a sharp growth in the number and use of motor vehicles and other transportation sources. As mentioned earlier, China and India, with a cumulative population of over 2.5 billion, are projected to continue their rapid economic

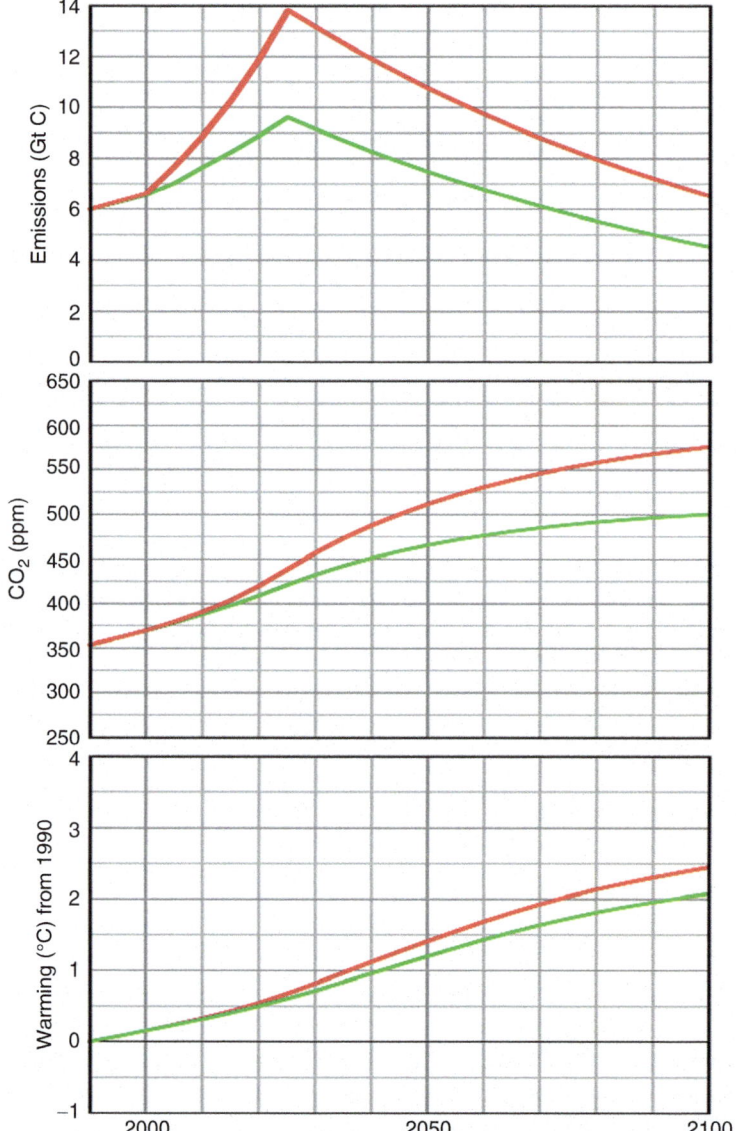

Fig. 1.19 Two mitigation scenarios starting in 2025: original (*green*) assumed 1.6% emission growth rate from 2000 to 2025, followed by an annual 1% reduction; revised (*red*) assumed a 3.0% growth rate from 2000 to 2025, followed by an annual 1% reduction

expansion with commensurate pressure on the power generation and transportation sectors. It should also be noted that the energy transformation category in Fig. 1.20 includes petroleum refining, natural gas, and coal conversion to liquids and biomass to alcohols, much of which will feed the transportation sector.

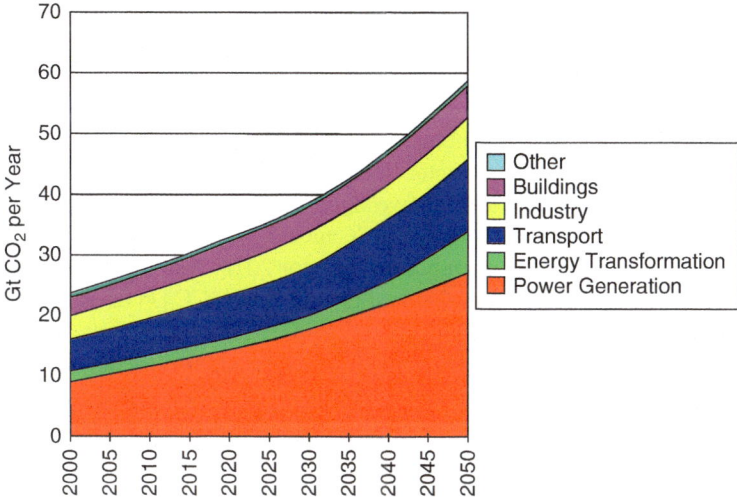

Fig. 1.20   Projected global CO_2 emission growth for key economic sectors, Gt per year

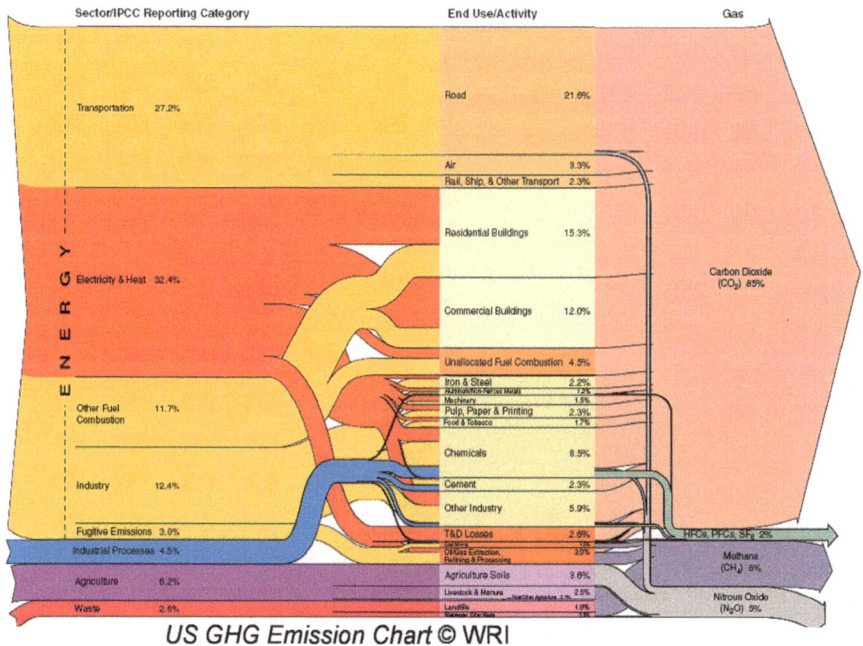

Fig. 1.21   U.S. energy and GHG emission flows by sector, end use, and gas in 2003

For the United States, the WRI [11] has generated Fig. 1.21, illustrating the relationship between sectors; end use/activities; and greenhouse gas emissions, including methane and nitrous oxide sources in CO_2 equivalents, for the year 2003. This graphic illustrates the relative importance and relationship of the power generation (electricity and associated waste heat in the figure), transportation, and industrial production, and the end use of energy in residential, commercial buildings, and industrial operations.

Gases other than CO_2 contribute significantly to warming. Figure 1.21 illustrates this for the United States. Although CO_2 is the dominant driver, methane and nitrous oxide are significant, together contributing 13% of the warming driving force. For the global view of the relative significance of the key greenhouse gases, Fig. 1.22 was generated using the MAGICC model. This figure illustrates the relative driving force of the key greenhouse gases for 2020, 2050, and 2100 assuming emissions per the modified IEA base case for CO_2 and IPCC [1] Scenario WRE750 for the other greenhouse gases. For this scenario, methane emissions are projected to grow at 0.5% per year until 2050, and remain constant for the next 50 years. For N_2O, emissions are assumed to grow at 0.4% per year until 2050 and the slow to a 0.1% growth rate until 2100. Also note for the forestry sector CO_2 net emissions are projected to decrease at about 2% per year to zero by 2075. Note that mitigating emissions of methane, a short-lived gas, allows for more near-term warming moderation, in contrast to a long-lived gas such as CO_2. Also note Fig. 1.22 projects that fine particles contribute a cooling effect in 2020 that transforms to a warming effect in later years. This is explained since emissions of sulfur dioxide are projected to increase until 2020, whereas the emissions will be reduced later in the century as countries install controls to mitigate that health and ecological impact of SO_2 and acidic sulfates. With such emission control, concentrations of sulfate particles, which form from SO_2 in the atmosphere and reflect incoming solar radiation, will consequently be reduced and their cooling effect reduced, yielding warming relative to 1990.

It is important to note that black carbon (BC), a component of fine particles, is a significant contributor to global warming even though the overall impact of fine

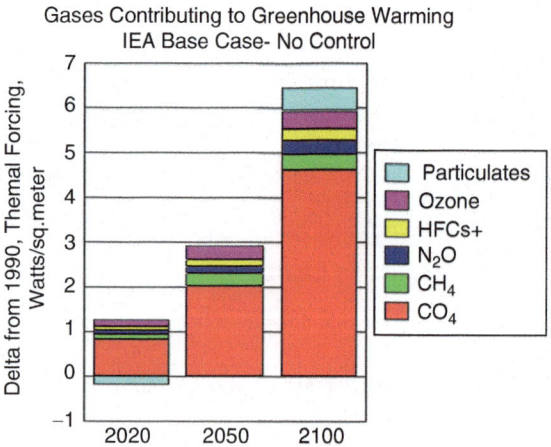

Fig. 1.22 Thermal driving forces (watts per square meter) of major GHGs relative to 1990

particles, dominated by reflective sulfates, is cooling. BC has a short atmospheric lifetime, is not well mixed in the atmosphere, and is a product of incomplete fuel combustion of fossil fuels and biomass. The sources of BC are widely dispersed and not well characterized, but appear dominated by mobile and stationary diesel engines, and residential fuel combustion in developing countries [12].

As mentioned earlier, this and subsequent chapters focus on energy technologies, and only CO_2 will be discussed, since it is the critical greenhouse gas and is growing at a fast rate. However, as noted earlier, an aggressive methane mitigation program could add about 0.3°C warming mitigation, to that achieved via CO_2 mitigation by 2100. A major mitigation program for ozone precursors and N_2O emissions could yield another 0.3°C warming reduction.

BC emission control could also contribute to warming mitigation, but the magnitude of this potential impact is difficult to quantify given the many uncertainties involved. However, Princeton [12] has recently estimated that a global BC mitigation program could yield a best guess value of 0.29 W/m^2 decrease in the global thermal driving force in 2100. This is comparable to the level achievable with an aggressive methane mitigation program. Given the short atmospheric lifetimes of BC, such benefits, if available, have the potential to be realized in the near term.

1.7 The Mitigation Challenge: What Can be Done and What Role Can Energy Technology Play?

One key question is, do we need new technology, or can we provide deep emission reductions with currently available generation and end use technologies? Three mitigation studies were analyzed to attempt to answer this important question. Enkvist [13] argues that the least expensive way to mitigate emissions in the short term will be to provide incentives to utilize existing technology, both on the end use efficiency side, buildings and mobile sources, and for low emission generation technologies, such as nuclear and wind. He also suggests that state of the art mitigation of non-CO_2 sources could be significant as well. The sum of the mitigation achievable with these state of the art technologies yields an annual savings of about 7.5 Gt CO_2 by 2030. However, assuming that the 3% global growth rate will continue until 2030 in the absence of such a mitigation program and that we wish to constrain warming to below about 2.5±0.7°C, it will be necessary to reduce emissions by about 30 Gt CO_2 in 2030. In the absence of fundamental cultural and lifestyle changes that dramatically reduce our energy usage, *new* energy technology will need to be developed and utilized if potentially catastrophic climate change is to be avoided. Based on the Enkvist analysis, such technology would need to be utilized to yield 74% of the required reduction in 2030. Less dependence on new technology could result if CO_2 emission growth rate would rapidly decelerate to about 1.6% annually, a typical growth rate in the 1990s. Barring an extended worldwide economic slowdown, this appears unlikely. In this case, available technologies could provide about 56% of the required mitigation.

Similar calculations have been made based on mitigation analyses conducted by Pacala [14] for the years 2004–2054 and IEA [6] for the years 2030–2050. For both references, when one calculates the role that existing technologies could play within the time frame of their assumed mitigation programs, it is estimated that such an aggressive utilization of existing technology could provide only about 25% of the required mitigation if the current 3% growth rate continues and about 45% of the needed mitigation if global emission growth decreases to a 1.6% CO_2 growth rate in the near term.

It should be noted that in the three studies described above, the estimate of the role that new technology must play is based on minimizing mitigation costs. It may be possible in some situations to push the use of existing technology to achieve greater carbon reductions. For example, earlier and more extensive use of current solar conversion technology could displace some coal utilizing carbon capture and storage (CCS), but at a much higher cost.

Therefore, it does not appear possible, in the absence of lifestyle/behavioral/ structural changes to mitigate the roughly 4 trillion tonnes of CO_2 that may be required to constrain warming below $2.0 \pm 0.7°C$ this century, without the extensive use of improved and in some cases breakthrough energy technologies. Such technologies are necessary for both energy production and to enhance end use efficiency (i.e., lower emission vehicles).

In order to understand the potential of various energy technologies to prevent CO_2 emissions, IEA [7] evaluated two key mitigation scenarios: the Accelerated Technology (ACT) scenario, which was formulated in their original Energy Technology Perspectives report in 2006 [6], and the new Blue Scenario formulated in the updated version of their analysis [7]. The recent scenario analysis was done at the request of G-8 Leaders and Energy Ministers in 2007. Of these, the Blue Map scenario is the most aggressive. The scenario *assumes an* aggressive and successful research, development, and demonstration program (RD&D) to develop and improve technologies and a comprehensive technology demonstration and deployment program. It also assumes policies in place that would encourage the use of these technologies in an accelerated time frame. These include CO_2 reduction incentives to encourage low-carbon technologies with costs up to US $200/metric ton CO_2. The incentives could take the form of regulation, pricing, tax breaks, voluntary programs, subsidies, or trading schemes.

Figure 1.23 illustrates the emission projections assumed for the two mitigation scenarios compared with the assumed baseline emission projection. The fundamental difference between the scenarios is that the ACT option aims at decreasing CO_2 emissions in 2050 to 2005 levels, while the more aggressive Blue scenario aims to reduce 2005 emissions in half by 2050. Also shown is the projected CO_2 concentrations in 2100 and the author's calculated values of 2100 and ultimate (equilibrium) warming for both scenarios. Included, is a plot depicting the implications of the current 3% emission growth rate if it would continue until 2030. As depicted on Fig. 1.23 for the ACT Map scenario extended to 2100, MAGICC calculations indicate best guess warming of 2.7°C relative to the pre-industrial era. *For the Blue scenario, warming in 2100 is projected to be 2.3°C.* Such significant warming is

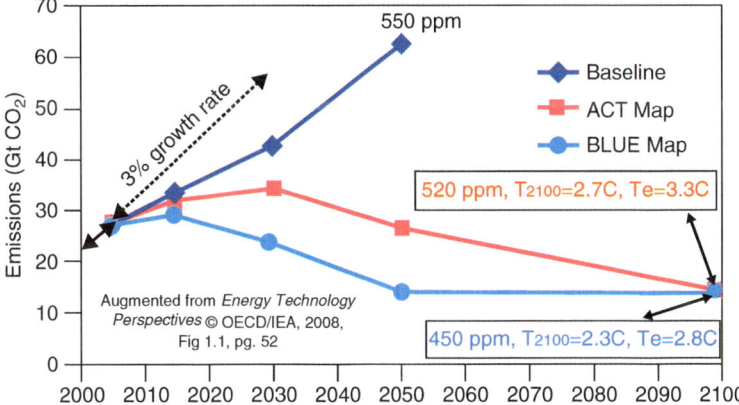

Fig. 1.23 The *ACT* and *Blue IEA* emission scenarios and their projected warming impacts. *Note: T_{2100}=best estimate warming in 2100; Te=best estimate equilibrium warming*

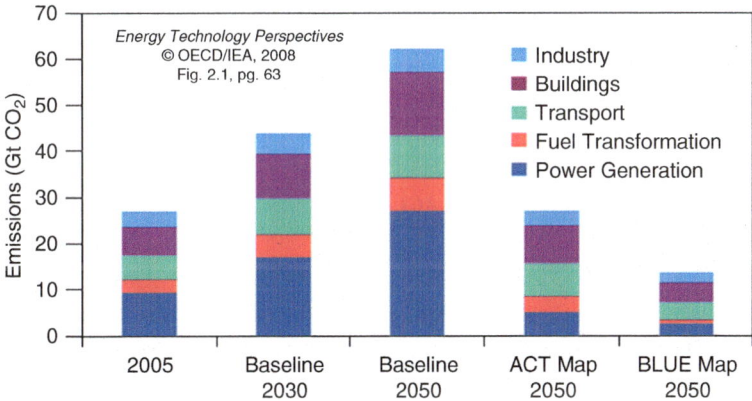

Fig. 1.24 Emissions by sector for *Baseline*, *ACT*, and *Blue scenarios* to 2050 in Gt CO_2

projected, despite the IEA assumption of an aggressive RD&D and deployment program, the optimistic assumption of a 1.7% growth rate in the near term compared to the recent 3% annual growth rate, and for the Blue scenario, the assumption that early and deep global reductions are implemented.

Figure 1.24 illustrates the energy sector implications of the ACT and Blue scenarios compared with projected baseline emissions up to the year 2050. For the less aggressive ACT scenario, major savings are achieved in the power generation sector. However, for the Blue scenario, major reductions are required in every energy sector.

Figure 1.25 summarizes the results of the IEA analysis by identifying technologies contributing to the CO_2 avoidance of both the ACT and Blue Map scenarios to 2050. The sum of all the bars yields the 35 and 48 Gt avoidance goals for the ACT and Blue scenarios, respectively. The figure illustrates the projected avoidance by

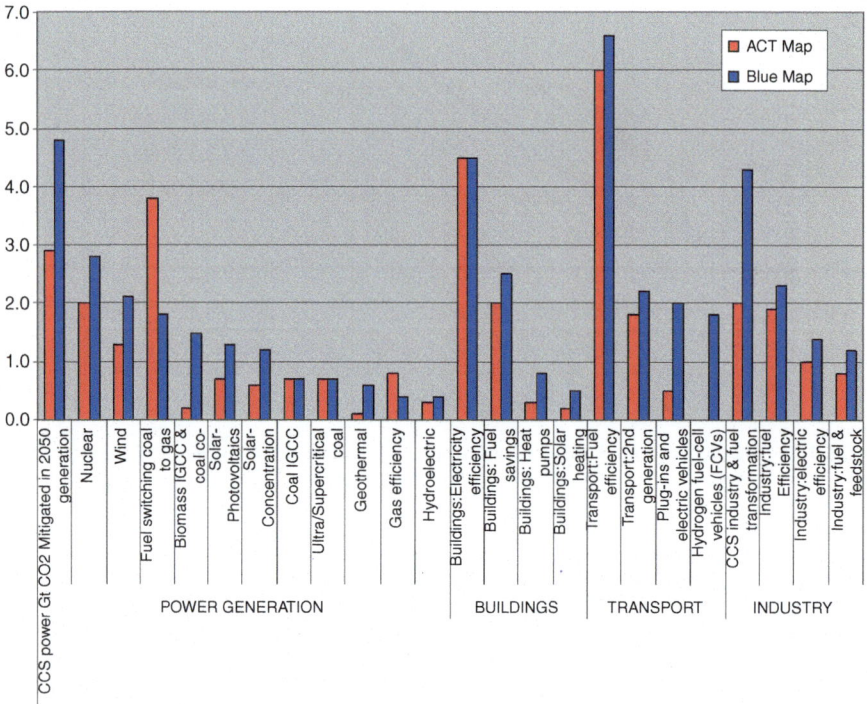

Fig. 1.25 Technologies needed to meet *ACT* and *Blue Map Scenarios* avoidance goal of 35 and 48 Gt CO$_2$ in 2050, respectively

technology in the key sectors. As can be seen, a diverse array of technologies in all energy sectors will be needed if these avoidance goals are to be met, especially for the Blue scenario. Of particular importance are end use technologies in the building, transport, and power generation sectors; and carbon storage technologies in the power generation and industrial sectors. It is important to note that the IEA [7] has characterized the technological changes that would be necessary to achieve carbon reductions consistent with these scenarios: as "A global revolution …. in ways that energy is supplied and used". For the more aggressive Blue scenario they concluded: "The Blue scenarios require urgent implementation of unprecedented and far reaching new policies in the energy sector."

1.8 What Are the Challenges of Early and Deep CO$_2$ Reductions?

It is instructive to examine the implications of an aggressive energy technology mitigation program. The Blue scenario is a useful option to examine, since it involves early and deep carbon reductions across all energy sectors, and since the

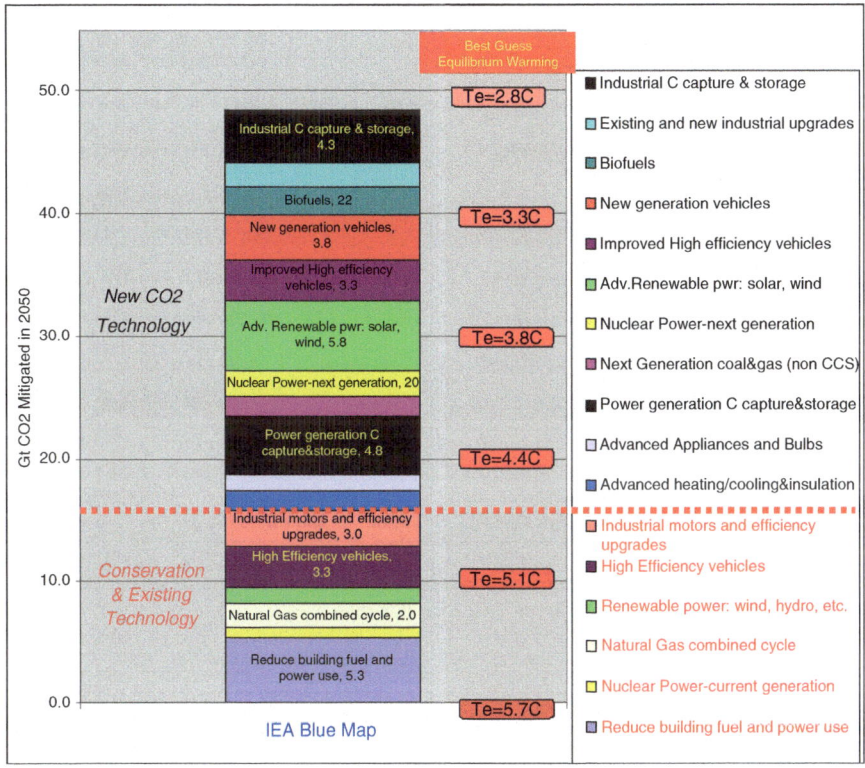

Fig. 1.26 Existing and new technologies needed for the *ACT* and *Blue Scenarios*

in-depth IEA analysis of this option offers us valuable insights regarding the research, development, demonstration, and deployment needs; the role that new technology must play and investment requirements. Figure 1.26 illustrates the role that new technology will have to play in order to control emissions consistent with the Blue scenario. The author has used engineering judgment to divide the technologies into *existing* and *new* categories. Also, best estimate equilibrium warming using the MAGICC model is included as a function of the Gt of CO_2 mitigated in 2050. As can be seen, new technology is projected to play a major role. Also note, in the absence of new technologies, it will be difficult to constrain ultimate warming below about 4°C, +/– the uncertainties!

In order to help quantify the technology requirements, IEA [7] generated Fig. 1.27. It attempts to quantify the **annual** need of low carbon power generation facilities in order to reduce emissions consistent with the two scenarios. As can be seen, a fundamental transformation of the power generation sector will be necessary. In addition to unprecedented construction of nuclear facilities and a fundamental shift of coal and gas facilities to incorporate carbon capture and storage, the Blue scenario will require a massive deployment of solar, wind, and geothermal plants.

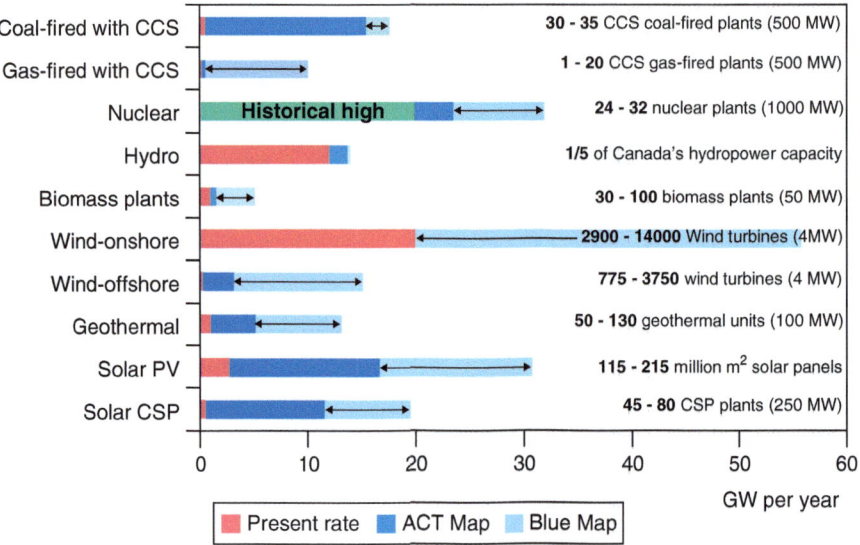

Fig. 1.27 Numbers of power generation plants and their GW per **year** production needed for *ACT* and *Blue scenarios* (note: *CSP* concentrated solar power)

A key question is: What are the research, development, demonstration, and deployment (RDD&D) requirements by technology for each energy sector? Fig. 1.28 has been derived from IEA's Blue scenario [7] to relate RDD&D resource needs compared with the quantity of CO_2 projected to be mitigated by technology. Note that the units are Gt per year, and for the costs, ***monthly*** expenditures in $ billions required over the assumed 40 year period (2010–2050). The monthly interval was used to allow the graphic to use the same ordinate values for mitigation and resource requirement quantification.

Note that when added together by technology, the monthly RDD&D requirements are estimated at $30 billion and total costs over the 40 year period at ***$14 trillion***. Of this amount about $11.9 trillion is the projected deployment costs. This suggests an RD&D requirement of about $2.1 trillion or about $52 billion per year, five times current funding levels. IEA [7] defines deployment costs as the total investment cost over time, needed to allow evolving technologies to improve to the point they are deemed to be cost competitive with existing technology, or if this is not possible, at least deemed affordable in the context of an aggressive mitigation program. As can be seen from the figure, most of the resources are required for mobile source technologies (electric, hybrid, and hydrogen/fuel cell vehicles) and for carbon capture and storage (coal generation, energy transformation, and industrial facilities). When these technologies are commercial and utilized per the Blue scenario, IEA estimates capital investment requirements over the baseline as $45 trillion. However, per IEA energy savings associated with these technologies could recover $43 trillion of that investment over time, assuming a 10% discount rate.

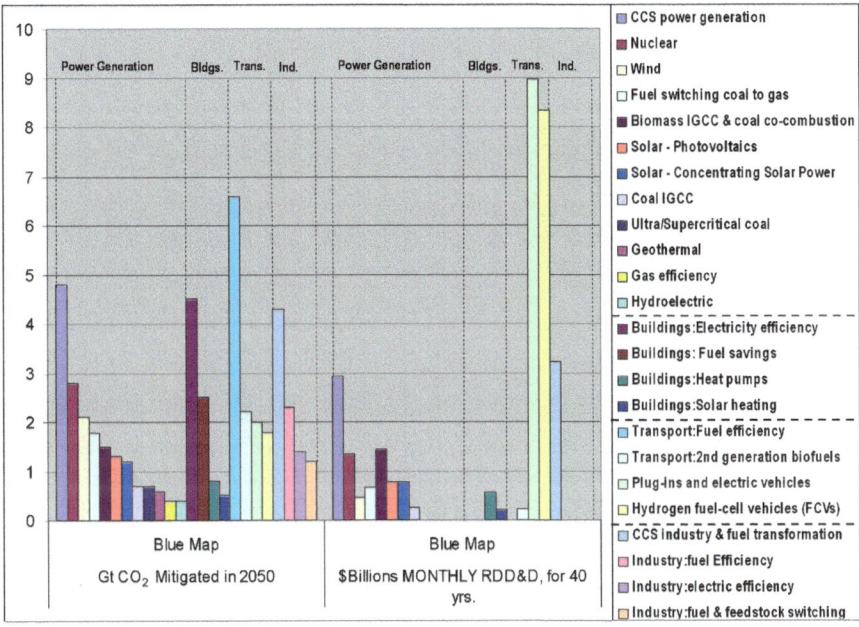

Fig. 1.28 CO_2 mitigated and corresponding RDD&D requirements by technology, for Blue scenario (Units, left, Gt CO_2 mitigated in 2050, right, $ billions of monthly RDD&D)

Let us now focus on these four critical sectors and examine the technology options available, their current state of the art, and the required RD&D for them to meet their potential to avoid CO_2 emissions. Tables 1.1–1.4 summarize the potential and status of key technologies based on the following recent energy technology assessments: IEA [6, 7], Hawksworth [8], Pacala [14], and Morgan [15]. Two additional references contained useful information relative to hydrogen/fuel cells, USEPA [16], and nuclear technologies, USEPA [17].

1.9 Power Generation Sector

According to the Center for Global Development [18], global power generation emissions have been growing at 3.7% annually, for the 2000–2008 period. It is important to note, that there are massive coal-fired power generation capacity expansions underway in China, India, and other countries. Since such plants have no CO_2 mitigation technology planned and can have lifetimes up to 50 years, the sooner such technology is ready for implementation and mandated, the sooner new plants can incorporate such technology and control emissions.

Table 1.1 Candidate technologies for power generation (CO$_2$ mitigation projected impact in Gt/year of CO$_2$)

Technology	Current state of the art	2050 Impact	Issues	Technology RD&D needs	Potential environmental impacts/R&D needs
Solar-photovoltaic and concentration (renewable)	First generation commercial, but very high costs	2.5	Costs unacceptably high, solar resource intermittent in many locations	High, breakthrough RD&D needed to develop and demo cells with higher efficiency and lower capital costs	Reduction in emissions of SOx, NOx, fine PM; fewer mining impacts and residues for disposal or use. Potential upstream emissions/effluents associate with manufacturing cells/*medium*
Wind power (renewable)	Commercial	2.1	Costs very dependent on strength of wind source, large turbines visually obtrusive, intermittent power	Medium, higher efficiencies, on-shore demonstrations	Reduction in emissions of SOx, NOx, fine PM; fewer mining impacts and residues for disposal or use; possible local impact on bird population/medium
Fuel Switching coal to gas	Commercial	1.8	Key issue is availability and affordability of natural gas	Medium, higher efficiencies with new materials desirable	Reduction in emissions of SOx, NOx, fine PM; fewer mining impacts and residues for disposal or use. Extraction R&D could enhance availability of CH4/*low*
Nuclear Power-next generation	Developmental, generation III+ and IV: e.g. Pebble bed modular and supercritical water cooled reactors	1.8	Deployment targeted by 2030 with a focus on lower cost, minimal waste, enhanced safety and resistance to proliferation	High, demonstrations of key technologies with complimentary research on important issues	Relative to coal, reduction in emissions of SOx, NOx, fine PM; fewer mining wastes. Small quantities of potent and long-lived waste, could contaminate small area/*high*

Coal IGCC with CO_2 capture and storage	1.6	*IGCC*: early commercialization, Underground storage (US): early development.	*IGCC*: High capital costs, questionable for low rank coals, potential reliability concerns; *US*: Cost, safety, efficacy	High, *IGCC*: Demos on a variety of coals, hot gas cleanup research; *US*: major program with long term demos at key geological formations to evaluate environment impact, efficacy, cost and safety	Lower power plant efficiency yields greater SOx, NOx, fine PM and coal mining impacts, including acid mine drainage. Sequestration could impact groundwater quality/*high*
Pulverized coal/oxygen combustion with CO_2 capture and storage	1.6	Developmental	Oxygen combustion allows lower cost CO_2 scrubbing, but oxygen production cost is high, *US* : Cost, safety and permanency	High, large pilot followed by full scale demos needed, low cost O_2 production needed, *US* requires major program (see write-up above)	Lower power plant efficiency yields greater SOx, NOx, fine PM and coal mining impacts, including acid mine drainage. Sequestration could impact groundwater quality/*high*
Pulverized coal with CO_2 capture and storage	1.6	Underground storage developmental; CO_2 scrubbing with MEA commercial but expensive	*US*: Cost, safety and efficacy issues, CO_2 scrubbing energy intensive: yielding high costs	High, *US* requires major program (see write-up above); affordable CO_2 removal technologies need to be developed and demonstrated	Lower power plant efficiency yields greater emissions of SOx, NOx, fine PM and coal mining impacts, including acid mine drainage. Sequestration could impact groundwater quality/*high*
Biomass as fuel gasified or co-fired with coal (renewable)	1.5	Commercial, steam cycles	Biomass dispersed source, limited to 20% when co-fired with coal	Medium, biomass/ GCC would enhance efficiency and CO_2 benefit; also genetic engineering to enhance biomass plantations	Reduction in emissions of SOx, NOx, fine PM; fewer mining impacts and residues for disposal or use; however potential eco impacts and excessive water use from biomass plantations/*medium*

(continued)

Table 1.1 (continued)

Technology	Current state of the art	2050 Impact	Issues	Technology RD&D needs	Potential environmental impacts/R&D needs
Nuclear power-current generation	Commercial, Pressurized Water Reactors and Boiling Water Reactors (generation III)	1.0	Plant siting, high capital costs, levelized cost 10–40% higher than coal, potential U shortages, safety, waste and proliferation	Medium, Waste disposal research	Relative to coal, reduction in emissions of SOx, NOx, fine PM; fewer mining wastes. Small quantities of potent and long-lived waste, could contaminate small area/*high*
More efficient coal fired power plants no CCS	Early commercialization of supercritical and ultra supercritical	0.7	Currently maximum efficiency of 45%, yielding 36% less CO_2 than current fleet	High, new affordable materials needed to enhance efficiency to 50–55%	Small reduction in emissions of SOx, NOx, fine PM; fewer mining impacts and residues for disposal or use/*low*
Coal IGCC with no CO_2 capture and storage	IGCC: early commercialization	0.7	IGCC: High capital costs, complexity and reliability concerns, only modest CO_2	High, demos on a variety of coals, hot gas cleanup research	Small reduction in SOx, NOx, fine PM; fewer mining impacts and residues for disposal or use/*medium*
Geothermal	Early commercialization	0.6	Cost of deep drilling and fracturing, distance from users	High, large number of demos in various geological formations	Potential for water and land pollution problems at geothermal site/*medium*
Natural gas combined cycle (new)	Commercial, 60% efficiency	0.4	Limited by natural gas availability, which is major constraint	Medium, higher efficiencies with new materials desirable	Reduction of SOx, NOx, fine PM; fewer mining impacts and residues. Extraction R&D could enhance availability of CH4/*low*
Hydroelectric (renewable)	Commercial	0.4	Capital costs high, potential eco disruption, siting challenges	Medium, minimize environmental footprint	Local ecological impacts/*low*

Table 1.2 Candidate technologies for CO_2 mitigation from buildings

	Technology	Current state of the art	Blue 2050 impact	Issues	Technology RD&D priority and needs	Potential environmental impacts/**R&D need**
Heating and cooling	Enhanced energy mgt. and high efficiency building envelope: insulation, sealants, windows, etc.	Commercial	2.5	Lack of incentive, high initial costs, long building lifetime	**Low/medium priority**, incremental improvements to lower cost and enhance performance	Less fossil fuel and nuclear power generation, and less on-site fossil fuel combustion, yield reduction in coal and natural gas emissions, and nuclear wastes/**low**
	High efficiency building heating and cooling, including heart pumps	Commercial	0.8	Lack of incentive, high initial costs	**Low/medium priority**, incremental improvements to lower cost and enhance performance	Same as above
	Solar heating and cooling	First generation commercial	0.5	High initial costs, availability of low cost efficient biomass heating systems	**Medium**, focus on development of advanced biomass stoves and solar heating technology in developing countries	Same as above
Appliances	More efficient electric appliances	Commercial	4.5	Higher initial costs and lack of information to the consumer	**Low/medium priority**, incremental improvements to lower cost and enhance performance	Less fossil fuel and nuclear power generation, yields reduction in coal and natural gas emissions, and nuclear wastes/**low**
	More efficient lighting systems	Commercial-fluorescent		Lack of incentive given higher initial costs	**Medium**, LED and OLED technology needs further development with aim of lowering initial cost	Same as above; however, mercury content of fluorescent bulbs could cause health and environmental problems/**med**
	Reduce stand-by losses from appliances, computer peripherals, etc.	Commercial		Lack of incentive from vendors and lack of knowledge from end-users	**Low**	Less fossil fuel and nuclear power generation, yields reduction in coal and natural gas emissions, and nuclear wastes/**low**

Table 1.3 Candidate technologies for CO_2 mitigation from mobile sources

	Technology	Current state of the art	Issues	Key enabling technologies	RD&D needs
Vehicles	Improvements: current internal combustion engine components	First generation: commercial	Lack of customer incentive major problem; trend to larger vehicles in US and recently Europe counter-productive		**Medium**, transmission and drive train improvements
	Non-engine Improvements: current vehicles; tires, A/C, light materials	First generation: commercial	Lack of customer incentive major problem; trend to larger vehicles in US and recently Europe counter-productive		**Medium**, lower weight construction, improved tires and more efficient A/Cs
	Hybrid electric vehicles (HEVs)	First generation: commercial	Higher costs (about $3000), "light" hybrids not as efficient as full hybrids, some newer models yield power over mileage benefits	**Batteries:** near term nickel metal hydride; longer Term: lithium Ion	**Medium/High**, minimize incremental cost, mostly battery related, and enhance efficiency
	Plug-in hybrid electric vehicles (PHEVs)	Developmental	Battery cost and lifetime key issues. Also requires low C electric generation to maximize carbon reduction benefits	**Batteries:** Near term nickel metal hydride; longer term: lithium ion	**High**, intensive R&D necessary to upgrade battery performance, lifetime and ability to allow deep cycling and rapid charging
	Full performance electric vehicles (FPEVs)	Developmental	Battery cost, storage capability and lifetime key issues. Also requires low C electric generation to maximize carbon reduction benefits	**Batteries:** longer term: lithium ion	**High**, intensive R&D necessary to upgrade battery performance, lifetime and ability to allow deep cycling and rapid charging

Fuels	Fuel cell electric vehicle (FCEV)	Developmental	Fuel cell costs and fuel cell stack life; also hydrogen production and storage, safety and lack of infrastructure	**H2 production:** lower cost low C processes **H2 Storage:** high pressure storage, and liquefied gas storage; both appear expensive **Fuel cells:** need to increase power per cell	**High**, breakthrough RD&D needed to develop cost competitive, long lived fuel cells. Hydrogen production and storage RD&D also needed
	Ethanol from sugar	Commercial	Limited by land capable of high sugar yields, e.g., sugar cane		**Medium**, develop sugar cane cultivars with higher yield and more frost tolerance
	Biodiesel and other fuels from biomass; thermo chemical processes	Developmental	Developmental, yet potentially high production and lower cost via gasification/Fischer-Tropsch synthesis		**High**, major RD&D needed to develop and demonstrate viable technology for biomass feedstock
	Biodiesel from vegetable oil	First generation: commercial	High costs, low yield from oil crops, limited waste cooking oils, low S a positive		**Low**
	Ethanol from grain/ starch, e.g., corn	Commercial	Limited by grain supply; high costs, energy intensive production		**Low**
	Ethanol from biomass/lignose cellulose;biochemical process	Early developmental	Inability to convert wide range of biomass types, high production costs, dispersed biomass source		**High**, breakthrough RD&D needed to develop lower cost generally applicable process(es)

Table 1.4 Candidate CO_2 mitigation technologies for industrial sources (impact in Gt/year)

Technology	Current state of the art	Blue 2050 impact	Issues	RD&D needs	Potential environmental impacts/**R&D need**
CO_2 capture and storage	Early development	4.3	Applicability limited to large energy-intensive industries, including fuel transformation processes; key questions: cost, safety, efficacy	High, major program with long term demos evaluating large number of geological formations to evaluate efficacy, cost and safety	Lower process efficiency yields greater air, water and land impacts per product produced, sequestration could impact groundwater quality/*high*
Motor systems	Commercial	1.4	For most industries not a major cost, lack of expertise for some industries	**Medium**: lower costs and higher efficiencies desirable	Reduction in coal emissions: Sox, NOx, PM and resides/*low*
Enhanced energy efficiency: existing basic material processes	Commercial	Enhanced fuel efficiency, total 2.3	Developing countries can have low energy efficiency due to lack of incentive and/or expertise	**Low**	Potential reduction in air emissions, water effluents and wastes/*low*
Steam systems (required for many industries)	Commercial		For most industries not a major cost; lack of expertise for some industries	**Low**	Reduction in coal emissions: Sox, NOx and PM and residues/*low*

Materials/product efficiency	First generation: commercial	Little incentive to minimize the CO$_2$ "content" of materials and products; life cycle analyses required	Medium, conduct life cycle analyses of key materials and products with the aim of minimizing CO$_2$ "content"	Potential reduction in air emissions, water effluents and wastes, depending on substitute material/*medium*
Cogeneration (combined heat and power)	Commercial	Limited by electric grid access that would allow the ability to feed electricity back to grid also high capital costs	*Low*	Reduction in coal emissions: Sox, NOx and PM and residues/*low*
Enhanced energy efficiency: new basic material processes	Developmental to Near-commercial depending on industry	New, innovative production processes require major RD&D and would need reasonable payback to replace more C intensive processes	**Medium/High**, Develop and demonstrate less carbon intensive production processes for key industries	Potential reduction in air emissions, water effluents and wastes, depending on new process/*high*
Fuel substitution in basic materials production	Commercial 1.2	Natural gas substitution for oil and coal can be expensive	Low	Unclear, environmental studies useful/*high*
Feedstock substitution in key industries	Commercial	Biomass and bioplastics can substitute for petroleum feedstocks and products, however cost high and availability low	**Medium**, develop affordable substitute feedstocks and products based on biomass	Unclear, environmental studies useful, depends on feedstock and process/*high*

Major reductions can result from lower emissions both on the generation side and on the user side as a result of lower usage via enhanced end use efficiency. Table 1.1 presents a summary of major generation options that offer significant opportunities for CO_2 mitigation. They are presented in the order of highest potential for CO_2 mitigation consistent with the IEA Blue scenario. Included in this and the subsequent tables are the IEA projected CO_2 savings for each technology in Gt of CO_2 in 2050 for the Blue scenario. Also included is information regarding potential environmental issues assuming wide scale deployment of the given technology, and the relative priority of environmental characterization and risk management research to understand and minimize these problems. Priority judgments were based on the potential magnitude of the environmental impacts and the relative availability of information on the magnitude and the mitigation potential of such impacts.

Key generation technologies include nuclear power, natural gas/combined cycle, and three coal combustion/capture technologies – Integrated Gasification Combined Cycle (IGCC), pulverized coal/oxygen combustion, and conventional pulverized coal – all with integrated CO_2 capture and underground storage. Figure 1.29 illustrates the major components of each capture technology. IGCC technology is the primary focus of the U.S. RD&D program. But this technology requires complex chemical processing and pure oxygen for the gasification process, and it cannot be readily retrofitted to existing plants. Oxy-combustion systems also require pure oxygen for combustion but are less complex and have the potential for retrofitting existing plants. CO_2 removal via scrubbing, adsorption, or membrane separation is conceptually simple and inherently retrofittable but is at

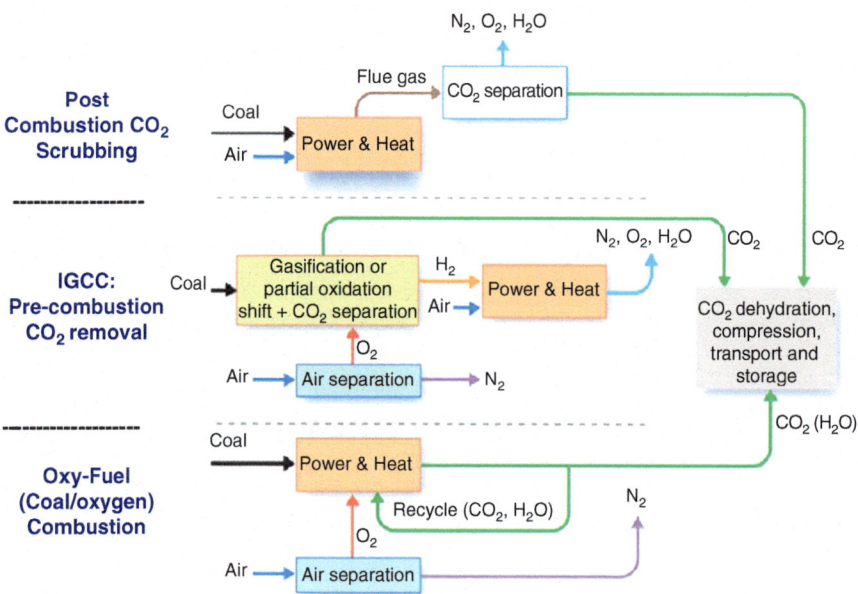

Fig. 1.29 Three key technologies capturing CO_2 from coal-fired power plants

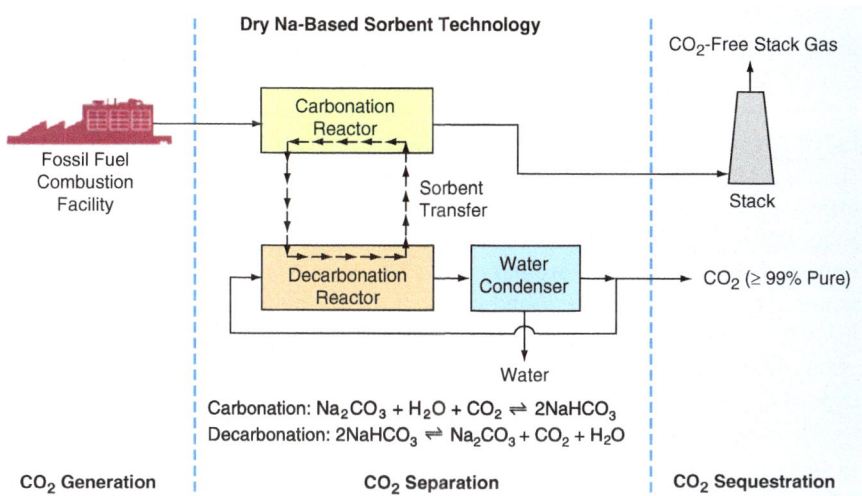

Fig. 1.30 RTI's Dry carbonate process for CO_2 capture

an early development stage; commercial amine scrubbers use large quantities of energy for sorbent regeneration and are expensive. Figure 1.30 schematically depicts a promising CO_2 capture technology under development by Research Triangle Institute (RTI). The Department of Energy has sponsored small pilot testing at EPA's Office of Research and Development's (ORD) Multi-pollutant Combustion Research facility. Early pilot testing results showed high CO_2 capture and efficient sorbent regeneration.

MIT [4] conducted an in-depth study of coal in a carbon-constrained world and concluded that: "... CO_2 capture and sequestration is the critical enabling technology that would reduce CO_2 emissions significantly while also allowing coal to meet the world's pressing energy needs." They concluded that current research funding is inadequate and "what is needed is to demonstrate an integrated system of capture, transportation, and storage of CO_2, at (appropriate) scale."

With the exception of wind power, renewable technologies are not projected by IEA [7] to have major mitigation impacts for the ACT scenario in the 2050 time frame. In the case of solar generation, both photovoltaic and concentrating technologies are currently prohibitively expensive. However, the Blue scenario assumes major improvements and cost reductions for both solar technologies, allowing them to play a major role in low carbon power generation before 2050. For biomass, major utilization is projected to be limited by its dispersed nature, its low energy density, and competition for the limited resource in the transportation sector.

An important factor that must be taking into account when considering a major restructuring of the power sector is the importance of water availability, needed for cooling and various unit operations and processes. Water supply in some regions could be compromised by climate change and scarce supplies would have to be shared with municipal, industrial, and agricultural users.

The author rates RD&D needs in the power generation sector critical, especially in the area of CCS and for the next generation of nuclear power plants. All three capture technologies described above warrant aggressive RD&D programs. The author concurs with MIT [4], that there are too many uncertainties with regard to IGCC to limit RD&D focus to that technology alone. Therefore, more emphasis should be placed on pulverized coal/oxygen (oxy-fuel) combustion, and high efficiency pulverized coal with CO_2 flue gas capture technology. Underground sequestration will be needed for each of these technologies and is in the developmental stage, with extraordinary potential. However, there are a host of economic, environmental, safety and efficacy questions that can only be resolved through a major program with a particular focus on demonstrations for the key geological formations most applicable to the greatest potential storage capacity.

An example of an important sequestration environmental issue is the potential of such operations to adversely impact drinking water sources. While CO_2 itself is not toxic, it could change subsurface geochemical conditions in such a way that toxic metals, such as arsenic, could be released into groundwater. In addition, impurities in the captured CO_2 stream could also impact drinking water quality. Because of these potential impacts and the likely large areas of the subsurface impacted by such sequestration if applied on a wide scale, this issue should be given a high research priority.

MIT [4] estimates that three full-scale CCS projects in the United States and ten worldwide are needed to cover the range of likely accessible geologies for large-scale storage.

For the next generation of nuclear reactors, the technology is quite promising and could start making a major impact by 2030. However, there needs to be a number of successful demonstrations to allow for resolution of remaining technical problems and to instill confidence in the utility industry that the technology is affordable and reliable, and to the public, that it is safe.

Given the resource, environmental and sustainability challenges associated with fossil fuel and nuclear power generation technologies, it would be highly desirable to generate all required electricity from affordable renewable resources. Therefore, major technological development efforts, should be focused on enhancing performance and reducing costs for wind power, both on-shore and offshore, and both solar generation technologies.

It should be noted that subsequent chapters discuss power generation mitigation in more detail: Chap. 2: Coal and Coal/Based Power Generation, Chap. 4: The Role of Nuclear Power: No Free Lunch, Chap. 5: Renewable Energy: Status and Prospects. Note that further consideration of the environmental implications of emerging technologies is discussed in Chap. 12: Potential Adverse Impacts of Greenhouse Gas Mitigation Strategies. Also, recognizing the importance of timely mitigation in the rapidly growing economies, Chap. 11 analyzes the climate change mitigation challenge unique to China, India, and Mexico. It is titled, The Role of Technology in Mitigating GHG Emissions from the Power Sector: the Case of China, India, and Mexico.

1.10   Building Sector

The building sector utilizes large quantities of electricity and fossil fuels directly and is expected to increase CO_2 emissions for the next several decades at about 2% per year [7]. Figure 1.21, illustrates the importance of this sector in the United States, with commercial and residential buildings contributing 27% to national greenhouse gas emissions via use of electricity and direct use of fossil fuels, mostly natural gas and oil. Table 1.2 summarizes major technologies capable of achieving significant reductions in CO_2 generation in the 2,050 time frame. The technologies are divided into two categories: (1) heating and cooling and (2) appliances, which include lighting.

For each of the two categories, the technologies are listed in the order of their potential impact in 2050 according to IEA for the Blue scenario. The technologies are either aimed at enhancing end use efficiency or are new alternative building heating/cooling technologies. It is important to note that those high-efficiency appliances and heating and cooling technologies are currently commercial, although there is potential for even higher efficiencies assuming a focused, successful research program. Lack of incentive and higher initial costs are the primary reasons for the slow rate of utilization. This is in contrast to the power generation sector, which is constrained by unavailable or undemonstrated technology.

Note that Chap. 9, Buildings Mitigation Opportunities with Health Issues Considered, analyzes major building efficiency options with a focus on energy efficiency/indoor air quality tradeoffs.

1.11   Transportation Sector

Emissions from the transportation sector are growing at a fast rate, and are projected to grow at 2% per year globally [7], driven by developing countries such as China and India, with a combined population of 2.5 billion. It is second only to the power generation sector in importance for the foreseeable future. There are two major technology categories: vehicles and fuels. Technology is currently commercially available capable of major reductions in CO_2 emissions per mile traveled, especially for light-duty vehicles. Table 1.3 summarizes the status of major technologies. Again, for each of the two categories, the technologies are listed in the order of their potential impact in 2050 according to IEA's Blue Scenario. The first two rows illustrate that major CO_2 reductions could be achieved by incorporating the most efficient internal combustion, chassis, A/C, and tire components. Also, hybrid technology, if optimized for efficiency and utilized with high-efficiency chassis components, can have a substantial positive impact. The main impediment to more robust utilization of these commercially available technologies appears to be higher initial costs for hybrids and buyer preferences that, in North America and more recently in Europe, are for larger, heavier, less-efficient vehicles. To the extent vehicle efficiency can be improved and renewable fuel options developed; major savings can be realized in the transportation sector.

IEA [6, 7] projected that substantial quantities of CO_2 will be emitted by gas and coal to liquid processes in what they refer to as the energy transformation sector as demand for oil exceeds global petroleum and natural gas extraction capability. Such processes would produce fuels primarily for the transportation sector. Such processes generating liquid fuels from tar sands and oil shale would be major emitters as well, unless the CO_2 is sequestered. In addition to concerns about large CO_2 emissions, such processes have the potential of generating large quantities of air and water pollutants and hazardous wastes, yielding serious environmental impacts. However, improvements in vehicle and engine technology to enhance conversion efficiency will lessen the need for such carbon intensive energy transformation processes.

Of all the biomass processes, thermo-chemical processes that can convert biomass to bio-diesel or other transportation fuels using gasification, pyrolysis, or Fischer-Tropsch technology, appear to have the most potential for CO_2 mitigation and should be considered for an aggressive RD&D program.

In addition, ethanol production by biochemical processing of biomass offers the potential for large-scale displacement of gasoline. However, breakthroughs will be necessary in the ability to chemically break down major biomass components to sugar for fermentation to produce ethanol.

Hydrogen/fuel cell vehicle technology is still in the development stage, since the fuel cell stack still has limitations in terms of cost and longevity, and hydrogen storage in vehicles remains problematical. Also, EPA [16] and IEA [7] assessments suggest that CO_2 savings would not be substantial unless or until the hydrogen could be generated from low-emission, renewable sources.

Despite the serious technical issues, in light of the ultimate potential of fuel cell/hydrogen and biochemical ethanol, the author believes both are also strong candidates for an aggressive RD&D focus with the aim of breakthrough technology.

Note that for biomass to make significant contributions to climate mitigation, thousands of square miles of dedicated plantations will be necessary. It will be important to ensure that such plantings are configured and maintained to minimize environmental damage by avoiding depletion of aquifers, pollution of surface and groundwater supplies, and degradation of soil quality. It is also necessary to understand the potential for excessive water utilization, especially in water stressed areas. Finally, there must be some level of confidence that such plantations will maintain their productivity as the climate changes in the decades ahead and that adverse impact on food production is avoided.

Note that Chap. 3: Coal and Biomass to Liquid Fuels, evaluates biofuels production technologies and Chap. 6: Mobile Source Mitigation Opportunities, discusses evolving low carbon vehicle propulsion technologies.

1.12   Industrial Sector

CO_2 emissions from the industrial sector are projected to grow at an annual rate of 1% per year over the next several decades (IEA, 7). Table 1.4 summarizes major technologies applicable to this sector. Although CO_2 emission control can be specific to a particular industry, there are a number of technologies that can be applied to a large fraction of the industrial sector. Technologies, which are generally applicable, include more efficient motors and steam generators and enhanced use of cogeneration technology; all are commercially available and offer the potential for major reductions. For the larger, more energy intensive industries such as cement kilns, ammonia production, and blast furnaces, CCS also offers the potential for mitigating large quantities of CO_2. However, as discussed earlier, CCS is in the early developmental stage with a host of questions that can only be resolved through a major program with a particular focus on demonstrations for key geological formations.

Developing and deploying new or modified industrial production processes can also yield important CO_2 emission mitigation potential. Processes can be modified to utilize more environmentally friendly feedstocks, or fundamentally new basic material processes can be introduced with inherently lower energy intensity.

Another approach that has potential is to encourage utilization of products which have lower CO_2 "content" (i.e., require less carbon intensive energy during the production, use, and disposal). These could be considered "climate-friendly" products. There is currently no incentive to use such products. Also, comprehensive life cycle analyses would be necessary to quantify product CO_2 "content."

More detail on the industrial mitigation challenges, with a focus on the cement industry, is available in Chap. 8, Reduction of Multi-pollutant Emissions from Industrial Sectors.

1.13   Geoengineering Options

Finally, there have been various geoengineering approaches suggested that could potentially slow warming until new energy technologies are developed and deployed. These options involve intentional, *direct manipulation of the earth's energy balance through interventions at the planetary scale.* For example, Wigley [19] suggested simulating volcanoes, which are known to cool the planet after high altitude eruptions, by purposely emitting large quantities of sulfate particles into the stratosphere. The objective would be to reflect incoming solar radiation. Such approaches are early in their conceptualization and would have to be carefully evaluated for their economic and environmental impacts.

Note that Chap. 9, Geoengineering: Direct Mitigation of Climate Warming, provides an overview of the more widely discussed project proposals, and highlights the critical uncertainties.

1.14   RD&D

IEA [6, 7], Hawksworth [8], Morgan [15], MIT [4], Princiotta [20], and the American Energy Innovation Council, AEIC, [21] have observed that RD&D funding in the energy area will need to be substantially increased to accelerate deployment and utilization of key technologies. As illustrated earlier, the later a mitigation program is initiated, the more severe emission cuts will need to be if CO_2 concentrations above 500 ppm are to be avoided. The Stern Report [22] concluded: "...support for energy R&D should at least double, and support for the deployment of new low-carbon technologies should increase up to fivefold." IEA [7] reviewed several references and concluded the range of increase for RD&D required was between 2 and 10 over current levels. IEA estimates a total of about $14 trillion of RD&D *plus deployment* would be required for their Blue scenario. Deployment costs are those costs that would allow construction and operation of near commercial technologies with the aim of improving performance and lowering the cost differential relative to the high carbon emission technology it would displace. Most recently, the AEIC [21], a council comprised of world class business leaders, called for US federal expenditures of "...$16 billion per year – an increase of $11 billion over current annual investments of about $5 billion ... the minimum level required."

It is important that such RD&D be conducted at both the federal and private sector levels. Federal funding is particularly relevant for those technologies that require substantial funding due to high capital costs and have a low probability of commercial impact and profitability in the near term. Examples include carbon capture and storage and next generation nuclear power technologies. Private sector funding for the developing lower cost, lower risk technologies could be encouraged by providing incentives, such as a significant price on carbon emissions.

Figure 1.31, generated from IEA data [23], depict IEA countries' public research expenditures in critical energy technology areas. It illustrates the relatively low

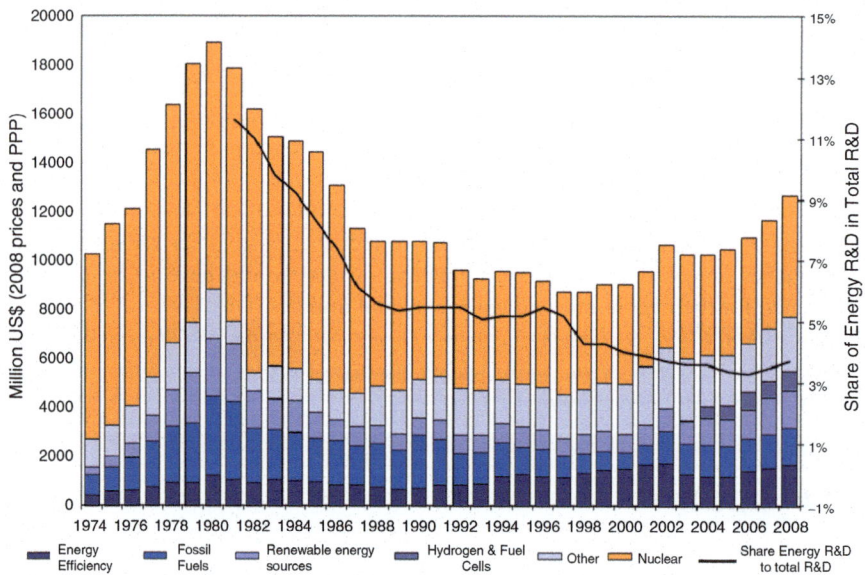

Fig. 1.31 IEA countries' public RD&D expenditures for key energy sectors, 2008 U.S. $ (millions)

funding in recent years and the major funding reductions since the major funding increases in the 1970s, which were motivated by the 1967 oil embargo. It should be recognized that, in the last few years, the United States has redirected some of its research resources to some key technologies, especially hydrogen/fuel cells, IGCC, carbon capture and storage, and biomass-to-ethanol technologies. The United States has coordinated its efforts in this area through the Climate Change Technology Program, CCTP [24]. Within the constraint of current budget priorities, the CCTP has coordinated a diversified portfolio of advanced technology research, development, demonstration, and deployment projects, focusing on energy efficiency enhancements; low-GHG-emission energy supply technologies; carbon capture, storage, and sequestration methods; and technologies to reduce emissions of non-CO_2 gases. The key agency responsible for CCTP related research is the Department of Energy, with about 86% of fiscal year 2008 CCTP funding. As part of this program, USEPA [25] is implementing a series of voluntary programs that encourage the reduction of greenhouse gas emissions, including Energy Star for the building sector, transportation programs, and non-CO_2 emission reduction programs in collaboration with industry. These programs, with their focus on conservation and low GHG technologies, could provide a foundation for an expanded program consistent with the mitigation challenge.

There have been two recent developments which have enhanced GHG mitigation technology R, D, D&D in the US. In FY 2009, stimulus funding through the American Recovery and Reinvestment Act supplemented funding in the energy area, with a one-time budget supplement of $6.8 billion. In addition, the creation of the Advanced Research Projects Agency-Energy (ARPA-E) has provided a significant boost to innovation in the energy technology area. ARPA-E focuses exclusively on high-risk, high payoff technologies that can change the way energy is generated, stored, and used, and has challenged innovators to come up with novel ideas with the aim of ultimately yielding breakthrough technologies. The program has the potential to accelerate development of breakthrough technologies, if funding levels are consistent with the challenge. The AEIC [21] has recommended annual funding levels for ARPA-E of $1billion.

Figure 1.32 depicts the same technologies as Fig. 1.25, with their contribution to CO_2 avoidance in 2050 for the Blue scenario, but characterizes each technology into high, medium, and low research priority categories. This is based on the author's judgment regarding the potential contribution to CO_2 avoidance each technology can achieve with an accelerated research, development, demonstration, and deployment program. It is noteworthy that for the coal generation sector, these priorities are consistent with MIT [4], which has conducted the most in-depth study of this critical energy source. Technologies earliest in their development cycle and having the greatest potential for major mitigation are ranked highest.

As indicated in the last column of Tables 1.1, 1.2 and 1.4, many of these technologies have the potential for significant environmental impacts via ecosystem damage and/or emissions/effluents to the air, water, and land. Therefore, a parallel research program to better understand such impacts for key technologies is indicated. Figure 1.33, which again is based on the IEA Blue technologies, indicates

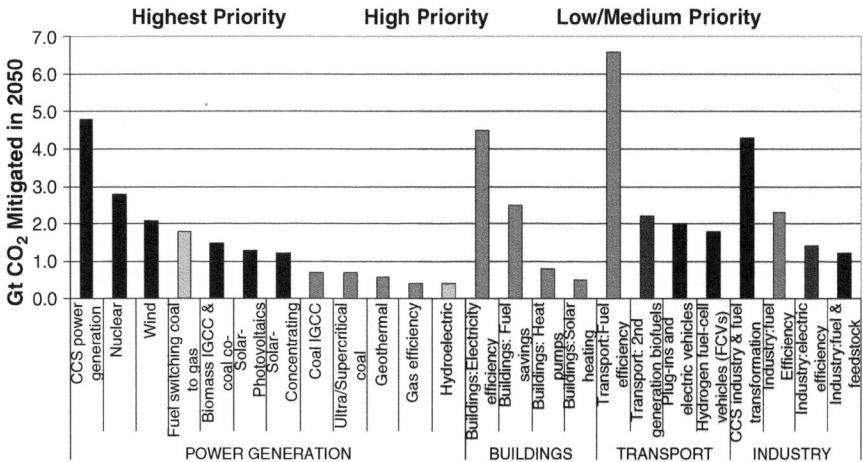

Fig. 1.32 Author's RD&D priorities to achieve Blue Scenario's CO_2 Avoidance Goal in 2050

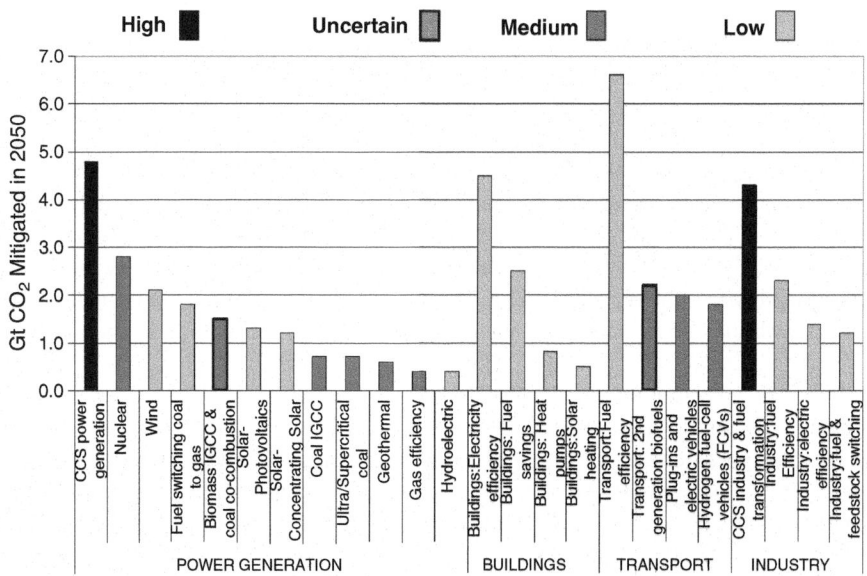

Fig. 1.33 Author's assessment of the potential environmental impacts of mitigation technologies for the Blue scenario

the author's judgment regarding the potential magnitude of environmental impacts, assuming wide scale utilization. Such a judgment involves consideration of the potential scope and impact of environmental/health impacts and the current knowledge on the quantification and potential mitigation of such impacts. As shown, advanced coal and biomass technologies are among those with the potential for major impacts and should be the focus of a comprehensive environmental assessment research program.

It should be noted that all the transportation technologies offer the potential for reducing U.S. dependence on foreign oil. Further, the countries that can bring these technologies to market first have the potential for major revenue streams from a multi-billion dollar international market.

As mentioned earlier, further consideration of the environmental implications of emerging technologies is discussed in Chap. 12: Potential Adverse Impacts of Greenhouse Gas Mitigation Strategies.

1.15 Summary and Conclusions

- **Greenhouse gas emissions** along with other deleterious anthropogenic impacts to Earth are **driven by an ever-increasing demand for human products and services.** This is further exacerbated by a growing population. Such trends challenge the long-term sustainability of modern civilization. **The monumental mitigation technology challenge would be substantially moderated, if human needs were downscaled and/or population growth was reduced**.
- Concentrations of CO_2 have increased to 383 ppm from a pre-industrial value of 278 ppm. This increase is due to anthropogenic emissions of CO_2 that can remain in the atmosphere more than 100 years. There is close to a scientific consensus that **much if not all of the nearly 0.8°C global warming seen since the pre-industrial era is a result of increased concentrations of CO_2 and other greenhouse gases**.
- **Global emissions of carbon dioxide have been accelerating at a rate of about 1.4% per year in the 1992–2002 time period. However, recent data suggests an acceleration of emission growth in recent years; 3.25% in the 2000–2008 period.** China's major expansion of its coal-fired power generation capacity has been the key factor in this acceleration in growth rate. **It will not be possible to have an effective global mitigation program without a serious commitment by the major emerging economies (e.g., China, India, and Brazil).**
- Projections of warming have been made on a credible business-as-usual case based on IEA [7] projections extended to 2100. This base case assumes a global annual growth rate of 1.6% in the next 25 years. Under this assumption, CO_2 concentration is projected to increase to 500 ppm in 2050 and 825 ppm by 2100. **Such concentrations will yield best estimate average warming since the pre-industrial era, of 1.9°C in 2050 and 3.9°C in 2100.**
- There **is still a large range of uncertainty associated with these warming projections**; the potential warming in 2100, since the pre-industrial era, could be as high as 4.9°C or as low as 2.5°C. Warming would continue into the next century, with equilibrium warming in the 2.7–10.5°C range, with the best estimate at 5.2°C above 1990 levels.

– If current worldwide emission trends continue and grow at 3% per year until 2030 before moderating, then projected warming, and potential consequences, would be substantially higher. This scenario will yield a **best-estimate average warming, relative to 1990, of 2.2°C in 2050 and 4.8°C in 2100.** Warming would continue into the next century, with equilibrium warming in the 3.4–13.2°C range, with the best estimate at 6.6°C above pre-industrial levels.

– **It is too late to prevent substantial additional warming; the most that can be achieved would be to moderate the projected warming.** The best result that appears achievable with CO_2 mitigation, assuming a major energy technology retooling, would be to constrain warming by about 2.4°C above pre-industrial levels (range 1.7°C and 3.1°C) by 2100. If other GHGs are aggressively mitigated, warming could be constrained to about 1.8°C. It is significant that an aggressive methane mitigation program could contribute in the order of an additional 0.3°C of warming avoidance. Aggressive mitigation of N_2O and ozone precursors can yield an additional 0.3°C warming reduction. Although less certain, control of black carbon emissions from key combustion sources might also provide comparable potential mitigation.

– Global impacts even for this constrained warming scenario are potentially serious. This suggests that the world community may have no remaining alternative other than to **pursue both mitigation and adaptation approaches aggressively**.

– In order to limit warming to about 2.5 ± 0.7°C utilizing CO_2 emission mitigation, it will be necessary for the world community to **decrease annual emissions at a rate of between 2% and 3% per year for the rest of the century**. The earlier the mitigation program starts, the less drastic the annual reductions would need to be. Since the base case assumes a roughly 1.6% positive growth rate, approximately one trillion tons of carbon (3.7 trillion tons of CO_2) will have to be mitigated by 2100 relative to the base case. This would be **an historic challenge**. Never has the world community had to face the prospects of fundamental energy production and utilization transformations to such an extent and at such a pace.

– Recent publications were used to relate the implications of a 4 trillion-ton CO2 mitigation program needed, along with aggressive mitigation of the other GHGs, to constrain warming below 2°C, to the key energy sectors and the technologies within those sectors that can contribute to the major mitigation challenge. It is concluded that an aggressive, cost effective mitigation program relying on *existing* **technologies is capable of mitigating only between about 25% and 45% of the required CO_2**, depending on projected business as usual CO_2 growth rates. Therefore, in the absence of fundamental lifestyle changes, new technologies are required for the key energy-related sectors: power generation, transportation, industrial production, and buildings. **The power-generation sector and transportation sectors are particularly important, since they are projected to grow at relatively high rates, driven especially by China and other actively developing countries.**

– The **power-generation sector, which has been growing at 3.7% annually in the 2000–2008 period.** Since the key source of emissions from this sector is coal combustion, it is critically important to develop affordable CO_2 mitigation technologies for such sources and to develop economical alternatives to coal-based power generation. CCS offers the potential to allow coal use while at the same time mitigating CO_2 emissions. The three major candidates for affordable CO_2 capture are: PC boilers with advanced CO_2 scrubbing, IGCC with carbon capture, and oxygen-fed (oxy-fuel) combustors. Of the three, only IGCC is being funded at levels approaching those needed. However, all three approaches rely on underground sequestration, an unproven technology at the scale required for coal-fired boilers, with many serious cost, efficacy, environmental, and safety issues. Nuclear power plants, natural gas/combined cycle plants, wind turbines and solar generators all have the potential to decrease dependence on coal use and make significant contributions to CO_2 avoidance. An accelerated RD&D program is particularly important for advanced nuclear reactors, since serious safety, proliferation, and waste-disposal concerns remain, and for solar power systems given their long-term potential if costs can be substantially reduced.

- The **building sector,** projected to grow globally at about 2% per year, is where much of the generated electricity is utilized and where there are many currently available technologies that can significantly reduce the use of electricity and other energy sources, with a corresponding decrease in CO_2 emissions. The constraints here are less technological and more socioeconomic. However, to the extent RD&D can lower cost and raise efficiency of building components, it can help provide extra incentive for building owners to invest in the most efficient heating and cooling systems, lighting, and appliances.
- Emissions from the **transportation sector** are projected to grow at 2% per year. The challenge in this sector is two-fold. The first challenge is that current propulsion systems all depend on fossil fuels with their associated CO_2 emissions, suggesting that technologies based on renewable sources, such as biomass, would be important. The second challenge is that the automobile industry, driven by consumer preferences (especially in North America), have offered heavy, inefficient vehicles such as sport utility vehicles. A review of developing technologies suggests that hybrid and plug-in hybrid vehicles and biomass-to-diesel fuel via thermo chemical processing are the most promising in the near term. However, cellulosic biomass-to-ethanol and hydrogen/fuel cell vehicles offer longer-term potential, if key technical, economic, and environmental issues are resolved and, in the case of hydrogen, renewable sources are developed.
- **Industrial sector** emissions are projected to grow at an annual rate of 1%. Although CO_2 emission avoidance approaches can be specific to a particular industry, the following key commercial technologies can be applied to a large fraction of the industrial sector: efficient motors, steam generators, and enhanced use of cogeneration technology. For the larger, more energy-intensive industries such as blast furnaces, CCS offers the potential for mitigating large quantities of CO_2. Developing and deploying new or modified industrial production processes can also yield important CO_2 emission mitigation potential. Another attractive

approach is to encourage utilization of products that have a lower life-cycle CO_2 content (i.e., require less carbon intensive energy during product production, use, and disposal).

– If near-term mitigation of four trillion tons of CO_2 is deemed a serious goal, **a major increase in RD&D resources will be needed. Current CO_2 mitigation research expenditures** in the United States and globally have been relatively flat in recent years, and the U.S. federal research expenditures on energy technologies are 70% lower than research expenditures in response to oil shortages in the mid-1970's. U.S. private sector research has fallen even more precipitously in recent years. It is important that such RD&D be conducted at both the federal and private sector levels. Federal funding is particularly relevant for those technologies that require substantial funding due to high capital costs and/ or have a low probability of near term commercial impact and profitability. Examples include carbon capture and storage, and the next generation nuclear power technologies. Private sector funding for the lower cost, lower risk technologies could be encouraged by providing incentives, such as regulatory drivers and meaningful carbon prices. Technology research, development, and demonstration are of particular importance for coal generation technologies: IGCC, oxygen coal combustion, and CO_2 capture technology for pulverized coal combustors. All of these technologies will have to be integrated with underground storage, a potentially breakthrough technology, but one which is at an early stage of development and faces environmental and cost issues. Also important are next generation nuclear power plants, solar technologies, biomass to diesel fuel processes, cellulosic biomass-to-ethanol production technology, and hydrogen production technology. Given their potential for wide scale utilization, all of these emerging technologies must evolve with full consideration of the tradeoffs associated with their unique environmental characteristics and their carbon mitigation potential, economics, safety, etc. Toward this end, concurrent research to assess potential environmental impacts and to identify risk management approaches is needed.

– Given the monumental challenge and uncertainties associated with a major mitigation program, it may be **prudent to consider all available and emerging technologies**. This suggests that fundamental research on energy technologies in addition to those currently in advanced stage of development, be part of the global research portfolio, since breakthroughs on today's leading edge technologies could yield tomorrow's alternatives. In addition, it is the author's opinion that it is prudent to consider geoengineering options, which although radical in concept, could potentially buy the time we may need to make the necessary adjustments in our energy and industrial infrastructure.

– Finally, availability of key technologies will be necessary but not sufficient to limit CO_2 emissions. Since many of these technologies have higher costs and/or greater operational uncertainties than currently available carbon intensive technologies, **robust regulatory/incentive programs will be necessary to encourage their utilization.**

References

1. Intergovernmental Panel on Climate Change (2007) Climate change 2007: the physical science basis, summary for policymakers, fourth assessment report of the Intergovernmental Panel on Climate Change (IPCC), Geneva, Switzerland. www.ipcc.ch/spm2feb07.pdf
2. World Resources Institute (2006) Climate Analysis Indicators Tool (CAIT) on-line database version 3.0. WRI, Washington, DC. http://cait.wri.org
3. Raupach MR et al (2007) Global and regional drivers of accelerating CO_2 emissions. In: Proceedings of the National Academy of Sciences of the United States of America, 22 May 2007. www.pnas.orgcgidoi10.1073pnas.0700609104
4. Massachusetts Institute of Technology (2007) The future of coal: an interdisciplinary MIT study. MIT, Cambridge
5. Le Quere M et al (2009) Trends in the sources and sinks of carbon dioxide. Nature Geosci 2:831–836
6. International Energy Agency (IEA) (2006) Energy technology perspectives 2006. Organization for Economic Cooperation and Development, IEA, Paris
7. International Energy Agency (IEA) (2008) Energy technology perspectives 2008. Organization for Economic Cooperation and Development, IEA, Paris
8. Hawksworth J (2006) The world in 2050: implications of global growth for carbon emissions and climate change policy. PriceWaterhouseCoopers, London
9. Wigley TML (2008) MAGICC/SCENGEN 5.3 user manual; MAGICC (Model for the Assessment of Greenhouse gas Induced Climate Change) can be downloaded from www.cgd. ucar.edu/cas/wigley/magicc/index.htm
10. Intergovernmental Panel on Climate Change(2007) Climate change 2007, impacts, adaptation and vulnerability, summary for policymakers, fourth assessment report of the Intergovernmental Panel on Climate Change (IPCC), Geneva. www.ipcc.ch/SPM13apr07.pdf
11. World Resources Institute (2007) U.S. GHG emissions flow chart. WWI, Washington, DC. www.cait.wri.org/figures/US-FlowChart.pdf
12. Princeton University Woodrow Wilson School of Public and International Affairs (2009) Black carbon a review and policy recommendations
13. Enkvist P, Naucler T, Rosander J (2007) A cost curve for greenhouse gas reduction. McKinsey Quarterly, http://www.epa.gov/oar/caaac/coaltech/2007_05_mckinsey.pdf
14. Pacala S, Socolow R (2004) Stabilization wedges: solving the climate problem for the next 50 years with current technologies. Science 305:968–972
15. Morgan G, Apt J, Lave L (June 2005) The U.S. electric power sector and climate change mitigation. The Pew Center on Climate Change, Arlington
16. U.S. Environmental Protection Agency, Yeh S, Loughlin D, Shay C, Gage C (2007) An integrated assessment of the impacts of hydrogen economy on transportation, energy use and air emissions. In: Proceeding of the IEEE Special Issue: Hydrogen Economy, 28 June 2007. http://pubs.its.ucdavis.edu/publication_detail.php?id=1110
17. U.S. Environmental Protection Agency, DeCarolis J, Shay C, Vijay S (2007), The potential mid-term role of nuclear power in the United States: a scenario analysis using MARKAL, ISBN 8188342815, In: Mathur J, Wagner H-J and Bansal NK (eds) Energy security, climate change and sustainable development. Anamaya, New Delhi 117–130
18. The Center for Global Action, Updated Carbon Monitoring for Action (CARMA) database. Available at: http://www.cgdev.org/content/article/detail/16578/, 2008
19. Wigley TM (2006) A combined mitigation/geoengineering approach to climate stabilization. Science 314(5798):452–454
20. Princiotta FT (1998) Renewable technologies and their role in mitigating greenhouse gas warming, US-Dutch symposium: facing the air pollution agenda for the 21st century, vol. 72. Elsevier Science Publishers
21. American Energy Innovation Council (2010) A business plan for America's energy future. http://www.americanenergyinnovation.org/full-report

22. Stern N (2006) Stern review on the economics of climate change, the Stern review, 2006. Pre-publication version. www.hmtreasury.gov.uk/independent_reviews/stern_review_ economics_climate_change/stern_review_report.cfm
23. International Energy Agency, RD&D budgets. www.iea.org/RDD/TableViewer/
24. Climate Change Technology Program. www.climatetechnology.gov/.
25. U.S. Environmental Protection Agency, Current, and near term greenhouse gas reduction initiative. www.epa.gov/climatechange/policy/neartermghgreduction.html

Chapter 2
Coal and Coal/Biomass-Based Power Generation

James R. Katzer

Abstract Coal is a key, growing component in power generation globally. It generates 50% of U.S. electricity, and criteria emissions from coal-based power generation are being reduced. However, CO_2 emissions management has become central to coal's future. To meet growing electricity demand, coal use is expected to increase in the foreseeable future because it is cheap and abundant. For this to happen CO_2 capture and geologic sequestration (CCS) is a critical technology. With CCS, coal-based power generation can be made much cleaner. Commercial demonstration of existing technologies, including CCS, with the resultant improvements that will accrue, is key to advancing coal-based power generation and addressing important environmental issues.

2.1 Introduction

Coal is used to generate 50% of U.S. electricity and about 40% of the electricity produced globally [1]. For China and India, the fraction of power that is based on coal is about 77% and 74% respectively [2], and it is growing. Because of its history, coal-based power generation in the absence of adequate controls can be a major emitter of air pollutants and thus is perceived as being dirty. CO_2 emissions from coal-based power generation have now also become a major concern. This chapter addresses both of these issues but focuses on CO_2 emissions and the implications to global climate change.

Total global CO_2 emissions from coal-based power generation totaled almost 7.5 billion tonnes in 2007; this is about 30% of the global fossil-related CO_2 emissions.

The findings included in this chapter do not necessarily reflect the view or policies of the Environmental Protection Agency. Mention of trade names or commercial products does not constitute Agency endorsement or recommendation for use.

J.R. Katzer (✉)
CBE, Iowa State University, Ames, IA 50012, USA
e-mail: jrksail@comcast.net

F.T. Princiotta (ed.), *Global Climate Change - The Technology Challenge*,
Advances in Global Change Research 38, DOI 10.1007/978-90-481-3153-2_2,
© Springer Science+Business Media B.V. 2011

Respectively, the U.S. and China emitted an estimated 5.9 and 6.7 billion tonnes of CO_2 from fossil fuel combustion and cement production, of which about 1.9 billion tonnes and 2.3 billion tonnes of CO_2 were from coal-based power generation in 2007 [3, 4]. Power plants are some of the largest single point source emitters; a typical 1,000 MW_e coal-fired power emits over 6 million tonnes of CO_2 per year. Examples of the annual CO_2 emissions for selected large power plants in several countries are given in Table 2.1 [5, 6].

Coal is a critical fuel for power generation because it is abundant and cheap – $1–$2 per million Btu, compared with $4–$12 per million Btu for natural gas and oil. Coal is also very abundant with estimated proven global reserves of about 900 billion tonnes which is equivalent to about 160 years at current production rates [7]. The three largest coal consumers – China, the U.S., and India – have about half the global reserves of coal and have limited reserves of other fossil fuels. The U.S., with about 250 billon tonnes of recoverable coal reserves, has 27% of the world total [1]. Global primary energy demand is projected to grow by just over 50% by 2030, and world electricity demand is projected to double by 2030. Given this situation, coal-based power could account for a significant portion of this growth, but that is not assured. This will require that growth continues and coal-based technologies improve significantly with respect to their environmental footprint. Figure 2.1 shows the increase

Table 2.1 CO_2 emissions from selected large coal-fired power plants

Plant name	Country	CO_2 emissions, million tonnes/year
Taichung	Taiwan	37.5
Poryong	South Korea	34.4
Tuoketuo	China	29.5
Vindhyachal	India	26.4
Hekinan	Japan	26.3
Janschwalde	Germany	24.9
Miller	USA, Alabama	18.7
Gibson	USA, Indiana	18.5
WA Parish	USA, Texas	18.3

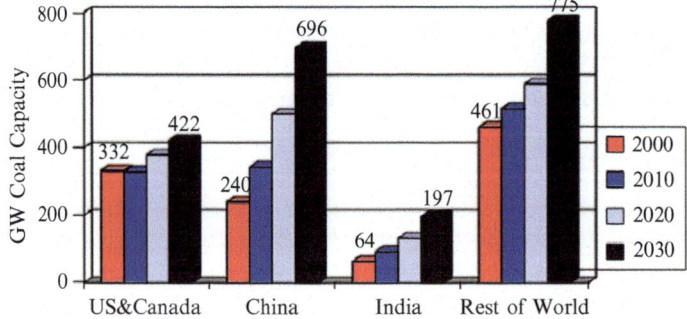

Fig. 2.1 Projected increase in coal-based power generation capacity by region to 2030 [1]

in coal generating capacity as projected by the IEA [1]. Internal Chinese projections for growth in coal-based power generation exceed IEA projections, reaching about 1,050 GW in 2030 and about 1,400 GW by 2050 [8]. Actual annual growth rate of coal-based power generation in China exceeded 20%/year from 2000 to 2007, and significant further growth is projected (Fig. 2.1). The criticality of reducing the environmental footprint of and controlling CO_2 emissions from coal-based power generation is obvious if these growth projections are to be realized. Further, most of this coal-based generation growth is in developing-world countries, which desperately need additional power generation to support economic growth, increase their standard of living, and reduce poverty.

This chapter focuses on the technologies for generating electricity from coal and on managing related emissions, particularly CO_2 emissions. It addresses the cost and performance of power generation from coal, of criteria emissions control, and of CO_2 capture and geologic sequestration (CCS) for different generating technologies. It is an update and expansion on a recent article by the author [9] and is based on a number of sources, particularly "The Future of Coal" [10], and recent comprehensive design work by Williams and co-workers at Princeton University, Princeton Environmental Institute [11]. Additional information is available in [12]. The impact of co-firing coal and biomass and of utilizing biomass alone on cost, performance, and CO_2 emissions associated with power generation is also considered. These considerations utilize the same cost and operational bases across all technologies, including those in Chap. 3, to make the comparisons as relevant as possible.

The overall approach used here involved picking a point set of design and operating conditions at which to compare technologies. The design bases for the comparisons include:

- Each unit was a Greenfield unit with 500 MW_e net generating capacity
- Each technology was designed to control criteria emissions to somewhat below today's best-demonstrated commercial performance.
- Costs were based on 2000–2006 detailed cost designs for the U.S. Gulf Coast; indexed to the mid-2007 construction cost environment as indicated by the Chemical Engineering Plant Cost Index. As indicated in Fig. 2.2, construction costs escalated rapidly from 2000, after a period of stability; mid-2007 CEPCI was a compromise level. Such rapid escalations as indicated by the HIS-CERA Index are likely not sustainable and could see self-correction when economic conditions change.
- Commercially demonstrated technologies were integrated, and cost estimates are for the Nth plant, where N is of order five to seven, for those technologies that are still evolving. This is meant to allow the learning's from the first couple of plants constructed to be engineered into future cycles of plants.
- Performance and costs are based on a single set of conditions for each technology and on the EPRI-recommended approach to calculate levelized cost of electricity (COE). Key economic and operating parameters are given in Table 2.2, and the properties of the feedstocks used, Illinois # 6, high-sulfur coal, and switch grass are given in Table 2.3.

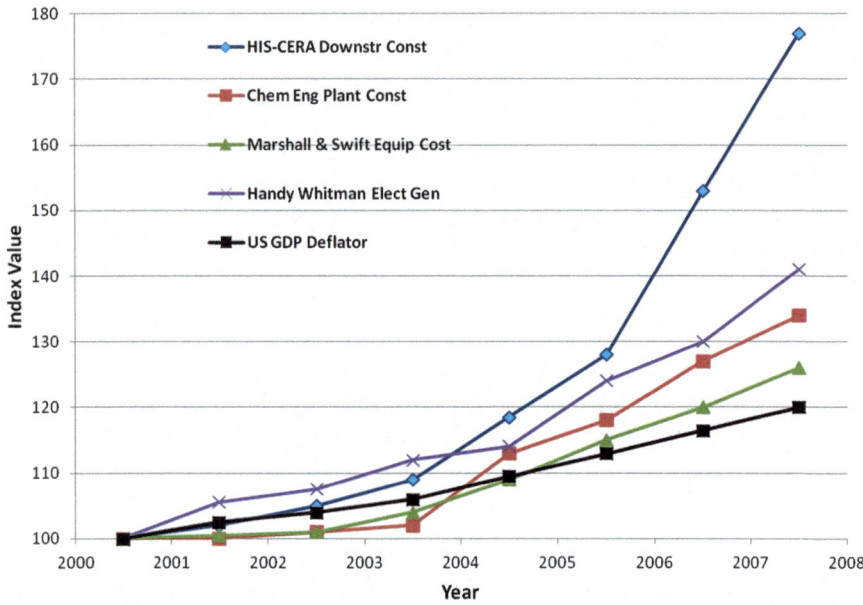

Fig. 2.2 Recent increases in construction cost as represented by several construction cost indices. Indices normalized to 100 in 2000 (Courtesy EPRI [13])

Table 2.2 Key economic and operating parameters used in developing cost comparisons [11]

Base year for capital costs, Gulf Coast	Mid-2007
Capital change rate, % of TPI per year	14.40
Interest during construction (3 year), % of TPC	7.16
O&M, % of TPC per year	4.00
Capacity factor of coal plans	85%

Table 2.3 Key parameters of feedstocks used in process analysis [11, 12]

Coal, Illinois #6, Herrin	
Coal price, $/GJ (HHV, as received)	1.71
Coal price, $/tonne, AR	46.4
HHV, MJ/kg (AR)	27.1
Wt% Carbon (AR)	63.7
Wt% Sulfur (AR)	2.51
Wt% Ash (AR)	9.7
Wt% Moisture (AR)	11.1
Biomass, Switchgrass	
Biomass price, $/GJ (HHV)	5
HHV, MJ/kg (AR)	15.9
Wt% Carbon (AR)	40
Wt% Sulfur (AR)	0.08
Wt% Ash (AR)	5.3
Wt% Moisture (AR)	15

This provides an indicative cost comparison from technology to technology. Obviously, coal type, plant site and location, dispatch strategy, and a myriad of design and operating parameter decisions will affect cost and operation but are not explored here [10]. The same comments apply to the estimates of Chap. 3 and will not be repeated there. In both chapters, production costs are presented by category so that the impact of capital cost, feedstock cost, etc., can be evaluated. The important issue is comparison among technologies, including without CO_2 capture and with CO_2 capture and geologic storage. Technology and costs for CO_2 transport and geologic storage are generation-technology independent, and costs are based on the CO_2 quantity stored.

2.2   Power Generating Technologies

2.2.1   Air-Blown Pulverized Coal (PC)

2.2.1.1   Without CO_2 Capture

A PC unit with a complete set of advanced criteria-emissions controls is shown in Fig. 2.3. It can be viewed as consisting of three blocks: the boiler block, the steam-cycle steam-turbine block, and the flue gas clean-up block as shown in Fig. 2.4. The design and operating conditions of the steam-cycle block largely determines the generating efficiency of the unit. For most existing PC units, the design and operating conditions of the steam cycle is below the critical point of water, which is referred to as subcritical operation. Operation above the critical point of water is referred to as supercritical operation. Ultra-supercritical is used to denote operation

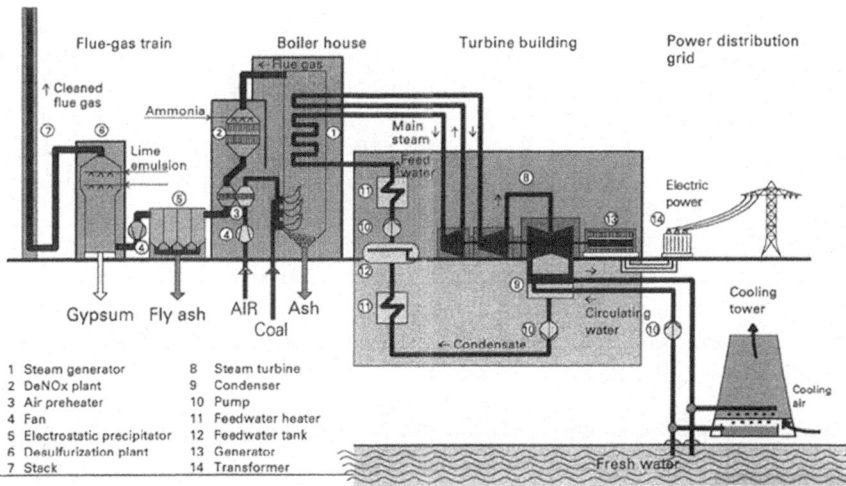

Fig. 2.3 Advanced pulverized coal unit (Courtesy ASME)

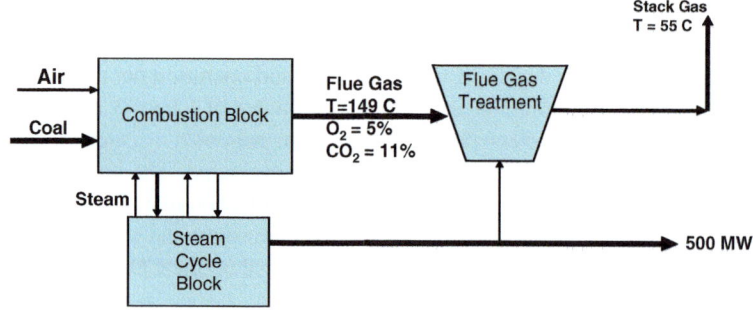

Fig. 2.4 Schematic of advanced pulverized coal unit

above a somewhat arbitrary set of operating parameters in the supercritical region. These ranges and typical generating efficiencies are summarized below.

For PC units, typical operating conditions and overall electrical generating efficiency are:

- Subcritical Unit

    - Steam-cycle operation to 550°C (1,025°F) and 2,400 psi
    - 33–37% overall generating efficiency (HHV)

- Supercritical Unit

    - Steam-cycle operation to 565°C (1,050°F) and 3,530 psi
    - 37–42% overall generating efficiency (HHV)

- Ultra-Supercritical Unit

    - Steam-cycle operation 600–620°C (1,110–1,150°F) and 4,650 psi
    - 42–45% overall generating efficiency (HHV)

Moving from subcritical to ultra-supercritical generating conditions reduces coal consumption by over 20% per kW_e-h of electricity generated. Moving from subcritical generating conditions to typical supercritical generating conditions can reduce coal consumption by over 10% per kW_e-h of electricity generated. Obviously, the higher the generating efficiency the lower the CO_2 emissions per kW_e-h of electricity generated. At a minimum, units need to be designed and operated at the highest efficiency that is economically justified to reduce CO_2 emissions. Current R&D programs are focusing on developing and proving materials and operating conditions above current ultra-supercritical conditions that could provide even higher PC generating efficiency. The next step in USCPC is 650°C (1,200°F) with generating efficiency exceeding 45%, with the next tranche being to 760°C (1,400°F) with efficiencies exceeding 48%. Materials properties, fabrication, and maintenance currently limit reaching these latter conditions.

The U.S. coal fleet consist largely subcritical generating units, with a limited number of supercritical units. Interest in supercritical technology in the U.S. has recently increased. India and China have built almost exclusively subcritical technology, but both countries have begun to construct a mix of sub and supercritical units. Meanwhile, Europe and Japan have built about a dozen ultra-supercritical units during the last decade [14]. Using modern materials technology, these units have reliability records equal to subcritical unit operation. The U.S. is behind in PC generating efficiency with a fleet average of about 33% [15].

2.2.1.2 With CO_2 Capture

A marked reduction of CO_2 emissions from PC power generation would require CO_2 capture from the flue gas, involving addition of another unit to the flue gas train. Today, the choice for CO_2 capture technology for PC generation would be amine absorption. Amine CO_2 capture is commercially proven in smaller-scale applications, including recovery of CO_2 from the flue gas of several smaller units for beverage, food and other industrial uses. The application of CO_2 capture from power plant flue gas is illustrated in Fig. 2.5. CO_2 is captured in the amine solution and then must be recovered from the solution. A large amount of energy is required to recover the CO_2 from the amine solution, regenerating the solution to capture more CO_2. A smaller amount of energy is needed to compress the CO_2 to a supercritical fluid. An energy diagram illustrating the parasitic energy requirements for CO_2 capture from a subcritical PC unit is shown in Fig. 2.6. For PC generating units that are designed for capture, the generating efficiency is reduced by about 9–11% points independent of steam cycle type. For subcritical, supercritical, and ultra-supercritical units estimated generating efficiency reductions are from 34% to 25%, from about 39% to 29%, and from 43% to 34% respectively for example. To maintain constant electrical output requires a 38–40% increase in coal consumption when a CO_2 capture designed unit is compared with a non-capture designed unit [10, 11, 13].

The energy comparison illustrated in Fig. 2.6 is for units that are designed specifically for capture or no-capture, and thus, all the components are of optimum

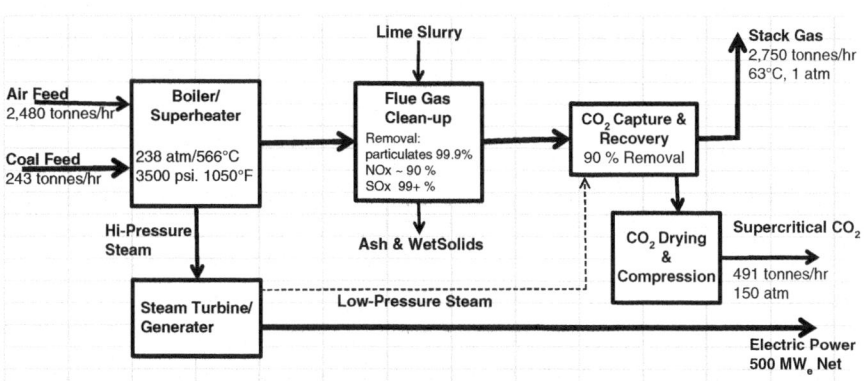

Fig. 2.5 Supercritical 500 MW$_e$ pulverized coal unit with CO_2 capture: projected generating efficiency is 29.3% vs. 38.5% for generation without CO_2 capture

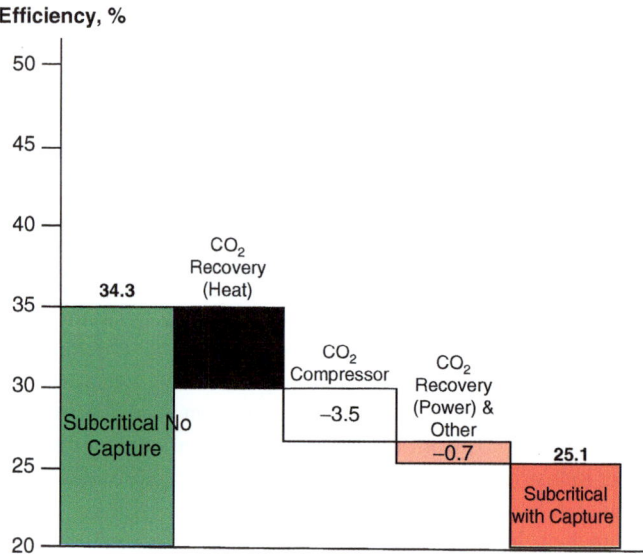

Fig. 2.6 Parasitic energy consumption associated with a subcritical PC unit with post-combustion CO_2 capture versus a subcritical PC unit without CO_2 capture [10]

size and performance to provide the maximum total unit efficiency. If a unit designed for no-capture is retrofitted for capture at a later date, the efficiency penalty for CO_2 capture is larger because some of the components become sub-optimum. This will be addressed further in the discussion of retrofitting.

Other approaches to CO_2 capture are being examined. For example, the use of chilled ammonia absorption is claimed to significantly reduce these energy require-ments and is being evaluated on a 1.7 MW_e system at a 1,224 MW_e commercial coal-fired generating station in Wisconsin [16, 17]. Additional approaches being pursued include unique framework solids, algal systems, frosting, and other adsor-bents. These are further from economic evaluation or demonstration. Improvements can be expected for absorption and adsorption systems, but there are physiochemical and thermodynamic limitations to how large these improvements will be.

2.2.2 Oxygen-Blown Power Generation

The main problem with CO_2 capture from air-blown units is the low CO_2 concen-tration in the flue gas due to nitrogen dilution. This can be solved by substituting oxygen for air. For PC combustion, this is Oxy-Fuel PC combustion. Another approach is to gasify the coal with oxygen and steam, and remove the CO_2 at high pressure prior to combustion of the syngas in a gas turbine. This approach is Integrated Gasification Combined Cycle (IGCC) power generation.

2.2.2.1 Oxy-Fuel PC Combustion

Oxy-fuel combustion, shown schematically in Fig. 2.7, addresses the high CO_2 capture and recovery costs, but it does so at the expense of an air-separation unit and its associated energy costs [18, 19]. The advantage is gained through being able to cool the flue gas, condensing out water, and leaving almost pure CO_2 which can then be compressed, with further drying, to produce supercritical CO_2 for geologic storage. A parasitic energy diagram for oxy-fuel is shown in Fig. 2.8.

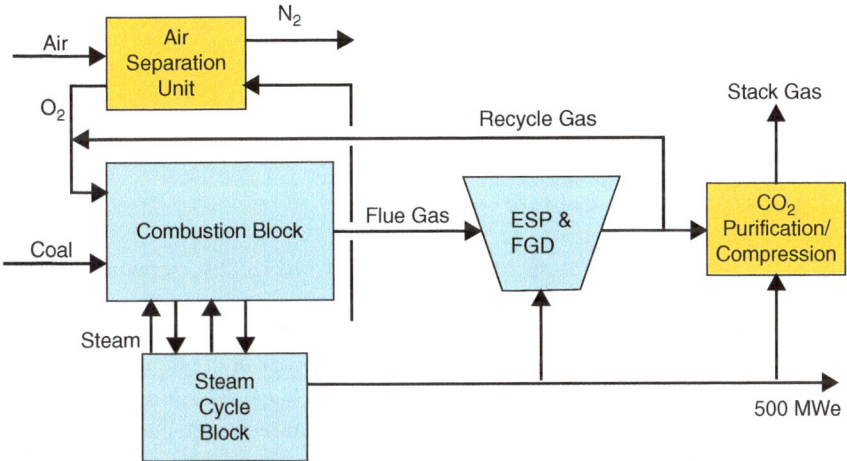

Fig. 2.7 Schematic of a pulverized coal oxy-fuel generating unit with CO_2 capture achieved by drying and compression. The volume of gas that goes up the stack is projected to be small

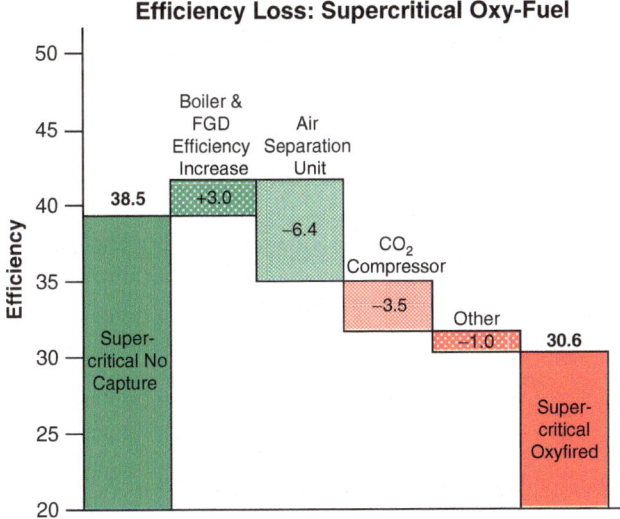

Fig. 2.8 Parasitic energy consumption associated with oxy-fuel combustion versus supercritical PC generation without CO_2 capture [10]

Boiler efficiency is improved somewhat, but this gain is more than offset by the power requirements of the oxygen separation unit.

The technology is in active pilot plant development, and the early stages of commercial demonstration. A 30 MW_{th} oxy-fuel pilot plant was commissioned in Schwarze Pumpe, Germany in mid-2008, with plans for a 300 MW demonstration plant followed by a 1,000 MW commercial plant [20, 21]. Because of the early state of commercial development, the performance and cost estimates are not as firm as those for PC or IGCC. Oxy-fuel PC has the potential for lower cost of electricity (COE) and lower CO_2 avoided cost than with PC capture. The development of this technology should be monitored.

2.2.3   IGCC

2.2.3.1   Without CO_2 Capture

IGCC power generation from coal is illustrated in Fig. 2.9, showing the main process components and stream flows. Oxygen from an air separation unit is used to combust sufficient carbon in the gasifier typically at 500–1,000 psig to increase the temperature to around 1,500°C (2,730°F). For typical coals, the ash melting point is between 1,200°C and 1,450°C. At 1,500°C, the coal ash melts and leaves the bottom of the gasifier as slag. At this temperature, water (steam), which is added with the coal or separately, reacts with the remaining carbon to convert it to

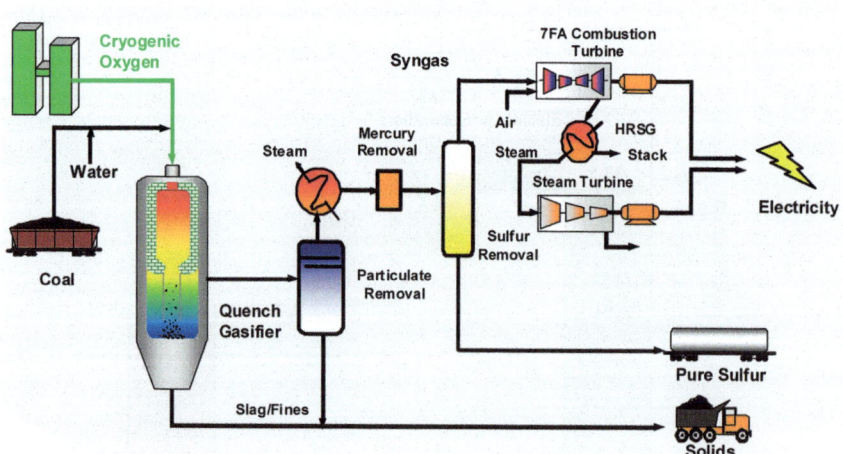

Fig. 2.9 A representative coal-based IGCC unit showing main process components and streams (Courtesy NETL)

syngas, a mixture mainly of carbon monoxide (CO) and hydrogen (H_2) and some CO_2, along with impurities such as H_2S, NH_3 & mercury. The syngas is quenched with water to remove particulate matter, cleaned of impurities, and then burned in a turbine in a combined-cycle power block that is very much like a natural gas combined-cycle (NGCC) unit (see Fig. 2.9). Because all the gases are contained at high pressure, high levels of particulate matter, sulfur, mercury, and other pollutant removal are possible. Air emissions levels from an IGCC unit should be similar to those from a NGCC unit. Coal mineral matter is removed from the gasifier as a solid, relatively dense vitreous slag.

The gasifier is the biggest variable in the system in terms of type (moving bed, fluid bed, and entrained flow), feed approach (water-slurry, dry feed), operating pressure, and the amount of heat removed from it. For IGCC units to date, entrained-flow gasifiers have been the primary choice. For electricity generation, without CO_2 capture, radiant and convective cooling sections behind the gasifier that produce high-pressure steam for additional power generation lead to efficiencies that can approach or exceed 40%. The additional heat removal options are illustrated in Fig. 2.10.

Figure 2.11 is a schematic of a 500 MW_e IGCC unit summarizing the operating conditions and giving the stream flows for no CO_2 capture. The unit is using Illinois #6 coal at rate of 185,000 kg/h or 4,400 tonnes of coal per day. This unit, which employs radiant cooling but not convective cooling, has a generating efficiency of 38% on an HHV basis [10, 13].

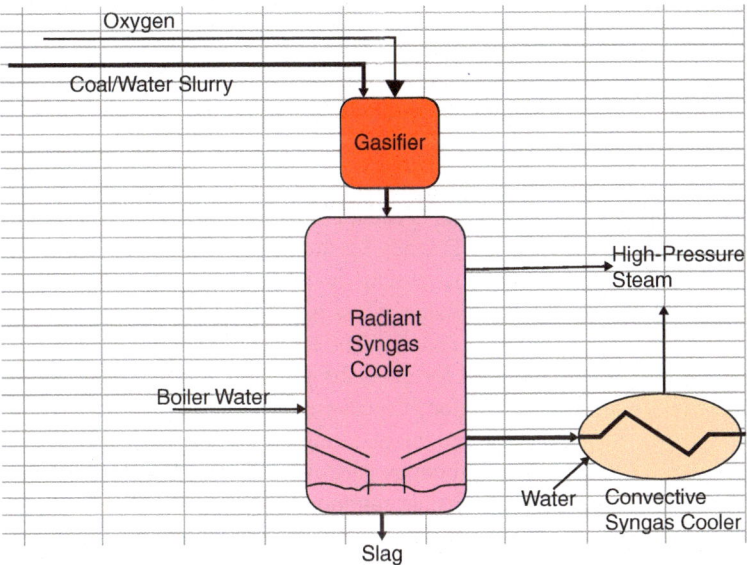

Fig. 2.10 Heat recovery options for an entrained-flow gasifier. Additional steam produced is used to generate electricity

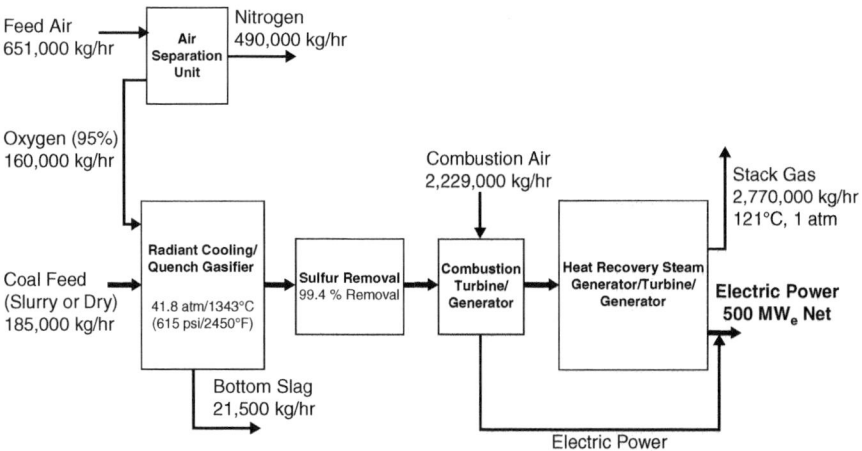

Fig. 2.11 Schematic of 500 MW$_e$ IGCC unit without CO_2 capture. Projected generating efficiency with radiant cooling is 38.4% [10]

2.2.3.2   With CO_2 Capture

A block diagram with key material flows for a 500 MW$_e$ IGCC unit designed for CO_2 capture is shown in Fig. 2.12. To achieve CO_2 capture with IGCC, the CO in the syngas must first be converted to CO_2 and H_2 via the water gas shift reaction ($CO + H_2O \rightarrow CO_2 + H_2$). To do this, two catalytic shift reactors are added just behind the quench to convert H_2O and CO to H_2 and CO_2. The gas clean-up train requires addition of a second gas-scrubbing unit located behind the sulfur-scrubbing unit to remove the CO_2. The CO_2 capture and recovery is done at high concentration and pressure, involves weak absorption, and recovery of the CO_2 is by pressure letdown. As such, it requires less energy and is cheaper than for dilute CO_2 capture from flue gas. The estimated generating efficiency for the design and operating parameters is 31–32% [10, 13]. Figure 2.13 illustrates the parasitic energy requirements to capture the CO_2 in an IGCC plant; the efficiency loss is about 7 percentage points vs. about 10–11% points for PC with CO_2 capture. The largest efficiency reduction is related to generating the steam required for the water gas shift reaction. CO_2 compression is second largest but is less than for PC because the compression begins at a higher pressure. After shift and clean up, the resulting gas stream is largely H_2, which is then burned in a combustion turbine as part of a combined cycle unit to generate power. Turbines that can burn high concentrations of H_2 have not yet been developed. All other technologies are commercial. For example, ammonia production from coal utilizes all the steps up to combustion and is practiced in the US, Europe, and particularly China. However, these technologies have yet to be integrated and

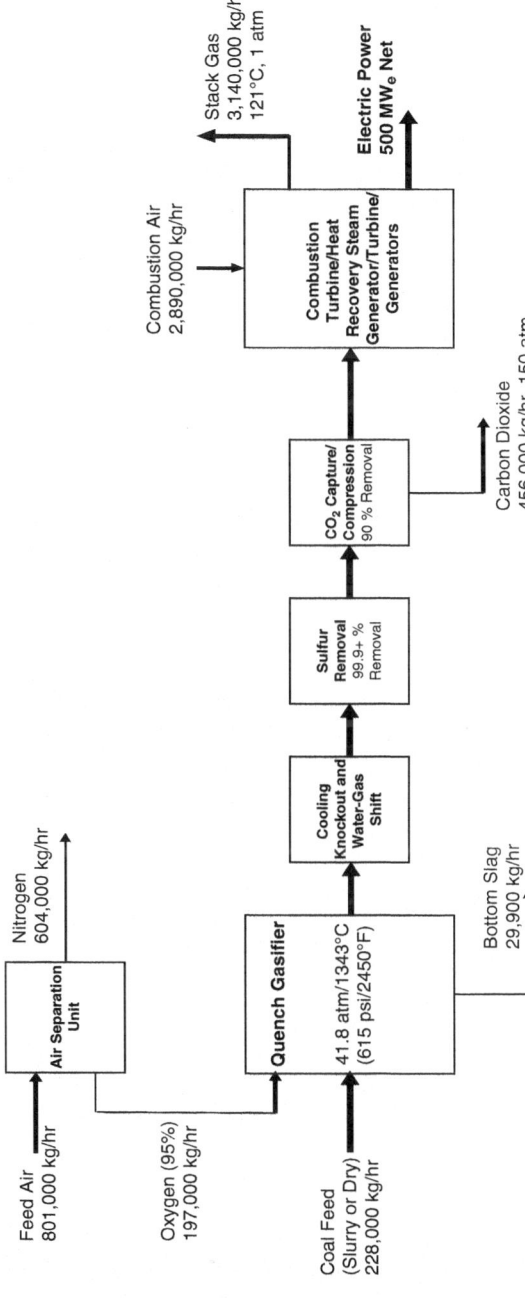

Fig. 2.12 500 MW$_e$ IGCC with CO$_2$ capture. Projected generating efficiency with quench gasifier is 31.2% [10]

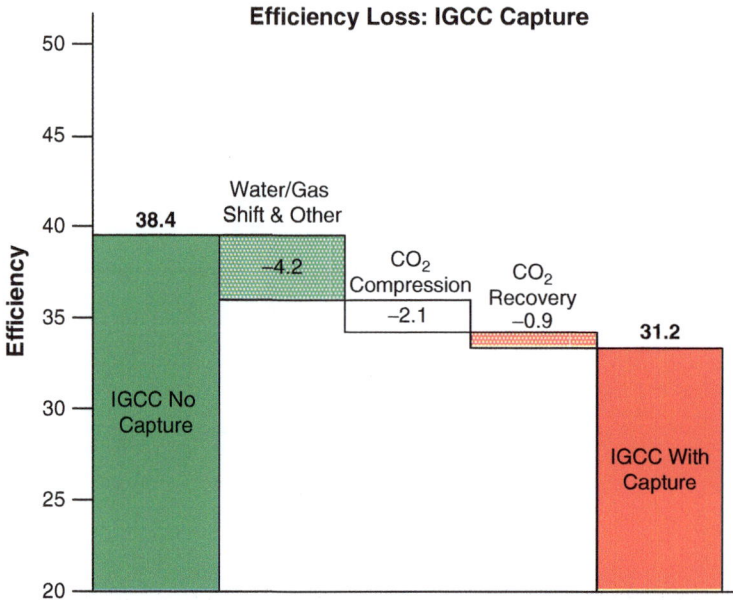

Fig. 2.13 Typical parasitic energy consumption associated with IGCC for pre-combustion CO_2 capture vs. IGCC designed for no CO_2 Capture [10]

demonstrated at the scale of operation required for large-scale power generation. Turbines that can burn very high H_2 concentrations are under development, but current turbines can only burn H_2 with appropriate dilution with N_2 from the air separation plant. Turbines designed for hydrogen combustion could provide an additional increase in generating efficiency.

2.3 Performance and Cost Summary

2.3.1 Cost

Table 2.4 summarizes the operating and cost parameters associated with the generating technologies discussed above. This indicative cost and performance data allow comparison among the generating technologies for mid-2007 Gulf-Coast construction costs.

Without CO_2 capture, PC has the lowest COE; the COE for IGCC is 10–15% higher. However with CO_2 capture, IGCC has the lowest COE. The cost of capture and compression for supercritical PC is about 4.8¢/kW$_e$-h; that for IGCC is about one half that or about 2.4¢/kW$_e$-h. The cost of transport and storage was estimated

Table 2.4 Performance and costs for coal-based power generating technologies [9–11]

	Subcritical PC		Supercritical PC		PC- Oxy	IGCC	
	w/o capture	w/ capture	w/o capture	w/ capture	w/ capture	w/o capture	w/ capture
PERFORMANCE							
Heat rate[a], Btu/kWₑ-h	9,950	13,600	8,710	12,600	11,200	8,910	10,500
Generating efficiency (HHV)	34.3%	25.1%	39.2%	27.2%	30.6%	38.3%	32.5%
CO_2 emitted, kg/h	453,000	61,900	396,000	57,100	25,400	400,000	45,600
CO_2 captured at 90%, kg/h[b]	0	557,000	0	514,000	482,000	0	422,000
CO_2 emitted, g/kWₑ-h	905	124	792	114	51	794	91
Life-cycle CO_2 emitted, g/kWₑ-h			831	170		833	138
COSTS							
Total plant cost, $/kWₑ	1,564	3,085	1,625	2,961	2,450	1,977	2,644
Cap. Ch ar.,¢/kWₑ-h @ 14.4%[c]	3.24	6.38	3.36	6.13	5.07	3.90	5.10
Fuel, ¢/kWₑ-h @ $1.50/MMBtu	1.80	2.45	1.57	2.26	2.01	1.63	1.89
O&M, ¢/kWₑ-h	0.84	1.66	0.87	1.59	1.32	1.06	1.33
COE, ¢/kWₑ-h	**5.87**	**10.50**	**5.81**	**9.98**	**8.40**	**6.60**	**8.32**
CO_2 Disposal Cost, ¢/kWₑ-h	0.00	0.72	0.00	0.67	0.63	0.00	0.57
COE[d], ¢/kWₑ-h	**5.87**	**11.22**	**5.81**	**10.65**	**9.03**	**6.60**	**8.89**
Cost of CO_2 avoided vs. same technology w/o capture[d], $/tonne		**$68**		**$71**	**$43**		**$37**

Basis: 500 MWₑ plant net output. Illinois #6 coal (63.7%wt C, 27.1 MJ/kg (HHV), 85% cap Fac)

[a]Efficiency=3,414 Btu/kWₑ-h/ (Heat rate)

[b]90% removal used for all capture cases, except PC-Oxy which was assumed 95%

[c]Annual capital charge rate of 14.4% from EPRI-TAG methodology, based on 55% debt @ 6.5%, 45% equity @ 11.5%, 38% tax rate, 2% inflation rate, 3 year construction period, 20 year book life, applied to total plant cost to calculate investment charge

[d]Includes the cost of CO_2 transport and geologic storage, details discussed in the text

to be \$6.5 per tonne of CO_2; the cost of transport and storage will be discussed later. The cost of CO_{2eq}-avoided for supercritical PC, including transport and geologic storage, is about \$71 per tonne of CO_{2eq}; that for IGCC is about \$37 per tonne. That for Oxy-fuel is estimated at \$43 per tonne of CO_{2eq}. These numbers include the cost of CO_2 capture and compression to a supercritical fluid, and CO_2 transport and injection. Its lower COE would appear to make IGCC the technology of choice for CO_2 management in power generation. However, Oxy-fuel has significant potential, and current research and demonstration activities will provide needed cost and engineering information. Further, for a lower rank coal and a higher plant elevation, the cost difference between IGCC and PC narrows. As such, significant reductions in the capture/recovery cost for PC could make it economically competitive with IGCC with capture in certain applications. In addition, the power industry still has concerns about IGCC operability and availability. Thus, we cannot declare any of these technologies superior to the others at this point.

2.3.2   Cleaning-up Coal

Coal has the reputation of being dirty, largely based on criteria air emissions. Table 2.5 gives the commercially demonstrated and projected emissions performance of PC and IGCC technologies [14, 22]. Electrostatic precipitators (ESP) or bag houses are employed on all U.S. PC units, and particulate matter (PM) emissions are typically low. Improved ESP or wet ESP can reduce PM emissions further, but at a cost. Flue gas desulfurization (FGD) is applied on about one-third of U.S. PC capacity, and thus "typical (average) U.S." SO_x emissions (Table 2.5) are quite high. "Best commercial" performance in Table 2.5 gives demonstrated, full commercial-scale levels of emissions reductions [14, 22, 23]. Additional reductions are possible. With CO_2 capture, SO_x emissions levels are expected to be even further reduced [24]. The best commercial emissions performance levels with IGCC is threefold to tenfold lower (Table 2.5). IGCC with CO_2 capture should have even lower emissions. In addition, IGCC produces a dense, vitreous slag that ties up most of the toxic components in the coal mineral matter so that they are not easily leached [25],

Table 2.5 Commercially demonstrated and projected emissions performance with CO_2 capture for PC and IGCC power generation [9, 10]

Technology	Case	Particulates Lb/MM Btu	SO_2 Lb/MM Btu	NO_x Lb/MM Btu	Mercury % removed
PC plant					
	Typical	0.02	0.22	0.11	
	Best commercial	0.015 (99.5%)	0.04 (99+%)	0.03 (90+%)	90
	Design w CO_2 cap.	0.01 (99.5+%)	0.0006 (99.99%)	0.03 (95+%)	75–85
IGCC plant					
	Best commercial	0.001	0.015 (99.8%)	0.01	95
	Design w CO_2 cap.	0.001	0.005 (99.9%)	0.01	>95

Table 2.6 Incremental cost of advanced PC generation emissions control vs. no emissions control [9]

	Capital cost[a] [$/kW$_e$]	O&M [¢/kW$_e$-h]	COE[b] [¢/kW$_e$-h]
PM control	55	0.20	**0.31**
NO$_x$	40	0.15	**0.23**
SO$_2$	200	0.30	**0.71**
Incremental cost vs. no control	295	0.65	1.25[c]

[a] Incremental capital costs are for a new-build plant
[b] Incremental COE impact for Illinois #6 coal with 99.5% PM reduction, 99.4% SO$_x$ reduction, and >90% NO$_x$ reduction
[c] When this is added to the "no-control" COE for SC PC, the total COE is 5.8¢/kW$_e$-h

and IGCC uses about 30% less water than supercritical PC. Although this discussion does not address the whole life-cycle for coal's environmental footprint, coal-use in the electricity generation step can, in fact, be very much cleaner than it is today, and CO$_2$ emissions can also be markedly reduced.

The estimated cost to achieve the emissions reductions used in the PC design basis, which is somewhat better than today's best demonstrated commercial performance vs. no emissions control is about 1.25¢/kW$_e$-h (see Table 2.6) out of about 5.8¢/kW$_e$-h total COE or about 20% [26–29]. CO$_2$ capture and recovery will increase the COE more than this, about 4¢/kW$_e$-h, based on today's PC technology. Cost reductions can be expected when this technology begins to be commercially practiced.

2.3.3 Biomass, and Coal Plus Biomass to Power

Burning fossil fuels for power generation and for transportation releases carbon that has been stored for millions of years as CO$_2$ into the atmosphere, resulting in the build-up of atmospheric CO$_2$. Using biomass as a fuel releases carbon removed as CO$_2$ from the atmosphere in a recent plant growth cycle and does not contribute to increasing atmospheric CO$_2$ concentration if done sustainably. Thus, using biomass in power generation reduces the life-cycle emissions of CO$_2$ per unit of power generated.

Biomass can be burned directly or co-fired with coal in a boiler. The major issues are effective size reduction of the biomass in order to feed it into the boiler and its lower energy density. Today, most biomass power plants burn demolition wood wastes, forest product wastes or agricultural wastes to produce steam for power generation. The U.S. has 11 GW of installed biomass-only plant capacity [30], with an average size of 20 MW$_e$. The industry average generating efficiency is of order 20%. Typically SO$_x$ emissions are low because biomass contains little sulfur, but NO$_x$ emissions can be quite high because the relatively high nitrogen content of many biofuels. These emissions can be controlled, at a cost, which, however, can be significant on small units. A generally more attractive approach is to co-fire biomass at levels of less than 25% (wt.%) with coal to gain the advantages of scale of a much larger generating facility and reduced CO_2 emissions per kW_e-h generated. The other option is to gasify the biomass and generate power in a combined-cycled configured power island. Gasification technology is considered below, and the economic and CO_2 impacts of using biomass to generate power are discussed.

2.3.3.1 Thermochemical Conversion of Biomass

Biomass gasification and/or pyrolysis involves the conversion of biomass to a mixture of carbon monoxide, hydrogen, carbon dioxide, methane, and other organics including bio-oils and tars, ash and small char particles. The concentration of these gases and other materials depends on the process design, and operating conditions. Gasification has the advantage that it can convert essentially any biomass material to syngas at sufficiently severe conditions (Fig. 2.14). This syngas can be burned in a boiler or can be cleaned and burned in a turbine in a combined cycle power island to produce electricity. It could also be cleaned and shifted to produce a synthesis gas from which a broad range of fuel and chemical products can be produced. This latter option is considered in Chap. 3. Biomass gasification exhibits many similarities to coal gasification, including a significant number of gasifier types and different approaches to gasification technology. Electricity or fuels produced via gasification of biomass should have low net CO_2 emissions; and if biomass gasification is combined with capture and geologic storage of CO_2, such processes have a negative CO_2 emission footprint.

Gasification can be carried out under a variety of pressure and temperature conditions. When relatively low pressures and temperatures are used, it is primarily a pyrolysis process. Under these less-severe conditions, the main products are a mixture of hydrogen, CO, and light hydrocarbons, bio-oil, tars, and char. For less-severe gasification (pyrolysis), the heating is usually indirect, avoiding the need for an expensive air separation unit, reducing the capital cost significantly. The mix of

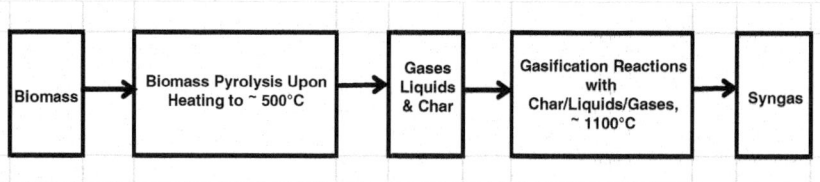

Fig. 2.14 Schematic of thermochemical conversion of biomass, flow from left to right. Pyrolysis produces a broad range of materials, including bio-oils and tars, which can undergoes gasification at higher temperatures to produce syngas, which is composed primarily of CO, H_2, and CO_2. One-step gasification at high temperature combines the pyrolysis and gasification stages to rapidly produce only syngas

primary products can be separated into several fractions for upgrading of for gasification or combustion. The gas stream can also be cleaned, compressed, the CO shifted to H_2 and CO_2, and the CO_2 removed for geologic storage.[1] If air is used directly as an oxidant in gasification, the nitrogen present results in a low Btu gas that is most easily used for steam or power generation via a boiler or combustion turbine but without CCS.

Biomass gasification using direct firing with oxygen at higher pressure and temperature produces a relatively pure syngas stream of CO and H_2, with some CO_2 and other gases. For temperatures greater than 1,100°C, little or no methane, higher hydrocarbons, or tar are present. The high oxygen content of biomass reduces the oxygen requirement and thus the air separation unit size and cost. With biomass there is almost no sulfur, limited ash, or few other contaminants to deal with, although there are issues with some feedstocks such as rice straw that contain silicon.

Several U.S. and European organizations are developing advanced biomass gasification technologies, and there are about ten different biomass gasifiers with a capacity greater than 100 tonnes per day operating in the U.S., Europe, and Japan. These units demonstrate a broad range of feedstocks, of feed capabilities, of gasifier characteristics, of product gas clean-up approaches, and of primary products. Biomass Technology Group (BTG) lists over 90 installations (most are small) and over 60 manufacturers of gasification technologies [31]. A recent NETL benchmarking report summarizes the status of larger scale biomass gasifiers [32] For example, at the McNeil Generating Station in Vermont, a low-pressure wood gasifier, which started operation in August 2000, converted 200 tonnes per day of wood chips into fuel gas for electricity generation [33].

Most of the gasification technologies have technical or operational challenges associated with them, but most of these issues are probably resolvable or manageable with commercial experience. Gasifier choice depends on the type of biomass feed and on the specific application of the gasification/pyrolysis products. The most persistent problem area appears to be biomass feed processing and handling, particularly if a gasifier must contend with different biomass feeds. DOE has funded five different advanced biomass R&D projects to advance the technology [33]. Although several of the available gasification technologies have been commercially demonstrated, biomass gasification technology has yet to be robustly demonstrated for commercial, integrated biomass gasification and power generation. The implication is that biomass gasification technology is still on a relatively steep learning curve, as is the integration of biomass gasification, gas clean up, and power generation or biofuel synthesis. A major characteristic of biomass gasification is that it will involve smaller units than coal gasification, and it will thus not benefit from the economies of scale of coal gasification. This is because of the dispersed nature

---

[1] Geologic storage of CO_2 (deep saline aquifer, depleted oil and gas reservoirs, and enhanced oil recovery) is considered most likely; other options such as deep ocean storage are considered unlikely.

of biomass and cost of biomass transport, which limit the area that could supply feedstock to a given plant to a relatively small radius near the plant site. This limits annual biomass feed availability to a given plant. This will increase the cost per unit product unless major process simplification and capital cost reductions can be achieved. A primary strategy has been to eliminate the air separation unit, which is typically required with most high-severity gasification technologies. This leads to gasification with air and involves nitrogen-diluted syngas or involves indirect heating to avoid nitrogen dilution which then typically produces a product stream containing more bio-oil, tar, and light hydrocarbon gases.

2.3.3.2 Power Generation

Next, the cost and performance, including CO_2 impacts, of biomass to power are examined. Because of the small scale of biomass-to-power plants and the higher cost of biomass vs. coal, biomass-combustion based power generation is generally not competitive with PC generation. However, with Renewable Energy Credits that are in effect in some locations, the technology can be profitable [34]. These cost issues can be reduced by co-firing with coal at a coal plant. In this case, there is additional expense associated with the higher cost of biomass vs. coal and with the facilities needed for biomass receiving, storage, preparation, and feeding into the boiler. These are in addition to the coal handling facilities. The rest of the PC unit remains essential the same. CO_2 emissions reductions are in direct proportion to the ratio of carbon per unit of biomass energy feed to the plant to the carbon per unit of coal energy feed.

The other approach to power generation involves gasification of biomass plus combined-cycle power generation. Using the approach outlined earlier for evaluating coal power plants, and utilizing a steam/oxygen blown fluidized-bed gasifier with gas cooling and gas cleaning for the biomass feed, a consistent set of cost and performance estimates were made [11, 35]. Table 2.7 summarizes these projections for biomass to power. It was assumed that the plants were sited such that 1 million tonnes of dry biomass per year is available for the plant (3,790 tonnes/day, 85% capacity factor).

Subcritical generation is assumed for conventional combustion, steam generation; CCS was not considered because of the high cost of flue gas CO_2 capture. Capital cost is higher primarily due to smaller unit size, and biomass fuel cost is more than twice that of coal. The resulting COE at 10¢/kW_e-h is about 70% higher than for a larger PC plant (5.8¢/kW_e-h) (Table 2.7 and Table 2.4). Although plant CO_2 emissions are similar to those of a coal-based plant, life-cycle CO_2 emissions are very small because the CO_2 was recently captured and will be recaptured in the next growth cycle. The small positive value (50 g CO_{2eq}/kW_e-h) is due to fossil-based emissions occurring in biomass production and transport.

Gasification-based generation [11] has higher efficiency with biomass, driven by lower utility costs for air separation and gas clean-up due to the high biomass oxygen content and low impurity levels. Estimated COE for biomass IGCC is about

Table 2.7 Projected cost and performance of power generation from biomass, and from coal plus biomass

	Biomass subcritical	Biomass IGCC		Coal/biomass IGCC	
	w/o capture	w/o capture	w/ capture	w/o capture	w/ capture
Performance					
Heat rate[a], Btu/kW_e-h	9,750	8,010	9,430	8,870	11,100
Generating efficiency (HHV), %	35.0	42.6	36.2	38.5	30.7
Coal feed, tonnes/day (AR)	0	0	0	3,480	3,480
Biomass feed, tonnes/day (AR)					
Carbon in feed, kg/h	63,100	63,100	63,100	155,000	155,000
CO_2 emitted, kg/h	229,000	209,000	29,200	530,000	12,800
CO_2 captured, kg/h[b]	0	0	179,000	0	132,000
Plant CO_2 emitted, g/kW_e-h	935	700	115	769	85
Life-Cycle CO_2 emitted, g/kW_e-h	50	−26	−740	478	−278
Costs					
Total plant cost, $/$kW_e$	1,910	1,768	2,529	1,920	2,620
Total capital required, $/$kW_e$	2,139	2,033	2,908	2,057	2,808
Inv. charge, ¢/kW_e-h @ 14.4%[c]	3.95	3.66	5.23	3.97	5.42
Fuel, ¢/kW_e-h	5.15	4.23	4.98	2.80	3.50
O&M, ¢/kW_e-h	1.03	0.95	1.36	1.03	1.40
COE, ¢/kW_e-h	**10.12**	**8.84**	**11.57**	**7.81**	**10.32**
CO_2 Disposal cost, ¢/kW_e-h	0.00	0.00	0.66	0.00	0.58
COE[d], ¢/kW_e-h total	**10.12**	**8.84**	**12.23**	**7.81**	**10.90**
Cost of CO_2 avoided vs. same technology w/o capture[d], $/tonne			47.6		41.5

Basis: 500 MW_e plant net output. Illinois #6 coal (63.7%wt C, 27.1 MJ/kg (HHV), 85% cap Fac)

[a]Efficiency =3,414 Btu/kW_e-h/ (heat rate)

[b]90% removal used for all capture cases, except PC-Oxy which was assumed 95%

[c]Annual capital charge rate of 14.4% from EPRI-TAG methodology, based on 55% debt @ 6.5%, 45% equity @ 11.5%, 38% tax rate, 2% inflation rate, 3 year construction period, 20 year book life, applied to total plant cost to calculate investment charge

[d]Includes the cost of CO_2 transport and geologic storage, details discussed in the text

50% higher than for PC generation and about 30% higher than coal IGCC, primarily because of the higher cost of biomass. Because biomass is the only feed, the electricity generated is essentially CO_2 neutral, because the CO_2 emitted from the plant was recently removed from the atmosphere and will be recaptured in the next growth cycle. However, it actually has a negative life-cycle CO_2 balance of minus 26 g CO_{2eq}/kW$_e$-h (without CCS) because the char from the gasification contains more carbon than fossil carbon consumed in the production and transportation of the biomass. This carbon is assumed permanently sequestered with the char. The COE for biomass IGCC is less than coal IGCC with CCS which still has a life-cycle CO_2 balance of +138 g CO_{2eq} /kW$_e$-h.

At zero life-cycle GHG (CO_{2eq}) price, biomass gasification with CCS has a COE that is about 25% higher than coal IGCC with CCS, but it has a large negative life-cycle CO_2 balance (−740 g CO_{2eq}/kW$_e$-h) associated with it. Therefore, it would receive a large CO_{2eq} credit in any carbon tax or carbon-trading regime. The cost of CO_{2eq} avoided is $48/tonne (Table 2.7). If coal-based PC with CCS is compared with biomass IGCC venting the CO_2 the avoided cost of CO_{2eq} is about $14/tonne. Comparing coal-based IGCC with CCS with biomass IGCC with venting, the cost of avoided CO_{2eq} is effectively zero (slightly negative) as is coal-based IGCC with CCS compared with biomass IGCC with CCS (slightly positive $/tonne CO_{2eq} avoided). These numbers should be considered to be zero within the ability to estimate costs at this point.

Co-feeding coal and biomass can provide improved generating economies-of-scale and reduced CO_2 emissions. To accommodate the different properties of biomass and coal, the process design estimates here are based on biomass gasification utilizing a steam/oxygen blown fluidized-bed gasifier and coal gasification via an entrained flow GE Texaco gasifier. Coal to biomass feeds were 60%/40% on an energy basis; 3,480 tonnes/day coal and 3,790 tonnes/day biomass, each on an as-received (AR) basis (Table 2.7) [11, 35]. The quenched syngas streams were combined to take advantage of available economies of scale downstream of gasification. With this configuration, without CCS, the COE for the coal/biomass case is about 20% higher (1.2¢/kW$_e$-h) than for the coal only case due mainly to the higher biomass fuel cost. Although the plant CO_2 emissions are similar, the life-cycle CO_2 emissions are 478 g CO_{2eq}/kW$_e$-h or about 40% less due to the 40% biomass feed, without CCS. With CCS, the COE is also about 20% higher (~2.0¢/kW$_e$-h) for the coal/biomass case than for the coal-only IGCC case. However, the life-cycle CO_{2eq} emissions are −278 g CO_{2eq}/kW$_e$-h for the coal plus biomass case vs. +138 g CO_{2eq}/kW$_e$h for the coal-only IGCC with CCS.

2.3.4 CCS Carbon Chain

The carbon chain from fossil fuel or biomass from its source through the process to carbon capture and sequestration (CCS) is shown in Fig. 2.15. We have already considered the first two steps (boxes) in the figure for power generation.

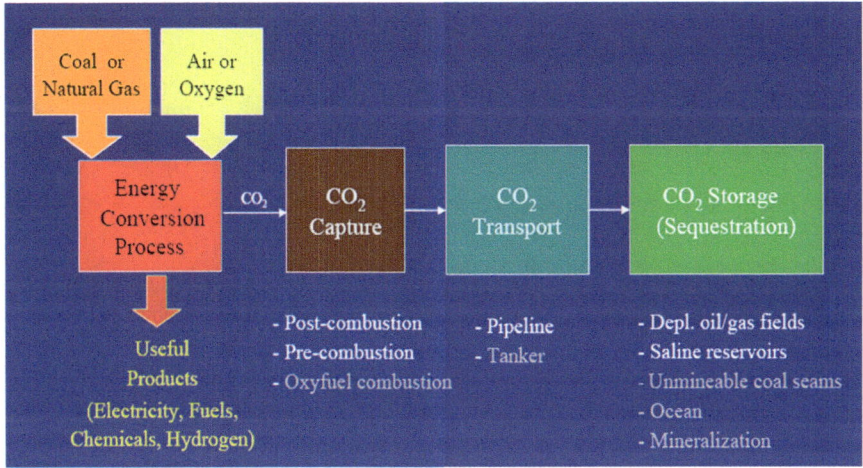

Fig. 2.15 Full carbon chain from fossil fuel input to carbon capture and sequestration

The last two steps in the carbon chain, pipeline transport and injection, will now be considered, as will their impact on COE estimated. These costs are already included in Tables 2.4 and 2.7; this section develops and discusses their cost basis. To estimate cost impact, a model of a typical CCS project is needed. Oil and gas reservoirs and enhanced oil recovery (EOR) are often discussed for geologic storage of CO_2. These storage sites-of-opportunity may play a role initially; but they have limited long-term potential because of the scale of CO_2 CCS that will be needed to make a major difference in managing CO_2 from coal-based power generation. Today, EOR uses 35–40 million tonnes of CO_2 per year which could be supplied by a few early CCS projects (see Table 2.1). Larger-volume, long-term storage for the U.S. will largely be in deep saline aquifers. These geologic formations underlie large portions of the U.S., particularly those areas that today have a lot of coal-based power generation and where additional coal-based generating capacity could be expected to be added as shown in Fig. 2.16. In fact, there is a high degree of coincidence between potential coal deposits, coal-based power generating sites, and potential geologic CO_2 storage sites.

The primary mode of CO_2 transport for sequestration operations will be via pipelines. There are over 2,500 km of CO_2 pipeline in the U.S. today, with a capacity in excess of 40 million tonnes CO_2/year. These pipelines were developed to support EOR operations, primarily in west Texas and Wyoming. In these pipelines, CO_2 is transported as a dense, single-phase fluid at ambient temperature and supercritical pressure. To avoid corrosion and hydrate formation, water levels are typically kept below 50 ppm. The pipeline technology is mature, and most costs can be estimated. The main unknowns are the costs of permitting, acquisition of right-of-way, and additional costs associated with local terrain (rivers, roads, high density inhabited areas).

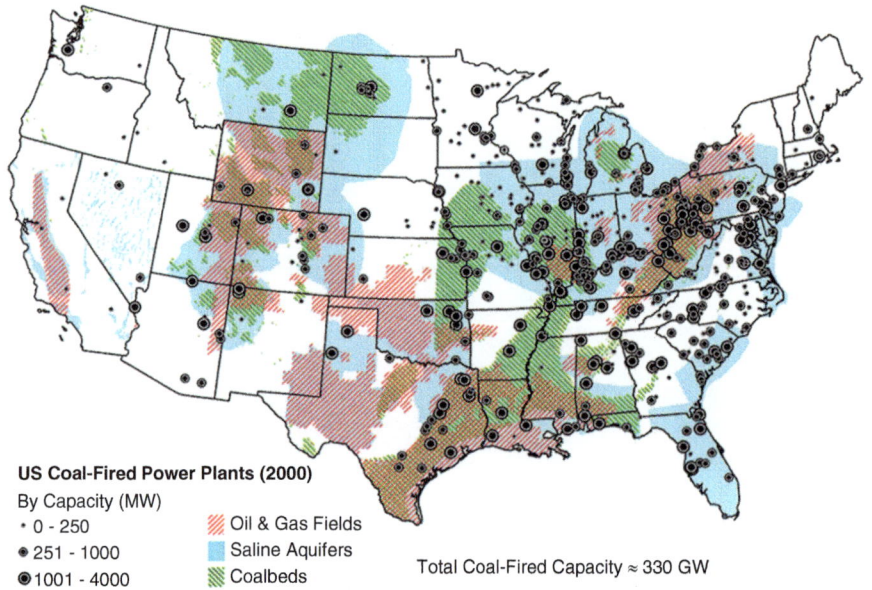

Fig. 2.16 Location of deep saline aquifers, oil and gas fields, coal beds and coal based power plants for U.S. [10]

However, rather than having long-distance CO_2 pipelines running across the country, a typical CCS power plant project could be expected to look something like that illustrated in Fig. 2.17. It is expected that sufficient capacity would be accessible within about a 100 km radius for a good location. Location is important, but once sited the CO_2 storage requirement for the lifetime of the power plant, which would be of order a billion barrels of liquid CO_2 should be within that area. The total reservoir CO_2 capacity must be sufficient so that by accessing different portions of the reservoir over the lifetime of the plant all the CO_2 captured can be safely stored.

The Transport and Storage (T&S) costs used here were updated to 2007 using recent reviews by McCollum and Ogden and by Tarka [36, 37]. Pipelining costs were updated using the Handy-Whitman Index of Public Utility Costs for Gas Transmission Line Pipe and Steel Distribution Pipe and operating costs updated using U.S. Bureau of Standards Producer Price Indices for the Oil and Gas Industry [37]. Capital costs were levelized over a 20-year period and include a 30% process contingency and a 20% project contingency. Monitoring costs are included and used the IEA Greenhouse Gas R&D Program [38]. This includes operational monitoring costs tracking the plume for 30 years and closure monitoring costs for the following 50 years. An operational fund is capitalized to cover the 80 year monitoring cycle. The storage site is chosen to be representative of a typical saline aquifer at a depth of 4,055 feet depth and 22 millidarcies permeability and 1,220-psi down-hole pressure. A $5 million initial site assessment was assumed. Costs were estimated from this basis.

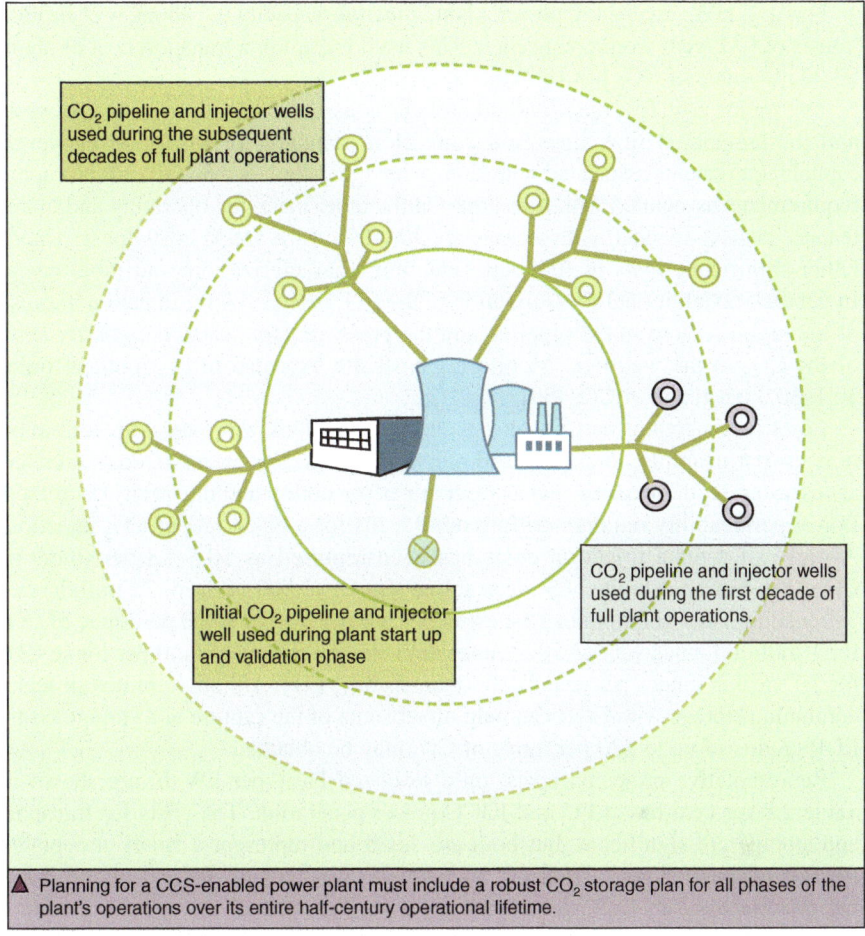

CO$_2$ pipeline and injector wells used during the subsequent decades of full plant operations

Initial CO$_2$ pipeline and injector well used during plant start up and validation phase

CO$_2$ pipeline and injector wells used during the first decade of full plant operations.

▲ Planning for a CCS-enabled power plant must include a robust CO$_2$ storage plan for all phases of the plant's operations over its entire half-century operational lifetime.

Fig. 2.17 Conceptual model of a typical CCS project (Courtesy Battelle, GTSP report)

The transportation costs dominate the CO$_2$ T&S costs and are highly non-linear with the amount of CO$_2$ transported. The economies-of-scale make transportation costs for large CCS projects much less expensive. For example, for 1 million tonnes of CO$_2$ per year (2,500 tonnes/day) the estimated transport cost is about $8.00 per tonne per 100 km; at 3.5 million tonnes CO$_2$ the estimated transport cost is about $5.00 and at 7 million tonnes of CO$_2$ per year the cost is about $3.00 per tonne per 100 km. These are typical values, but costs are dependent on pipeline costs which can be highly variable from project to project due to both physical (e.g., terrain the pipeline must traverse) and political considerations. In addition, there are a number of other issues that can have a significant impact on potential CCS projects. These include state and federal laws and regulations, and political, public, and environmental concerns.

For a 1 GW_e coal-fired power plant, pipeline capacity of about 6–7 million tonnes of CO_2/year would be needed. This would result in a transport cost of about $3.00 per tonne of CO_2 per 100 km.

The major cost for injection and storage is associated with drilling the wells and the associated flow lines and connectors required for injection. However, capital requirements associated with storage are typically less than 20% of the capital requirements associated with transport. On the other hand, the operating and maintenance costs associated with storage are 30–40% of the O&M costs for transport. Other significant costs include site selection, characterization, and monitoring. In general, no additional pressurization of the CO_2 is required for injection because of the high pressure in the pipeline and the pressure gain due to the gravity head of the CO_2 in the wellbore. Monitoring costs are expected to be small, of order $0.1–$0.3 per tonne of CO_2 [39].

Costs for injecting the CO_2 into geologic formations will vary with formation type, its permeability, thickness, and other properties. For example, costs increase as reservoir depth increases and as reservoir permeability and injectivity decreases. Lower permeability requires drilling more wells for a given rate of CO_2 injection. A range of typical injection costs has been reported as $0.5–$8 per tonne of CO_2 [39]. For an average U.S. deep saline aquifer (1,000 m deep, 22 millidarcies permeability, and 160 m thick) the estimated storage cost is $1.60 per tonne of CO_2 for 1 million tonnes/year (2,500 tonnes/day) storage rate and $0.50 per tonne CO_2 for 3.5 million tonnes per year (10,000 tonnes/day) [37]. Although limited in scale, combining storage with EOR can help offset some of the capture and storage costs. EOR credits of up to $20 per tonne of CO_2 may be obtained.

Representative projected costs, on a levelized basis per kW_e-h, are shown in Table 2.8 for coal-based PC and IGCC power generation. The costs for transport and storage are significant, but both are small and represent a small, acceptable fraction of the total cost. Transport and storage costs include the cost of constructing pipelines and of drilling the injection wells, as well as the system operating costs. The numbers used were updated to 2007 using typical terrain and saline reservoir properties [36, 37]. The largest cost is in CO_2 capture and compression (Table 2.8). For IGCC, the projected cost of CCS would increase the bus bar cost of electricity by about 40%, from 6.8 to about 9.4¢/kW_e-h. IGCC with CCS vs. PC venting would represent about a 50% increase in the bus bar COE. This electricity

Table 2.8 Cost of CCS projected for PC and IGCC generation with CO_2 capture

Technology	PC	IGCC
CCS Step	¢/kW_e-h	¢/kW_e-h
Capture	3.3	1.3
Compression	0.9	0.5
Transport	0.5	0.5
Injection	0.1	0.1
Totals	**4.8**	**2.4**

would be very low emissions electricity, including low CO_2 emissions. Furthermore, it is economically competitive with electricity generated by wind power and by new nuclear power plants [1].

Comprehensive geological reviews suggest that for carefully selected storage sites, there are no irresolvable technical issues for CO_2 injection and storage with respect to its efficacy and safety[10]. However, there are technical issues that require better understanding. We have 30 years of successful CO_2 injection experience from which we have found no critical issues. The Sleipner Project in Norway [40] has been injecting 1 million tonnes/year of CO_2 into the Utsira Saline Formation since 1996 using a single well bore. Weyburn in Canada [41] has injected 0.85 million tonnes/year of CO_2 into the Midvale reservoir for EOR since 2000, and In Salah [42] has been injecting 1 million tonnes/year of CO_2 into the water leg of the gas field for several years also. None of these projects have encountered any problems, and there is no sign of CO_2 leakage.

2.4 A Forward View

2.4.1 Retrofitting Existing Plants

Addressing the future of coal-based power generation raises many key issues and challenges. A number of these are analyzed next. To achieve really significant reductions, CO_2 emissions from the existing coal fleet will have to be reduced. This will require retrofitting CO_2 capture on existing units, or repowering them with high-efficiency technology with CO_2 capture such as IGCC-CCS or replacing them with other technology.

Retrofitting existing units involves several factors that significantly affect the economics and viability of the unit. These include unit age, size, and operating efficiency, as well as land availability or other space constraints at the plant site. Existing units are frequently smaller, have low generating efficiency, and may not have highly efficient emissions control systems relative to large, new builds. The energy requirement for CO_2 capture is usually higher for retrofits because of less efficient heat integration for sorbent regeneration in an existing plant. For power generation, plant output reduction approaches 40% vs. the 30% reduction for purpose-built plants [39, 43–45]. Existing plants that are not equipped for adequate NO_x control or with a flue gas desulfurization (FGD) system for SO_2 control must be retrofitted or upgraded for high-efficiency sulfur capture in addition to the CO_2 capture and recovery system. All these factors lead to higher overall costs for retrofits. Figure 2.18 illustrates the retrofit of a subcritical PC unit with MEA (monoethanolamine) flue gas scrubbing. The original unit had a generating efficiency of 35% (HHV) without CO_2 capture; after retrofit with CO_2 capture the original 500 MW_e unit produces only 294 MW_e and has a generating efficiency of 20.5% (HHV) or a 41.5% derating. The efficiency reduction for a

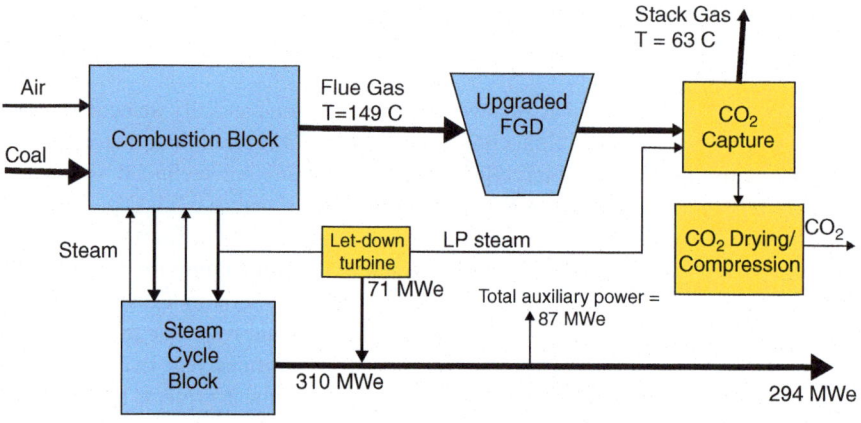

Fig. 2.18 Retrofit of a subcritical PC unit with amine CO_2 capture

CO_2 capture purpose-built unit would be about 28% (HHV) or a 28% derating. For the purpose-built unit, everything is optimally sized; for the retrofit unit, steam is diverted from the turbine for sorbent regeneration, and the turbine is operating at about 58% of design loading, far from its conditions for optimum performance.

If the original unit is fully paid off, the cost of electricity after retrofit could be slightly less to somewhat more than that for a new purpose-built PC plant with CO_2 capture based on the new capital required [43, 44]. However, an operating plant will usually have some residual value, particularly if flue gas clean-up technology has recently been added; and the reduction in plant efficiency and output, increased on-site space requirements, and unit downtime are all complex factors not fully accounted for in this analysis. For smaller, older units, rebuilding the entire boiler and power generation sections or replacing them with IGCC (repowering) may be the best alternatives [44, 45]. Generally, the cost of CO_2 avoided is expected to be 30–40% higher than for a purpose-built capture-plant. For example, an MEA retrofit of a supercritical PC is projected to cost almost as much as a new unit on a $/kW$_e$ basis from an Alstom retrofit design study [43]. Retrofit capture costs have been projected to range from 2 to 7¢/kWe-h from best to worst case scenarios considered with 90% CO_2 capture in a feasibility study by Alstom [46]. Further, retrofits require case-by-case detailed design-based examination. Although there is no one answer for existing subcritical PC units, CO_2 capture will likely be achieved through repowering with a supercritical PC unit with CO_2 capture or with oxyfuel or with IGCC-CCS or other technology, rather than retrofitting. A recent MIT symposium on retrofitting PC plants for CO_2 control addressed all of these issues but found no easy, cheap solutions [45]. The option of producing fuels and power from biomass and coal is a new option that is discussed in Chap. 3.

2.4.2   *Electricity and CO$_2$ Avoided Costs*

Figure 2.19 shows the levelized COE for coal- and biomass-based power generation and the components that make up the total cost. The COE for natural gas combined-cycle power generation with venting and with CCS for two natural gas prices is also included in Fig. 2.19 [11, 35, 47, 48]. As discussed above, with CO$_2$ venting the COE is lowest for conventional PC and about 10–15% higher with IGCC. The COE for NGCC is about the same as for PC for a gas price of $6/GJ, but at a gas price of $16/GJ it is more than twice that. The fuel cost is the largest COE driver in NGCC and is also large for biomass to power. Adding CCS increases the cost for all generating technologies but less so for technologies that involve gasification, for example IGCC vs. PC. As a result, coal IGCC with CCS is the most attractive technology, as is biomass to power (BTP) with CO$_2$ venting, their COEs are essentially the same.

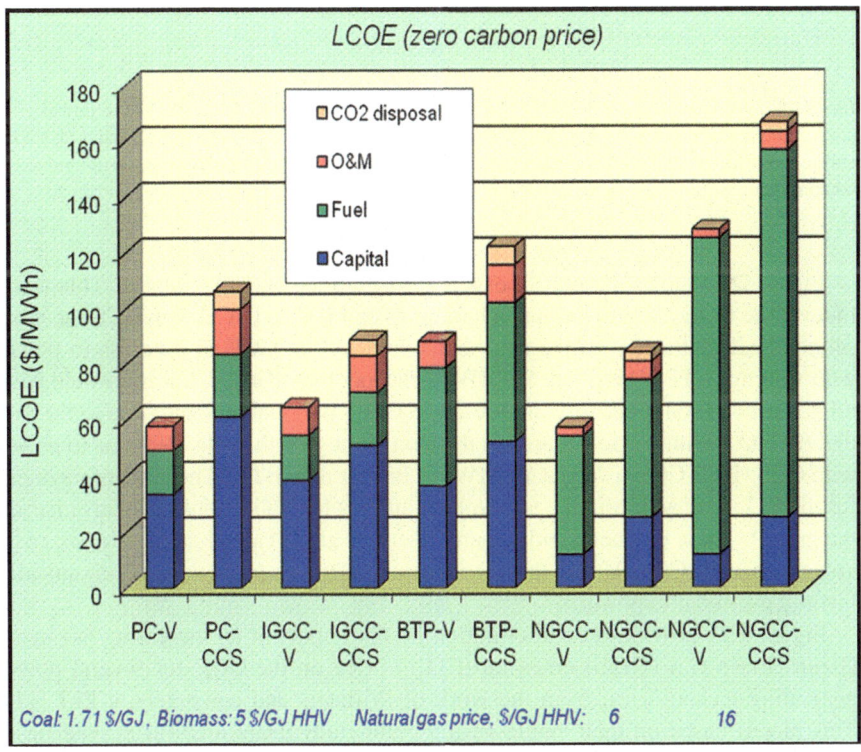

Fig. 2.19 Levelized cost of electricity for newly-built coal-based and biomass-based power generation technologies at study point-design conditions and zero price on CO$_2$ emission. Natural gas combined cycle generation with two natural gas prices included ($100/MW$_e$-h equals 10¢/ kW$_e$-h.) (Courtesy Williams et al. [11])

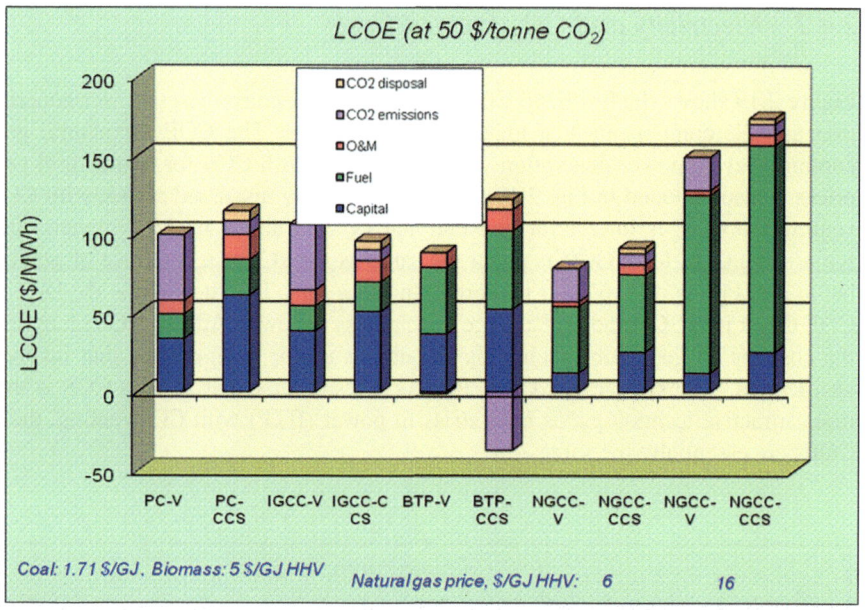

Fig. 2.20 Levelized cost of electricity for newly-built coal-based and biomass-based power generation technologies at the study point-design conditions with a $50 per tonne price on CO_{2eq} emitted to the atmosphere. Natural gas combined cycle generation with natural gas is included for comparison ($100/MW$_e$-h equals 10¢/kW$_e$-h) (Courtesy Williams et al. [11])

Figure 2.20 shows the impact of a $50/tonne price on CO_{2eq} vented. This tends to level the COE of many of the technologies and makes IGCC with CCS the most economically attractive of the coal technologies. The COE of biomass to power with venting (BTP-V) is about $9/MW$_e$-h cheaper then IGCC-CCS and would have no tax imposed on it because the life-cycle CO_{2eq} is essentially zero due to the fact that the CO_2 emitted is recaptured in the next plant growth cycle. Biomass to power with CCS (BTP-CCS) is about $2/MW$_e$-h cheaper than BTP-V because of payment for the CO_2 removed from the atmosphere and geologically stored (negative bar on the graph). These payments more than offset the added capital and feedstock costs associated with CCS. Under these conditions, biomass to power is economically favored over coal to power.

Figure 2.21 provides key information on the impact of an increasing life-cycle Green House Gas (GHG) emission (CO_{2eq}) price on the COE for several power generating technologies, from the work of Williams and coworkers at PEI [11]. This plot is based on their single design-point study using a consistent database. Included is the impact of GHG emissions price on the cost of average grid power and on the cost of power from existing, fully paid-off, coal plants. Crossover points are the CO_{2eq} price at which economics would induce a shift from one technology to the other for new power plants. For example, the CO_{2eq} cost that would induce a shift from IGCC venting (CTP-V) to IGCC with CCS (CTP-CCS) is $38/tonne

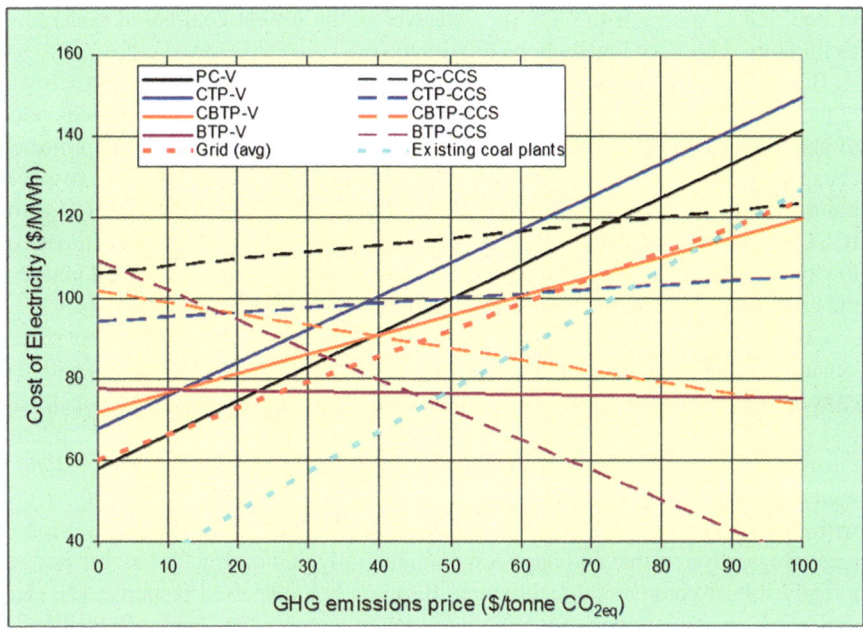

Fig. 2.21 Cost of Electricity (COE) as a function of GHG emissions price. Crossover points are the CO_2 prices required to economically induce a shift from one technology to the other. Also indicated is the impact of CO_2 price on the average cost of grid power today and the cost of power generated by existing coal plants (Courtesy Williams et al. [11])

CO_{2eq} (Fig. 2.21). The CO_{2eq} emissions price required to induce a shift from newly designed PC venting (PC-V) to newly designed IGCC with CCS (CTP-CCS) is about \$50/tonne CO_{2eq} and to drive a shift from PC-V to PC-CCS requires a CO_{2eq} price of about \$73/tonne. This is the situation which would exist when the demand for electric power is growing and new coal-based power plants are being designed and built. It would also be the situation for repowering old, obsolete power plants.

In the case of existing coal-based plants that are fully operational where there is insufficient growth in electricity demand to warrant new plants, as might be the case in the U.S., the relevant crossover point is between existing, venting PC plants and new IGCC with CCS (CTP-CCS). Under these conditions, a CO_{2eq} price of over \$75/tonne would be required to induce the construction of a new IGCC plant with CCS (CTP-CCS). This approach would also apply to repowering existing PC plants with IGCC with CCS.

These CO_{2eq} price cross-over points suggest that significant shifting to IGCC or other coal-based power plants with CCS would occur at a relatively low CO_{2eq} price (less than ~\$40/tonne) in economies that have growing electricity demand, i.e. are building new plants. In economies with stagnant electricity demand, because of conservation efforts, etc., the CO_{2eq} price to induce a shift would have to be much higher (more like \$75/tonne), to induce a switch from existing, venting PC plants to new IGCC plants with CCS (or whatever is the lowest coal-based generating technology with very low carbon emissions).

Biomass to power plants (BTP) using biomass gasification both with CCS and without CCS would economically replace existing coal plants at an emissions price of about \$50/tonne CO_{2eq}. For new plants, biomass to power (BTP-V) is projected cheaper than IGCC with CCS (IGCC-CCS) for all CO_{2eq} prices, and the crossover point for biomass to power with CCS (BTP-CCS) is less than \$30/tonne CO_{2eq} for IGCC-CCS. If the estimated COE is low because of a low capital cost estimate or low biomass cost estimate, the appropriate curve shifts upward by that amount, but the crossover points remain within a relatively small CO_{2eq} price range.

Combined coal and biomass (~60%/40% on an energy basis) – based power generation (CBTP) without and with CCS have low crossover CO_{2eq} prices with the conventionally considered all-coal based power generation. These are basically all less than \$40/tonne CO_{2eq} for new plants. However, to replace existing PC plants, the existing emissions price would have to exceed \$55/tonne CO_{2eq}. The challenge with the biomass and the combined coal and biomass cases is the lack of experience with biomass gasification, and the availability of biomass. Biomass gasification is technologically feasible and has been commercially demonstrated, but it is not yet a really robust commercial technology. Biomass is a dispersed resource, and thus supplying large quantities of it to a given site on a continuous basis is a challenge. Because it is less dense and typically contains significant water, collection and transport over long distances is not economically attractive. This limits the size of potential plants. This makes coal plus biomass configurations more attractive because it provides economies of scale, and reduces CO_{2eq} emissions, while coal supplements available biomass. In addition, small amounts of biomass with coal (around 10% on an energy basis) in IGCC with CCS (CTP-CCS) can produce zero life-cycle GHG electricity.

The estimates developed here are all based on bituminous coal, for which the COE favors IGCC with CCS in a CO_2-constrained environment. Although, about 50% of U.S. coal reserves are bituminous, the remaining 50% are sub bituminous coal and lignite. Lower rank coals and higher elevation plant locations narrow the cost difference between IGCC and PC with CO_2 capture [10, 49]. Cost improvements for PC capture could make it economically competitive with IGCC in certain applications, and Oxy-fuel PC looks potentially competitive also. Thus, it is too early to decide which technology will be cheapest for coal-based power generation with CO_2 capture. All technologies need to remain under development and demonstration until there is sufficient commercial-scale experience to decide.

There is always a need for innovative technology in coal-based power generation to improve operations, increase generating efficiency, and to reduce emissions and CO_2 capture costs. A number of technologies are being developed to reduce cost and improve performance at the bench and pilot scale. However, it is important to note that conventional coal-based power generation is a mature technology, and PC units have been highly optimized. Technology already exists to capture CO_2 from PC and IGCC units, although it is typically applied at smaller scale in other applications. These technologies need to be commercially demonstrated, integrated, and optimized on the scale of power generation. Waiting for research to provide that "unique solution" is not a rational approach if there is any urgency to the CO_2 emissions issue. The rational approach is to put available commercial technology into practice, integrate it into the full generating and emissions control system, and begin to move along the learning-by-doing curve. This typically results in significant cost reductions, improved effectiveness and efficiency, and increased operability, reliability, and robustness. Rubin and coworkers at Carnegie Mellon University have studied the impact of learning-by-doing on cost for a number of technologies, including the power industry (e.g. see [50]). From the historical experience curves for a range of power generation technologies, LNG plants and oxygen and hydrogen production, Rubin et al. [50] estimated that for the generating technologies considered above, the CO_2 capture cost could undergo a 13–15% capital cost reduction and a 13–26% total cost reduction with 100 GW_e of new installed capacity. This is in addition to the cost reductions that will also be taking place in the base plant, such as IGCC which will be undergoing learning-by-doing cost reductions with increasing commercial applications. The same can be said for the CO_2 transport and storage component of the total generating and CCS chain.

Commercial technologies exist that can be utilized and integrated to achieve effective power generation with CCS today. These have been applied in commercial operation but frequently at a smaller scale than required for power generation. Application of these technologies at commercial power-plant scale will ultimately result in significant improvements in them and in significant cost reductions. Similarly, CO_2 sequestration (geologic storage) is commercially demonstrated at the 1 million tonnes per year at several locations in the world, and more demonstrations are planned internationally. The DOE Regional Partnership program has started to develop a geologic database but needs to accelerate and expand in scale.

Geologic storage still needs full-scale, well-monitored demonstrations at several locations and in different geologies in the U.S. to develop the needed site choice, permitting, monitoring and closure procedures and to gain needed public and political support for the more widespread application of the technology. These are large-scale, expensive activities, which if successfully demonstrated and applied are mainly aimed at benefiting society, and thus, society has a stake in supporting them. The can also be said in support for combined demonstration programs among several countries that depend heavily on coal-based power generation (such as India, China, and the US) and are likely to remain heavily dependent on coal-based power generation for the foreseeable future. Such demonstrations can take up to a decade to plan, build, and operate to gain desired learning. Thus, there is urgency to start down the path.

In addition to CO_2 emissions, criteria emissions from coal-based power generation can be very low if the available control technologies are applied. When CO_2 capture is applied, these criteria emissions can be expected to be even lower, resulting in a small environmental footprint for clean coal technology. With CO_2 capture and sequestration, "clean coal" can provide base-load electricity that is cost competitive with wind and new nuclear and can continue to help maintain our energy diversity. Thus, "clean coal" would appear to continue to be an economic choice for base-load power generation of very low emissions electricity, including low CO_2 emissions.

In summary with respect to the path forward:

- The technologies required for CO_2 capture with power generation are commercial and can be expected to improve in cost and performance from operation at scale and learning-by-doing. Major R&D developments are not needed to begin applying them now. However, major R&D will be needed to support their application and to help drive improvements in them and the development of new and improved technologies. The order of commercial readiness is: (1) IGCC-CCS, (2) PC with post capture, and (3) oxy-fuel.
- It is technically feasible to safely and effectively store large quantities of CO_2 in deep saline aquifers, and the U. S. storage capacity in such reservoirs appears very large. This needs to be clearly demonstrated on a commercial scale and some technical issues need resolution. Improved storage capacity estimates need to be made by country. The U.S. appears to have storage capacity potential in excess of several hundred to over a 1,000 gigatonnes of CO_2. China appears to have large storage capacity close to with much its coal use, but India may have a more limited storage potential [10].
- A broad range of regulatory issues, including: permitting guidelines and procedures, liability and ownership, monitoring and certification, site closure, remediation, require resolution so that projects can proceed forward in a smooth, efficient manner.
- For CCS to be available to apply on a large scale, it is critical to gain political and public confidence in the safety and efficacy of geologic storage.

To resolve these issues and establish CCS as a viable technology for managing CO_2 emissions, it is necessary to carry out 3–5 large-scale CCS demonstration projects in the U.S. and 7–12 globally at the 1 million tonnes CO_2 per year scale, using different generation technologies, focusing on different geologies, and operated for several years [10]. Effective demonstration of technical, economic, and institutional features of CCS at commercial scale with coal combustion and conversion plants, will: (1) give policymakers and the public confidence that a practical carbon mitigation option exists, (2) shorten the deployment time and reduce the cost for carbon capture and sequestration when a carbon emission control policy is adopted, and (3) maintain opportunities for the lowest cost and most widely available energy form to be used to meet the electricity needs of the U.S. and the developing world in an environmentally acceptable manner. If completed expeditiously, this program can provide the U. S. and the rest of the world with robust technical options for addressing CO_2 emissions from power generation and for liquid transportation fuels production from coal as discussed in Chap. 3. If the U.S. took the lead in these activities, it could also provide a broad technology base for U.S. companies to apply globally and would also strengthen our engineering and technology base to deal with other energy/technology issues in the future.

With a robust set of technology options, it is in theory feasible to markedly reduce CO_2 emissions from coal-based electric power, but to drive this, the price set on CO_{2eq} emissions will have to be high. This is particularly the case if power demand does not grow and the activity focuses on replacing units in the existing fleet. With growth in power demand and with the end-of-life retirement of existing units the emissions reduction will occur at a lower emissions price. IEA projects in the World Energy Outlook 2008 that the GHG emissions price will have to be about \$90/tonne CO_{2eq} by 2030 to realize the 550 stabilization trajectory [51]. EPRI gives a detailed analysis of reductions potential in the generating portfolio including the role of coal with CCS [52]. Chap. 3 presents an important route for achieving significant reduction in emissions associated with coal-based power generation at a significantly lower cost.

References

1. IEA (2006) World energy outlook. In: OECD/IEA (ed) World energy outlook 2006. OECD, Paris
2. EIA (2008) International energy outlook 2008. http://www.eia.doe.gov/oiaf/ieo/electricity.html
3. MNP (2008) Global CO_2 emissions increased by 3% in 2007. China contributing two-thirds to increase. Netherlands Environmental Assessment Agency
4. EIA (2007) Emissions from energy consumption for electricity production and useful thermal output at combined-heat-and-power plants. Electric Power Annual
5. Tollefson J (2007) Countries with highest CO_2-emitting power sectors. Nature 450:1
6. CARMA (2007) Carbon monitoring for action, Power plant CO_2 emissions, www.carma.org
7. BP (2006) Statistical review of world energy 2006. British Petroleum Ltd, London, www.bp.com/productlanding.do
8. Yi W (2008) Clean coal-based power generation technology option In: China, in China-India-U.S. science, technology, and innovation workshop, Bangalore
9. Katzer J (2008) The future of coal-based power generation. Chemical Engineering Progress
10. MIT (2007) In: Katzer J (ed) The future of coal, options for a carbon-constrained world. MIT, Cambridge
11. Kreutz TG, Larson ED, Williams RH (2008) Personal communication, Princeton University, PEI, Princeton, NJ, USA
12. Williams RH, Larson ED, Liu G, Kreutz TG (2008) Fischer-Tropsch fuels from coal and biomass: strategic advantages of once-through ('Polygeneration') configurations. In: Proceedings of the 9th international conference on Greenhouse Gas Control Technologies, Washington, DC, 16–20 Nov 2008
13. Booras G (2008) Economic assessment of advanced coal-based power plants with CO_2 capture. In: MIT carbon sequestration forum IX, Cambridge, 16–17 Sept 2008.
14. PowerClean TN (2004) In: Network PT (ed) Fossil fuel power generation state-of-the-art. University of Ulster, Coleraine, UK, pp 9–10
15. CURC (2006) CURC/EPRI technology roadmap update. Coal Utilization Research Council, Washington, DC, p 10
16. Fraser J (2008) Chilled ammonia carbon capture process to be demonstrated. The Energy Blog, http//thefraserdomain.typepad.com/energy/2008/03/chilled-ammonia.html
17. Alstom (2008) Clean power today, chilled ammonia carbon capture. http://www.power.alstom.com/home/about_us/clean_power_today/chilled_ammonia_carbon_capture/28478. EN.php?languageId=EN&dir=/home/about_us/clean_power_today/chilled_ammonia_carbon_capture/

18. Dillon DJ, Panesar RS, Wall RA, Allam RJ, White V, Gibbins J, Haines MR (2004) Oxy-combustion processes for CO_2 capture from advanced supercritical PF and NGCC power plant, in Greenhouse Gas Technologies Conference 7, Vancouver
19. Jordal K et-al. (2004) Oxyfuel combustion of coal-fired power generation with CO_2 capture – opportunities and challenges
20. Vattenfall (2008) Vattenfall's project on CCS. http://www.vattenfall.com/www/co2_en/co2_en/399862newsx/1226829repor/1221774test/index.jsp
21. Reuters-News-Service (2005) Vattenfall plans CO_2-free power plant in Germany. www.planetard.com
22. USEPA (2005) Continuous Emissions Monitoring System (CEMS) Data base of 2005 power plant emissions data. EPA
23. Thompson J (2005) Integrated Gasification Combined Cycle (IGCC): environmental performance. In: Platts IGCC symposium, Pittsburgh
24. Holt N (2007) Preliminary economics of SCPC & IGCC with CO_2 capture and storage. In: 2nd IGCC and Xtl conference, Freiberg
25. EPRI (1989) Long-term leaching tests with coal gasification slag. In: GS-6439. Electric Power Research Institute, Palo Alto
26. Oskarsson K, Berglund A, Rolf D, Snellman U, Stenback O, Fritz J (1997) A planner's guide for selecting clean-coal technologies for power plants. In: World Bank (ed) World bank technical paper no. 387. World Bank, Washington
27. Tavoulareas ES, Charpentier JP (1995) Clean coal technologies for developing countries. In: World bank technical paper no. 286. World Bank, Washington, DC
28. Chaisson J (2005) EPA flue gas emission control cost information, personal communication, Clean Air Taskforce, Boston
29. Rutkowski MD, Klett MG, Maxwell RC (2002) The cost of mercury removal in an IGCC plant. In: 2002 Parsons
30. PowerScorecard (2008) Electricity from biomass, http://www.powerscorecard.org/Tech_detail.cfm?resource_id=1
31. BTG (2004) Biomass gasification. http://www.btgworld.com/
32. Ciferno JP, Marano JJ (2002) Benchmarking biomass gasification technologies for fuels, chemicals and hydrogen production. NETL, DOE
33. EIA (2007) Biomass for electricity generation. DOE (ed), Washington
34. Brooks D (2006) PSNH's wood-burning plant makes power practical. Nashua Telegraph, Nashua
35. Kreutz TG, Larson ED, Liu G, Williams RH (2008) Fischer-Tropsch fuels from coal and biomass (Paper). http://www.princeton.edu/pei/energy/publications/texts/Kreutz-et-al-PCC-2008-10-7-08.pdf
36. McCollum DL, Ogden JM (2006) In: I.o.T. Studies (ed) Techno-economic models for carbon dioxide compression, transport, and storage. University of California, Davis
37. Tarka T (2008) In: NETL (ed) Systems analysis technical note, CO_2 transport and storage costs. Department of Energy, Pittsburgh
38. IEA (2004) Overview of monitoring requirements for geologic storage projects, I.G.G.R.D. Programme (ed), IEA
39. IPCC, Metz Bea (eds) (2005) IPCC special report on carbon dioxide capture and storage. Cambridge University, Cambridge
40. Arts R, Eiken O, Chadwick A, Zweigel P, van der Meer L, Zinszner B (2004) Monitoring of CO_2 injected at Sleipner using time-lapse seismic data. Energy 29:1383
41. Wilson M, Monea M (eds) (2004) IEA GHG Weyburn CO_2 monitoring & storage project summary report 2000–2004. IEA, p 273
42. Riddiford F, Wright I, Espie T, Torqui A (2004) Monitoring geological storage: in Salah gas CO_2 storage project. In: GJHGT-7, Vancouver
43. Bozzuto CR, Nsakala N, Liljedahl GN, Palkes M, Marion JL, Vogel D, Gupta JC, Fugate M, Guha MK (2001) In: N. U. S Department of Energy (ed) Engineering feasibility and economics of CO_2 capture on an existing coal-fired power plant. Alstom Power Inc., Richmond

44. Simbeck D, (2001) CO_2 mitigation economics for existing coal-fired power plants. In: First national conference on carbon sequestration, Washington, DC, 14–17 May 2001
45. MIT (2009) Retrofitting of coal-fired power plants for CO_2 emissions reductions. web.mit. ecu/mitei/docs/reports/meeting-report.pdf. Accessed 23 March 2009
46. Alstom Power Co (2006) Sequestration for existing power plants feasibility study Report No. ME-AM26-04NT41817.401.01.01.003, Windsor, CT
47. Kreutz TG, Larson ED, Liu G, Williams RH (2008) Fischer-Tropsch fuels from coal and biomass. In: 25th annual international Pittsburgh coal conference, Pittsburgh, 2008
48. NRC (2009) America's energy future. National Research Council, Washington
49. NCC (2004) Opportunities to expedite the construction of new coal-based power plants. National Coal Council
50. Rubin ES, Yeh S, Antes M, Berkenpas M, Davidson J (2007) Use of experience curves to estimate the future cost of power plants with CO_2 capture. Int J Greenhouse Gas Control 1:188–197
51. IEA (2008) World energy outlook for 2008. OECD, Paris
52. EPRI (2008) The power to reduce CO_2 emissions: the full portfolio. Electric Power Research Institute, Palo Alto

Chapter 3
Coal and Biomass to Liquid Fuels

James R. Katzer

Abstract Demand for liquid transportation fuels has been increasing by over 2%/year over the last two decades and is accelerating in the emerging economies which are moving to automobile ownership. Almost all liquid transportation fuels are derived from petroleum, which at the same time is coming under increasing demand pressure and price instability. A high degree of dependence on petroleum brings concerns about diversity and security as well as issues of decreasing CO_2 emissions associated with the transportation sector. This chapter examines the potential to use coal and biomass to replace petroleum-derived liquid fuels and thereby to address the concerns that are associated with near total dependence on petroleum-based liquid transportation fuels. The evaluation centers on the U.S. but is easily expandable to other developed countries and the developing world.

3.1 Introduction

Global transportation depends on liquid fuels, and these are almost entirely derived from crude oil. For a number of countries, including the U.S., China, India, and Europe, there is rising concern over access to sufficient crude supply, over energy and national security, over the cost of that supply, and over diversity of the energy sources for liquid transportation fuels. In addition, there is growing concern over reducing CO_2 emissions from transportation. The transportation sector contributes one third of the U.S. Greenhouse Gas (GHG) emissions, and it is likely that the transportation sector will have to do its share in reducing GHG emissions in the

The findings included in this chapter do not necessarily reflect the view or policies of the Environmental Protection Agency. Mention of trade names or commercial products does not constitute Agency endorsement or recommendation for use.

J.R. Katzer (✉)
Chemical and Biological Engineering, Iowa State University, Ames, IA 50011, USA
e-mail: jrksail@comcast.net

future. The conversion of coal to liquid fuels requires large energy inputs which in turn result in additional production of CO_2. Therefore, coal to liquids (CTL) without aggressively applying carbon capture and sequestration (CCS) would have a negative impact on the greenhouse gas emissions balance. However, CTL offers the opportunity to reduce petroleum imports and to diversify the sources of our transportation fuel supply. As the price of petroleum and natural gas increase relative to that of coal, there will be increasing interest in the commercial potential of synthetic liquids produced from coal and also from biomass.

In 1979, the United States, anticipating increases in the price of oil to $100 per barrel, embarked on a major synthetic fuels program intended to produce up to two million barrels of oil equivalent per day of natural gas from coal and of synthetic liquids from shale and coal. A quasi-independent government corporation, "The Synthetic Fuels Corporation" (SFC), was formed for this purpose. The SFC undertook to finance approximately six synfuels projects using a combination of indirect incentives. When the price of oil fell to about $20 per barrel in the early 1980s, the need for and particularly the economics of a government-supported synfuels program disappeared; and the SFC was terminated in 1985. The SFC experience suggests the dangers of building a program around a single issue or assumption such as the future world oil price or to base it on a single goal. Today, concerns about greenhouse gas emissions, and energy security, and diversity of supply have been added to the issue of crude oil price.

This chapter focuses primarily on the thermochemical conversion of coal and/or biomass to liquid fuels. Figure 3.1 illustrates the routes for coal conversion to liquid transportation fuels. If biomass is substituted for coal, the top and bottom routes in Fig. 3.1 still apply. For biomass, liquid fuels can be produced using biochemical conversions, such as the conversion of cellulose to ethanol. This chapter addresses the biochemical routes in less detail. The conversion of coal to liquid transportation fuels will most likely involve one of two main "routes". The "indirect route" involves breaking down the coal to form low molecular weight gases which can then be catalytically converted to synthesize liquid fuels. The other is the so-called "direct route" which converts the coal to liquid products without going through low molecular weight gases as a primary intermediate. The "indirect route" gasifies the coal as the first step. This chapter evaluates the cost and performance of technologies for the production of 50,000 bpd of Fischer-Tropsch based diesel and gasoline, and 50,000 bpd of gasoline from coal produced via methanol synthesis followed by the conversion of the methanol to gasoline (MTG). The conversion of biomass to liquid fuels, and combined biomass and coal conversion to liquid transport fuels in smaller-scale plants, limited by biomass availability, is also evaluated. All of these conversions are based on gasification, and the technology cost and performance evaluations use the same cost and operational bases and assumptions used in Chap. 2 on power generation. Thus, the technology background developed for use of coal and biomass in power generation will not be repeated here. Similarly, the issues associated with CCS are the same as those associated with power generation and will not be discussed in this chapter. Because the estimates in this chapter use the same consistent cost basis, comparisons should be quite accurate.

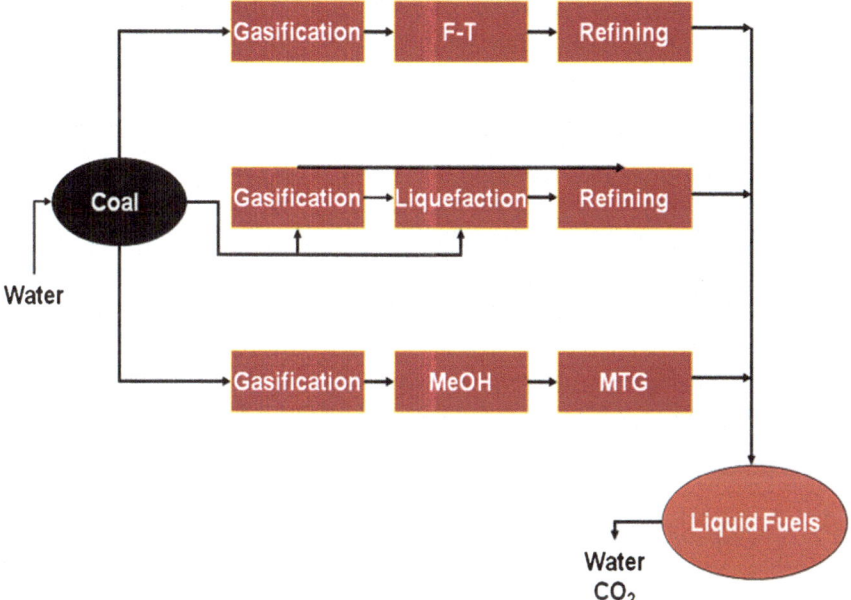

Fig. 3.1 Schematic for conversion of coal to liquid fuels by direct liquefaction (*center*) and by indirect liquefaction routes (*top* and *bottom*)

3.2 Indirect Routes to Liquid Fuels

3.2.1 *Gasification*

The initial step in the production of liquid fuels, such as diesel fuel, gasoline, or methanol from coal is the gasification of coal to produce syngas. Coal gasification is commercially deployed today. Several different types of gasifier may be selected and used, but complete gasification is required for downstream synthesis operations. An entrained-flow gasifier (GE/Texaco) was used for the coal-gasification evaluations here, as in coal to electricity (Chap. 2). For biomass gasification, a steam/oxygen-blown fluidized-bed gasifier was used. Biomass gasification is technically ready for aggressive commercial demonstration but is not commercially robust enough to assure efficient, effective commercial deployment today and will evolve with further development.

In gasification sufficient carbon is burned using oxygen to increase the temperature to above $1,200 - 1,400°C$. At this temperature the char remaining reacts with water (steam) to produce an equilibrium mixture of CO, H_2, and CO_2, which is referred to as syngas. The overall reaction is:

$$C + H_2O \rightarrow CO + H_2 + CO_2$$

The syngas produced is about 55% CO, 34% H_2, and 10% CO_2 on a volume or molar basis, along with low concentrations of impurities. The syngas composition will depend on the gasifier and the feedstock, and the impurities must be removed.

After gasification, the hydrogen to carbon monoxide ratio is adjusted to that required for the desired synthesis reaction by the water-gas shift reaction:

$$CO + H_2O \rightarrow CO_2 + H_2$$

The CO_2 is then removed from the gas stream, and the remaining synthesis gas can be catalytically converted to synthetic liquid transportation fuel. The CO_2 that is removed can either be vented to the atmosphere as a stream of pure CO_2, or can be further compressed and transported to a geologic storage site and sequestered. In either case, this CO_2 removal is a required step for the synthesis process.

3.2.2   Liquid Fuels Synthesis

Two, commercially proven technologies for converting the synthesis gas to liquid transportation fuels are considered. One of the technologies, Fischer-Tropsch synthesis, was developed in Germany in the 1920s, and several commercial plants were constructed to produce transportation fuels during World War II. The technology was commercialized by Sasol in South Africa in the 1950s and expanded in the early 1980s to produce about 150,000 bpd of transportation fuels from coal. Early versions of the technology used iron-based catalysts, which produce a broad range of products from methane to high molecular-weight waxes, but also produce a range of oxygenated materials. Figure 3.2 schematically shows the process layout of a Fischer-Tropsch process.

The primary Fischer-Tropsch synthesis reaction is illustrated by:

$$(2n+1)H_2 + nCO \rightarrow C_nH_{(2n+2)} + nH_2O$$

The hydrocarbon distribution follows an Anderson-Schultz-Flory distribution given by:

$$W_n / n = (1-\alpha)^2 \alpha^{(n-1)}$$

- W_n is weight fraction of hydrocarbon with n carbons
- α is chain growth probability

On a molecular basis, methane is most prevalent product. Higher α increases the molecular weight of the product. α increases with decreasing temperature and is also affected by catalyst type.

Because of the broad molecular weight range of products, their normal-paraffinic nature, and the presence of oxygenates, Fischer-Tropsch product requires a large

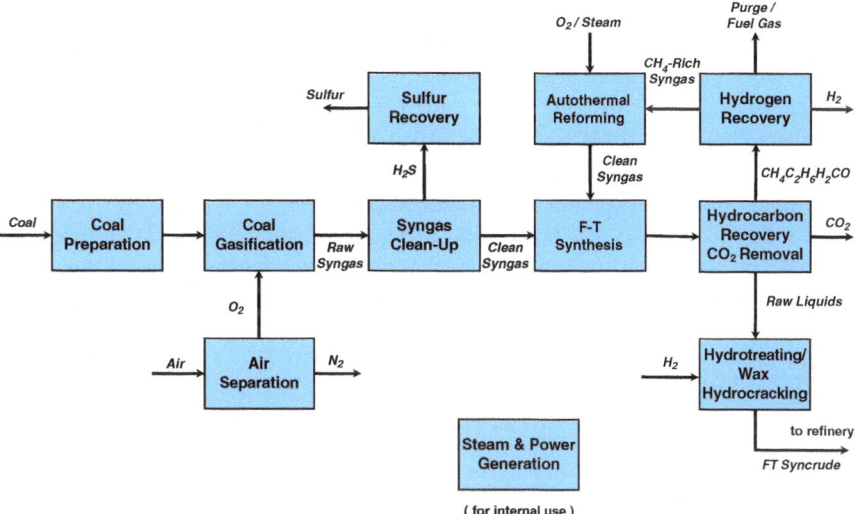

Fig. 3.2 Schematic of Fischer-Tropsch process producing liquid transportation fuels from coal [1]

Table 3.1 Typical product distributions in wt% for iron Fischer-Tropsch catalysts operating at low temperature (LTFT) and at high temperature (HTFT)

Product	LTFT	HTFT
CH_4	4	7
C_2-C_4 olefins	4	24
C_2-C_4 paraffins	4	6
Gasoline	18	36
Middle distillate	19	12
Heavy oils and waxes	48	9
Water soluble oxygenates	3	6

amount of refining to produce finished transportation fuels. Recently developed catalysts, based on cobalt, are more selective to hydrocarbon products but still produce a broad molecular-weight range of normal-paraffinic compounds that require upgrading. The cobalt-based product distribution requires somewhat less refining to produce final transportation fuels. Cobalt catalysts are very sensitive to poisoning by sulfur and other impurities. Synthesis temperature markedly affects product distribution as indicated in Table 3.1. A number of reactor types have been applied to Fischer-Tropsch synthesis. Each has advantages and disadvantages. The best reactor type and catalyst for commercial-scale coal-to-fuels is yet to be clearly demonstrated. The two primary contenders appear to be ebulated, slurry-bed (fixed fluid-bed) or fixed-bed reactors.

FT technology continues to be improved. The large Sasol plants in South Africa originally used circulating fluid-bed Synthol reactors, but these have been replaced with fixed fluid-bed reactors (SAS reactors). These form the basis for the large (36 ft diameter) slurry reactors, which produce ~17,000 bpd of fuels each, installed at the Oryx gas-to-liquids plant in Qatar. Shell has developed improved fixed-bed Fischer-Tropsch reactor technology which forms the basis of their Shell Middle Distillate Synthesis (SMDS) process. They built a 12,000 bpd plant in Bintulu, Malaysia, which has been operating since 1980. Others have developed or are developing their own version of FT technology.

Another technology for production of liquid transportation fuels from coal or biomass involves the production of methanol from the synthesis gas, followed by its conversion to gasoline by methanol-to-gasoline (MTG) technology developed by Mobil Oil. The main reactions include:

• Syngas of methanol

$$CO + 2H_2 \rightarrow CH_3OH$$

• Methanol dehydration to dimethyl ether

$$2CH_3OH \rightarrow CH_3OCH_3 + H_2O$$

• MTG Reactions over a shape selective zeolite catalyst

$$CH_3OCH_3 \rightarrow \left(n - CH_y -\right) + H_2O$$

where (n–CH$_y$–) represents gasoline, which is a complex hydrocarbon mixture.

Figure 3.3 illustrates the main processing steps in coal to gasoline using MTG. Methanol synthesis is large-scale commercial technology that can be supplied by several technology licensors and is used to produce methanol from coal today. It is well-developed and highly selective to methanol [2]. Single-train methane-based methanol plants up to 5,500 tonnes of methanol per day have been built.

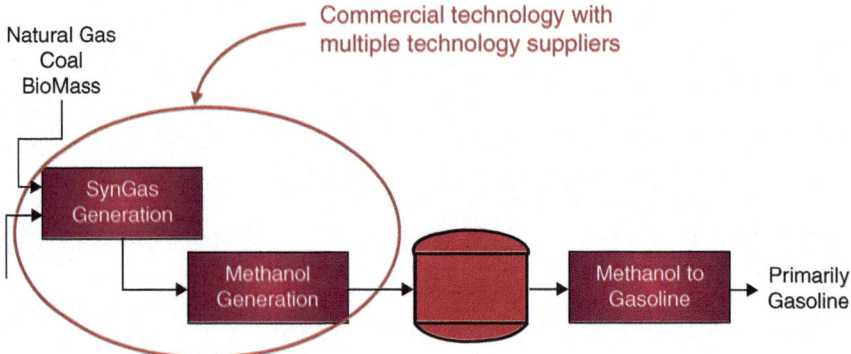

Fig. 3.3 Gasoline production from coal, biomass, or natural gas using methanol to gasoline (MTG) technology. Feeds are natural gas, coal, or biomass, water and oxygen

Table 3.2 Typical gross (wt.) hydrocarbon product distribution from methanol conversion to gasoline by MTG [5]. Given in tons of product per 1,000 tonnes of methanol feed

Products	Production (1,000 tonnes MeOH)
Gasoline (88% of HC)	387
Fuel gas	7
LPG	46

MTG, developed by Mobil Oil in the 1970s, is based on shape-selective zeolite catalysts that produce almost-exclusively hydrocarbon molecules in the gasoline range [3, 4]. The principle product is high-octane gasoline. Methanol synthesis is highly selective, and methanol is almost the only product. Methanol is then converted to gasoline in another highly selective process. Table 3.2 gives a typical gross (wt.) product distribution for the hydrocarbon fraction in the conversion of methanol by MTG [5]. The gasoline fraction has an average research octane number of 92.2 and an average motor octane number of 82.6 and meets all other specifications for U.S. commercial gasoline [5]. A 14,500 bpd gasoline plant was started-up in New Zealand in 1985 and operated for about 10 years using natural gas as feed to produce methanol, which was then converted to gasoline [5]. The MTG portion of the plant was shut down when it became economically attractive to sell methanol rather than gasoline because of decreasing crude price. A second-generation, 100,000 tonnes/ year MTG plant was constructed in Shanxi Province, China and started up in June 2009 [6]. Plans to increase the scale to one million tonnes/year have been announced [6]. Two MTG projects have been under consideration for the U.S. although lower crude prices have slowed progress on them. A variant of MTG involves conversion of methanol to olefins and the conversion of these olefins to gasoline and diesel. It is referred to as methanol to olefins, gasoline, and diesel.

3.3 Direct Coal Liquefaction

The direct liquefaction of coal involves partially deconstructing the coal structure thermally under conditions of high temperature and pressure and adding hydrogen either directly or indirectly via a donor-solvent to stabilize the fragments formed. Otherwise these fragments would recombine (repolymerize) to form heavy tar. The technology goes back to Germany in the mid-1920s and was put into large-scale production there in 1939 to produce motor fuels. From the 1970s to the 1990s, U. S. DOE conducted R&D on direct liquefaction of coal and supported construction of two large-scale pilot plants to demonstrate the technology. Plans to build a large-scale plant in the U.S. were canceled due to decreasing crude cost, increasing capital cost projections, concern over technical risk, and increasingly tight specifications on transportation fuels, which would lead to significantly increased refining cost of the coal-derived liquids for the products to meet required fuel specifications.

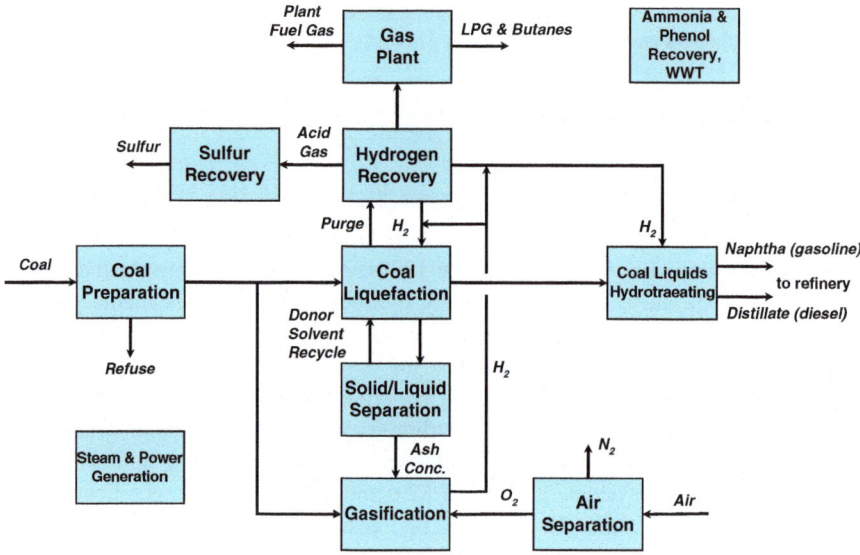

Fig. 3.4 Schematic of a direct coal liquefaction process [1]

Figure 3.4 is a schematic flow diagram illustrating direct liquefaction of coal. Process conditions require temperature between 400°C and 430°C (750–800°F) and pressures between 200 and 250 atm (3,000–3,500 psi). The coal conversion occurs in a two-stage ebulating-bed reactor system with additional hydrogen or donor solvent added between stages. Because coal has a carbon to hydrogen ratio less than 1.0 and transportation fuels have a carbon to hydrogen ratio of about 2.0, large quantities of hydrogen need to be added in the process. This hydrogen is typically provided by the gasification of coal as shown in Fig. 3.4. The resulting "liquid" products from the coal are very heavy, very aromatic, and contain large amounts of sulfur, oxygen, and nitrogen. Thus, they require a lot of upgrading in relatively high-severity refinery-type hydro processing processes.

The Shenhua Group Corp. constructed a first train (one million tonnes per year liquid product) direct-liquefaction plant in Inner Mongolia, which started up in 2009 [7]. Expansion to five million tonnes/year is planned if economics are positive. India is also evaluating direct liquefaction, but has not committed to a plant.

3.4 Liquid Fuels Production: Cost and Performance

Scale is an important factor when considering coal (or biomass) to liquid transportation fuels. Larger scale plants mean lower costs per barrel of liquid fuel product. A 50,000 bpd liquid transportation fuels plant consumes over five times as much coal and emits over three times as much CO_2 as does a 500 MW_e IGCC power plant.

Yet, a plant of this size would produce less than ~0.3% of the U.S. daily liquid transportation fuels demand. The total life-cycle emission of CO_2 from the coal to fuels process is markedly larger than for that for fuels derived from petroleum if carbon capture and sequestration (CCS) is not employed. Without CCS, FT synthesis of liquid transportation fuels emits about twice as much CO_2, including the combustion of the fuel, as compared with crude oil-derived fuels. With CCS, the full life-cycle CO_2 emissions for liquid transportation fuels produced from coal can be comparable with the total emissions from petroleum-based liquid transportation fuels. However, for liquid transportation fuels produced from coal or biomass, CCS does not require costly separation of CO_2 as is required in power generation. Instead, the CO_2 separation is a required, integral part of the process and thus, is included in the process and the cost of producing the fuel, independent of whether the CO_2 is vented or compressed further, transported and geologically stored.

The cost and performance of various technologies for converting coal, biomass, and coal/biomass into liquid transportation fuels are summarized below. Several recent studies have evaluated the economics of both FT synthesis of fuels, and synthetic natural gas production [8–12]. For FT synthesis of fuels, reported capital costs ranged from $42,000 to $63,000/bpd capacity, of which the FT reactor and associated equipment accounted for from $15,000 to $35,000/bpd of the costs. These costs are for different dates in time and have different bases and assumptions, and different levels of performance. NETL recently expanded its estimates to include thermochemical production of liquid fuels from coal and biomass [13].

To obtain consistent updated estimates for liquid transportation fuels synthesis from coal and biomass, Williams and coworkers at the Princeton Environmental Institute [14] carried out an extensive design and evaluation study using the same cost bases and assumptions used for the estimation of power generation costs from coal and biomass summarized in Chap. 2. Tables 3.3 and 3.4 summarize the key parameters. Details of the work are covered in the paper on "Fischer-Tropsch Fuels from Coal and Biomass" [15]. Since the routes considered here involved gasification of the coal or biomass feedstock, the front part of the plant is the same as used for power generation from coal or biomass by gasification with the exception of size. This difference was typically accounted for by a different numbers of trains.

Table 3.3 Key economic and operating parameters used in estimates of conversion of coal and biomass to liquid transport fuels

Base year for capital costs, Gulf Coast	Mid-2007
Capital charge rate, % of total TPI[a] per year	14.4
Interest during construction (3 years), % of TPC[b]	7.16
O&M, % of TPC per year	4.00
Capacity factor of fuel plants	90%

[a]TPI is total plant investment = TPC plus interest during construction
[b]TPC is Total Plant Cost = "overmight" capital investment to construct plant

Table 3.4 Key analytical data on feedstocks used in estimates of conversion of coal and biomass to liquid transport fuels

Coal, Illinois #6, Herrin	
Coal price, $/GJ (HHV, as received)	1.71
Coal price, $/tonne, AR	46.4
HHV, MJ/kg (AR)	27.1
Wt% carbon (AR)	63.7
Wt% sulfur (AR)	2.51
Wt% ash (AR)	9.7
Wt% moisture (AR)	11.1
Biomass, switchgrass	
Biomass price, $/GJ (HHV)	5
HHV, MJ/kg (AR)	15.9
Wt% carbon (AR)	40
Wt% sulfur (AR)	0.08
Wt% ash (AR)	5.3
Wt% moisture (AR)	15

Gasification of coal utilized a GE/Texaco entrained-flow gasifier, and gasification of switch grass (or other biomass) utilized an oxygen-steam blown fluid-bed gasifier. Costs were updated to a 2007-cost basis using the Chemical Engineering Plant Construction Cost Index. Technology required for gas clean-up, for water-gas shift to achieve the desired H_2 to CO ratio for synthesis, and that for the separation of CO_2 from the H_2 plus CO stream was then integrated into the technology chain. The required synthesis gas conversion technology (FT or methanol/MTG) was then integrated with the front-end of the process (gasification, clean-up, etc.), and the needed refining was added to the back-end to produce liquid transportation fuels that meet current fuel standards. Power generation technology was added to the back end to produce electricity from purge gas and other fuel gas streams. Mass and energy balances were carried out using Aspen Plus to allow sizing of the equipment. The typical process configuration involved recycle of the unconverted synthesis gas exiting the reactor, back to the reactor to increase conversion to liquid transportation fuels, referred to as recycle (RC). These RC cases included designs without geologic storage (venting the separated CO_2, (V)) and with geologic storage of CO_2 (CCS). Light-end hydrocarbon products and purge synthesis gas were burned in a power block to generate electricity. Some of the design options involved use of the unconverted synthesis gas after passing through the reactor once (no recycle) and gaseous hydrocarbons for enhanced power generation in a combined-cycle power block. These are referred to as "once through" or OT options. These OT configurations produce larger quantities of electric power (35+% for OT vs. ~ 10% for RC), in addition to liquid transportation fuels. The results for RC configurations are summarized in Table 3.5.

Table 3.5 Summary of cost and performance of coal and biomass to liquid transportation fuels process technologies: recycle cases producing maximum liquid fuels [14] key economic and operating parameters are given in Tables 3.3 and 3.4

	Coal to liquids Via FT		Coal to liquids via MTG		Biomass to liquids via FT		Coal/biomass to liquids via FT	
	CO_2 Vented	w CCS	CO_2 vented	w CCS	CO_2 Vented	w CCS	CO_2 Vented	w CCS
Performance								
Coal feed, tonnes/day (AR)	24,300	24,300	20,800	21,100	0	0	2,750	2,750
Biomass feed, tonnes/day (AR)	0	0	0	0	3,580	3,580	3,580	3,580
Plant efficiency, %(LHV)	49.0%	47.6%	52.4%	51.2%	51.9%	50.2%	51.1%	49.5%
Carbon in feed, kg/s	179	179	154		17	17	37	37
Dissel, bbl/day	28,700	28,700	0	0	2,540	2,550	5,740	5,740
Gasoline, bbl/day	21,300	21,300	50,000	50,000	1,870	1,870	4,260	4,260
Power exported, MWs	427	317	147	79	340	24.0	97	74
CO_2 captured and stored, kg/s	0	338	0	114	0.0	31.2	0	73
Total CO_2 emitted per gal fuel, kg/gal	25.4	11.5			25.30	2.90	26.2	11.2
Life-cycle CO_2 emitted, kg/gal	26.5	12.6			-4.8	-15.9	10.6	0.3
Total fuel life-cycle CO_2/LC petrCO_2	2.18	1.03	2.08	1.17	-0.14	-1.35	1.23	-0.02
COSTS								
Total plant Cost, m $	$4,900	$4,950	$3,940	$4,020	$636	$650	$1,320	$1,340
Specific plant cost, $bbl/day	$97,600	$98,900	$78,800	$80,400	$144,000	$147,000	$132,000	$134,000

(continued)

Table 3.5 (continued)

	Coal to liquids Via FT		Coal to liquids via MTG		Biomass to liquids via FT		Coal/biomass to liquids via FT	
	CO_2 Vented	w CCS	CO_2 vented	w CCS	CO_2 Vented	w CCS	CO_2 Vented	w CCS
Inv. Charge, 2007 $/ C3J Fuel (LHV)	8.42	8.53	7.37	7.49	12.43	12.65	11.34	11.52
Cost cost @ $1.71/CJ (HHV)	4.14	4.14	3.83	3.88	0.00	0.00	3.34	2.34
Biomass cost @ $5/GJ (HHV)	0.00	0.00	0.00	0.00	11.88	11.88	5.24	5.24
Oper&Main @ 496 of TPC/year	2.18	2.21	1.91	1.94	3.23	3.28	2.94	2.99
Co-Product electricity @ 6.0 c/kWh	−2.26	−1.68	−1.08	−0.89	−2.07	−1.45	−2.57	−1.97
CO_2 disposal Cost	0.00	0.49	0.00	0.43	0.00	1.29	0.00	0.89
Total fuel cost, $/GJ (LHV)	12.48	13.71	12.00	13.86	25.47	27.66	19.30	21.01
Total fuel cost, $/gge	**$1.50**	**$1.64**	**$1.44**	**$1.54**	**$3.05**	**$3.32**	**$2.31**	**$2.52**
Breakeven oil price, $/bbl	**$56**	**$68**	**$47**	**$51**	**$127**	**$139**	**$93**	**$103**
Cost of CO_2 avoided[4] vs. Same technology w/o capture, $/tonne		**11**		**10**		**20**		**15**

3.4.1 Coal-to-Liquid Fuels

Figures 3.5, 3.6, and 3.7 show the process configurations for coal-to-liquid fuels using FT and MTG technologies in the recycle configuration with CO_2 vented and with CCS. The only difference in the process configuration for the CCS case vs. the CO_2-venting case is the addition of CO_2 compression. For a facility producing 50,000 bbl/day of liquid transportation fuels, with the CO_2 being vented, the total plant cost (TPC) is estimated at $4.9 billion or $97,600 per stream-day-barrel, which equals TPC/50,000 barrels per day. This results in a projected fuel cost of $1.50 per gallon of gasoline equivalent, gge[1]. The capital charge is the largest component in this cost, and coal cost is next largest component at about one-half the investment charge per gallon of gasoline equivalent (Table 3.5).

This results in a breakeven crude oil price[2] of $56 per barrel for these cost and economic assumptions [16]. The addition of geologic CO_2 storage (CCS) adds slightly to the total plant cost, the cost of the fuels produced, and the breakeven crude oil price, but these impacts are not large because the CO_2 capture and separation is a required integral component of the fuels synthesis process whether or not CCS is practiced. The transport and geologic storage costs decrease with increasing amounts of CO_2 transported and geologically stored using the latest cost estimates, and CTL plants produce very large amounts of CO_2. These costs are significant but not a major cost of the fuels production. These costs could also change substantially with location. Thus, the cost of CO_{2eq} avoided[3] is low, and is estimated to be about $11/tonne of CO_{2eq}. This is much less than the avoided cost for power generation which is in the $40/tonne of CO_{2eq} range (IGCC).

Without CO_2 compression, transport, and geologic storage, the CTL plant emits 1.8 times as much CO_2 as is emitted in the combustion of the FT fuels produced in the plant. When the life-cycle greenhouse gas (LC GHG) emissions are estimated for petroleum-based diesel and gasoline produced from crude oil using the GREET model, and the LC GHG emissions are estimated for the FT fuels produced here, including GHG emissions in coal mining and in transportation[4] [15, 18], the ratio

---

[1] For mixed liquid products such as gasoline and diesel and for uniform comparison, all production costs were expressed in terms of $ per gallon of gasoline equivalent, gge.

[2] The breakeven crude oil price (BEOP) is the crude oil price in $/bbl at which the wholesale price of the petroleum-derived products would equal the calculated cost of the production of the FT fuels on a $/GJ basis. The BEOP was determined by using the average difference between the crude acquisition cost and the wholesale price of the products between 1990 and 2003, in 2007 $. This averaged 32.3 ¢/gal for gasoline and 24.9 ¢/gal for diesel fuel (an additional 5 ¢/gal was added to account for the cost of producing low sulfur diesel ([16], EIA, 2008)).

[3] Cost of CO_{2eq} avoided is estimated as {[levelized FT liquid product cost (in $/GJ) for CCS design] minus [levelized FT liquid product cost for vent design]} divided by {[greenhouse gas emissions (in tonnes CO_{2eq} per GJ of liquid product) for vent design] minus [greenhouse gas emissions (in tonnes CO_{2eq} per GJ of liquid product) for CCS design]}

[4] The non-conversion plant components of greenhouse gas emissions were estimated from GREET version 1.8. Production, transportation and refinery green house gas emissions for petroleum-derived fuels were also estimated from GREET. (Argonne National Laboratory, Transportation Technology R&D Center, 2008. *The Greenhouse Gases, Regulated Emissions, and Energy Use in Transportation (GREET) Model* [17]).

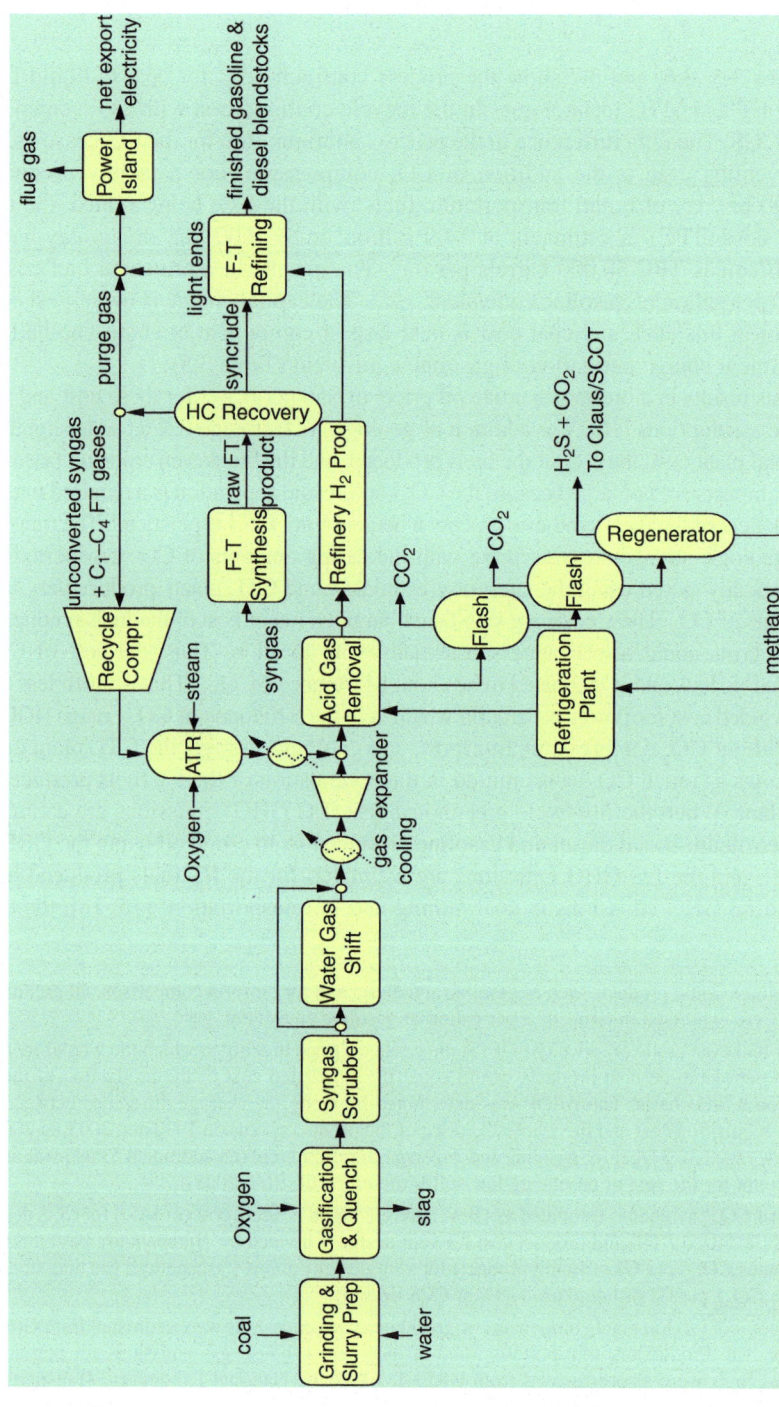

Fig. 3.5 Schematic of a coal-to-liquid fuels plant using the Fischer-Tropsch synthesis process, with CO_2 venting [15, 17]

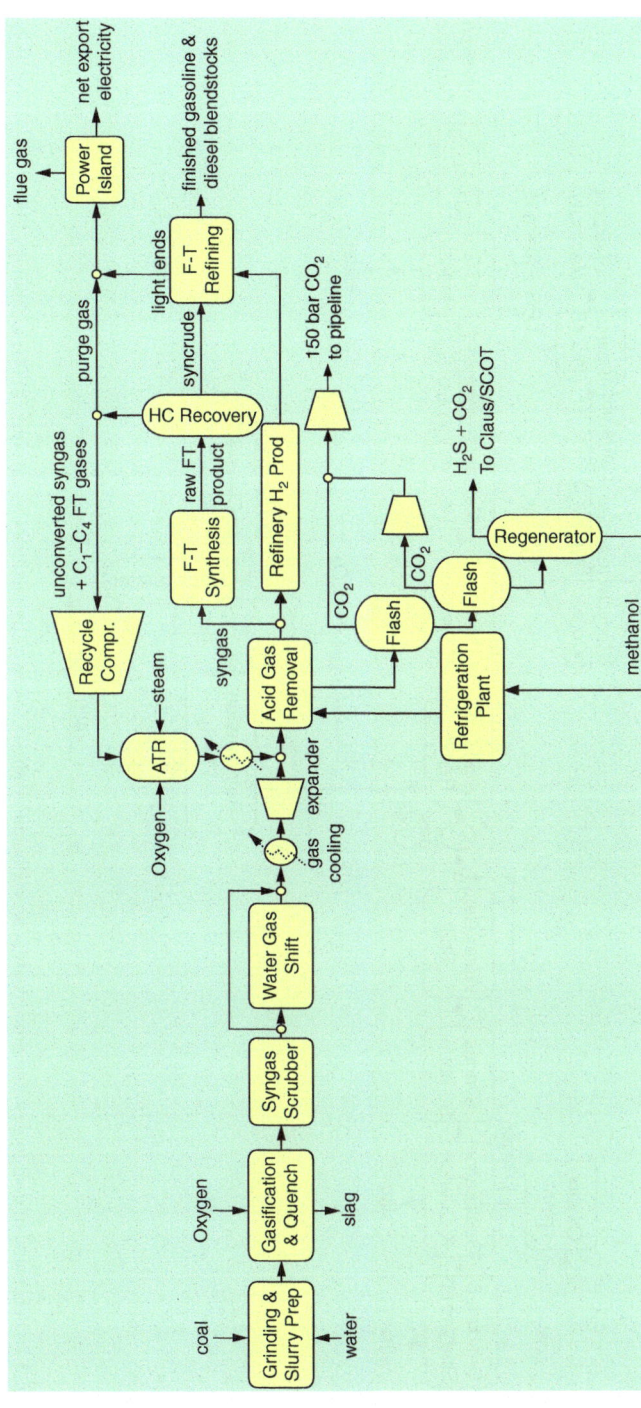

Fig. 3.6 Schematic of coal-to-liquid fuels plant using the Fischer-Tropsch synthesis process, with compression of CO_2 for transport and geologic storage [15, 17]

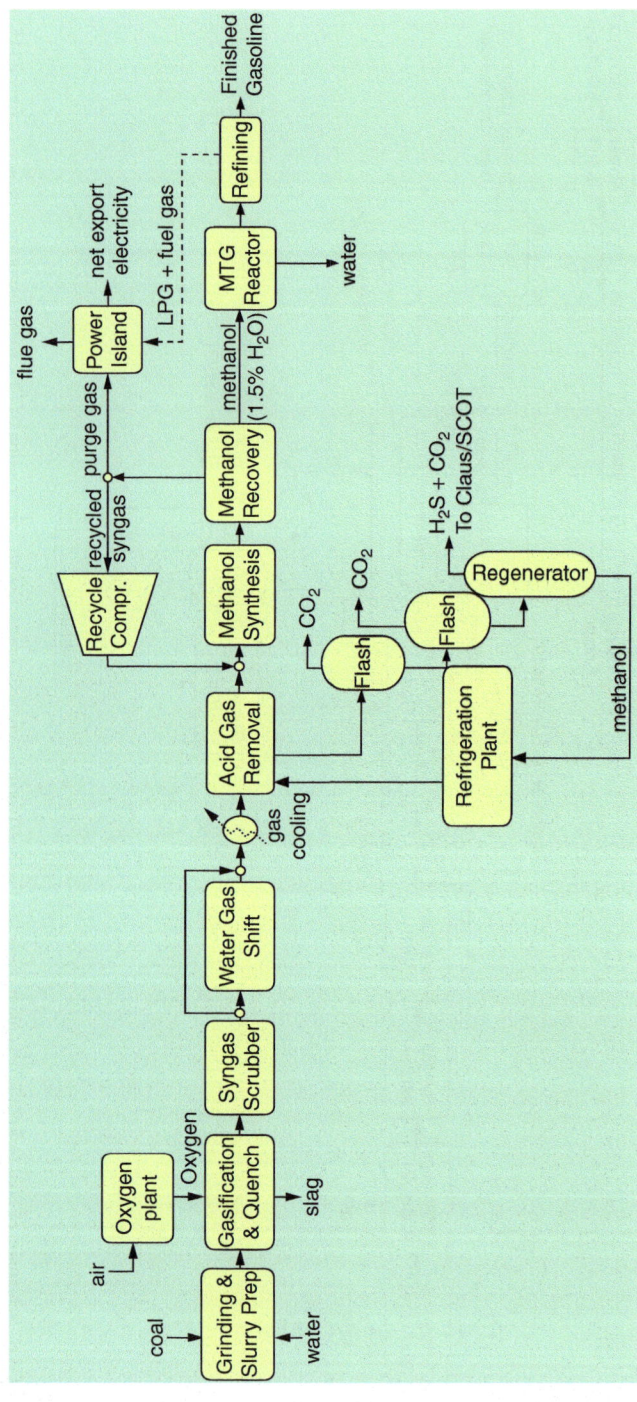

Fig. 3.7 Schematic of coal to gasoline process utilizing methanol synthesis followed by methanol to gasoline (MTG) technology [18]

of LC GHG emissions for the CTL fuels to that for the crude oil-based fuels is 2.18 (Table 3.5). If the CO_2 that is already separated from the synthesis gas, as part of the synthesis process, is compressed, transported, and geologically stored, the ratio of the LC GHG emissions for the FT produced fuels to that for the petroleum-based fuels becomes 1.03. Thus, with geologic CO_2 storage, coal-based liquid transportation fuels can be essentially equivalent to petroleum-based transportation fuels with respect to their LC GHG emissions.

This estimate is based on the assumption that the electricity sold to the grid displaces electricity, and its associated CO_2 "content", generated by an IGCC power plant with CO_2 venting in the case of an FT plant with venting; and displaces electricity generated by an IGCC power plant with CCS (90% removal) in the case of an FT plant with CCS. The CO_2 accounting approach used for the electricity displaced can obviously change the ratio of LC GHG emissions for the FT fuels relative to the same amount of petroleum-derived fuels; but because the electricity output represents only about 12% of the product out on an energy basis, the impact is not large. Based on the plant configuration used here, the coal to liquid fuels process captures and sequesters about 85% of the CO_2 that the plant emits without CCS. The ratio of LC GHG emissions for FT-derived fuels from CTL to that for petroleum-derived fuels could be less than 1.0 if changes in process configuration and operation were made to capture and sequester more CO_2, but this would incur additional cost. The detailed design choices would be driven by economic analysis if there were no policy requirements on relative emissions for synthetic fuels production.

3.4.2 Biomass to Liquid Fuels

For biomass gasification, the process configuration to produce liquid fuels using FT synthesis is similar to that shown in Fig. 3.5 except that gasification involves an oxygen-steam blown fluid-bed gasifier with quench rather than the GE/Texaco entrained-flow gasifier used for coal gasification. In addition, sulfur removal requirements are much lower because of the very low sulfur content of biomass. Because of the dispersed nature of biomass supply, the annual amount of biomass available for any one plant site was assumed to be one million tonnes(dry)/year which results in a feed rate of 3,580 tonnes (AR)/day [2,370 tonnes(dry)/day]. Switchgrass was used for these evaluations, but the estimates for corn stover were essentially the same.

Because of the smaller scale, about one-tenth the size as for CTL, and the higher cost of biomass vs. coal, the product cost is estimated at $3.05/gge, and the break-even crude oil price is $127/bbl with CO_2 venting (Table 3.5). The scale-economics effect shows up in the cost per stream-day-barrel, which is over $144,000 vs. the larger CTL plant where it is less than $100,000 (Table 3.5). If CCS is added to the scheme, the fuel cost increases to $3.32/gge; and the crude oil breakeven price is $139/bbl. From a GHG perspective, the LC GHG emissions for biomass based FT fuels/LC GHG emissions from equivalent petroleum-based fuels is −0.14 for the CO_2 vented case, and the ratio is −1.35 in the CCS case (Table 3.5). With CO_2 venting, the LC GHG emissions for the biomass FT fuels are slightly negative because of unconverted carbon in the char which is returned to the land and effectively stored there, offsetting fossil-based emissions associated with growing, harvesting and transporting the biomass. With CCS, the LC GHG emissions are highly negative because CO_2 is being removed from the atmosphere with each growth cycle and geologically stored. If methanol/MTG is used, instead of FT, the economics are more favorable, and the LC GHG emissions are similar. The cost of CO_{2eq} avoided is about \$20/tonne CO_{2eq} for biomass to liquid fuels. It is higher because of the higher cost per stream-day-barrel of the smaller plant.

3.4.3 Coal plus Biomass to Liquid Fuels

The benefit of producing liquid transportation fuels from biomass is that the LC GHG emissions are essentially zero (carbon-free liquid fuels); and with geologic storage of CO_2, the LC GHG emissions can be highly negative. The diffuse nature of biomass and its limited availability surrounding a given plant site dictate smaller plant scale. Smaller plant size combined with higher biomass cost result in a high liquid fuel cost. Converting coal plus biomass together to liquid fuels, referred to here as CBTL, can provide benefits of the economies of plant scale, and of lower average feedstock cost due to coal use, and can thus produce lower-cost liquid fuels with reduced LC GHG emissions. There are many ratios of coal to biomass that could be considered; but for illustrative purposes, a ratio that offsets the positive CO_2 emissions from coal-derived liquid fuels with the negative emissions potential of liquid transportation fuels produced from biomass is very illustrative. A liquid transportation fuel so-produced would have LC GHG emissions essentially equivalent to those of petroleum fuels if process CO_2 is vented; and it would be essentially carbon-free liquid fuel if CCS is practiced.

To achieve this would require a biomass/coal ratio of about 40% biomass to 60% coal on an energy basis. A CBTL plant consuming about one million tonnes(dry)/year of biomass would feed about 3,580 tonnes(AR)/day of biomass and 2,750 tonnes(AR)/day of coal. From a LC GHG perspective, it makes no difference whether both the biomass and coal are processed in the same plant (co-processing) or are processed in plants that are entirely separate in processing or in geography. However, integrating the plants and co-processing takes advantage of the economics-of-scale and reduces the cost of the liquid fuels produced.

Figure 3.8 illustrates a process configuration for the joint conversion of biomass and coal. Because of the different handling and gasification behaviors of coal and biomass, a different gasifier was used for each in this evaluation. The synthesis gas streams were combined after the gasification and gas clean-up to gain economies of scale downstream. The plant employs recycle around the reactor to maximize liquid fuel production, and the separated CO_2 is either vented or geologically stored.

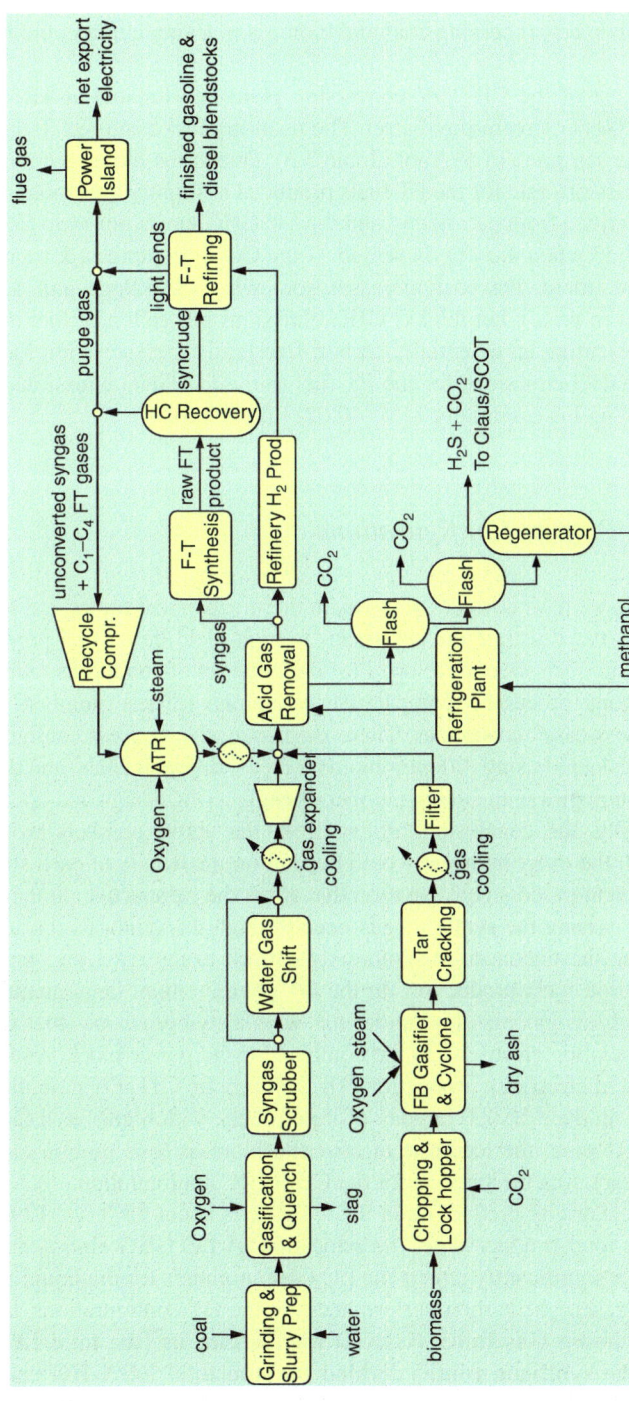

Fig. 3.8 Process configuration for the conversion of coal and biomass to liquid transportation fuels using Fischer-Tropsch technology with the venting of CO_2. The schematic for CCS is not shown here; if CO_2 CCS is to be included, the CO_2 streams are compressed as shown in Fig. 3.6 [15, 17]

Total liquid transportation fuels production is 10,000 bbl/day for this plant scale. As gasifiers evolve, co-processing coal and biomass in a single unit could become the norm.

The capital cost of the CBTL-V conversion plant is estimated at $1.3 billion (TPC) or $132,000 per stream-day-barrel. The resulting fuel cost is $2.31/gge with a crude oil breakeven price of $93/bbl (Table 3.5). Due to the biomass component, the LC GHG emissions rate for the FT fuels produced is slightly lower than that for the same fuels derived from petroleum, and the LC GHG emissions ratio (FT fuels/ petrol fuels) is 1.23 when the CO_2 is vented. If the CO_2 is compressed, transported and geologically stored, the cost increases somewhat ($2.52/gge and $103/bbl crude oil breakeven price), but the LC GHG emissions per gallon of fuel is essentially zero (representing an essentially carbon-free liquid transportation fuel), and the ratio of LC GHG emissions for the FT fuel to the LC GHG emissions for the petroleum-based fuel is essentially zero (−0.02) (Table 3.5).

3.4.4   Other Process Configurations

For clarity above, we have considered the main thermochemical routes and simplest cases to evaluate and illustrate the key issues in high-yield liquid fuel production. Design and cost numbers were developed for only one set of conditions to facilitate comparisons among processes (comparative economics). These numbers do not represent "business-case" economics. There are many other process configurations that could be considered using different biomass types, different coals, and different gasification technologies with a large number of engineering design and operational variants. Generally, the conclusions drawn from the above numbers would not change, although the exact numbers would depend on the details of each situation. However, a different process configuration that alters the perspective on the overall picture involves passing the synthesis gas once through the synthesis reactor (OT) and not recycling the unconverted synthesis gas back to the reactor to maximize liquid transportation fuels production. In the OT configuration, large quantities of unconverted synthesis gas plus low molecular weight hydrocarbons (fuel gas) are available for generating electricity in the combined-cycle power block, resulting in markedly increased electricity production [18–21]. For the CTL-OT case, the plant was designed to produce 50,000 bbl/day of liquid fuels, with higher coal feed rate. For the CBTL-OT case, the feed-rate of coal and biomass (one million tonnes of biomass (dry)/year), was the same as for the CBTL-RC configuration.

Because the electricity production in the OT case is about 35% (vs. 12% in the RC case) of the total product slate, the allocation of LC GHG emissions to the electricity portion significantly affects the LC GHG numbers for the liquid fuel. To more-clearly represent the carbon performance for the OT configurations, a global life-cycle **Greenhouse Gas Index (GHGI) was defined as {the total LC GHG emissions for the synthetic route} divided by {the total LC GHG emissions reduction resulting from the displacement of products produced by the**

Table 3.6 Cost and performance of once-through Fischer-Tropsch systems with increased electricity production [14, 19]

	CTL-OT-V	CTL-OT-CCS	CBTL-OT-V	CBTL-OT-CCS
Coal feed, tonnes/ day (AR)	33,140	33,140	3,420	3,420
Biomass feed, tonnes/ day (AR)	0	0	3,580	3,580
Liquid FT products, barrels per day	50,000	50,000	8,100	8,100
Power exported, kW$_e$	1,745	1,470	315	276
Cost of fuels, $/gge	1.08	1.39	2.10	2.48
Crude oil equivalent price, $/bbl	37	51	84	108
LC GGH ratio, synthetic/current	1.35	0.68	0.79	0.10

current route} (petroleum-derived gasoline and diesel and electricity from a supercritical PC plant venting CO_2 {831 g CO_2/kW$_e$-h}, Table 2.4, Chap. 2) [19].

A 50,000 bpd CTL-FT-OT plant requires 33,140 tonnes of coal/day (increased from 26,700 tonnes/day) and exports an estimated 1,740 kW$_e$ of electricity to the grid, compared with 427 kW$_e$ for the CTL-FT-RC plant, a four-fold increase (Table 3.6 vs. Table 3.5). If the co-product electricity is priced at 6.0 ¢/kW$_e$-h, the 2007 U. S. average electricity generating cost, the cost of the liquid transportation fuels produced is projected to be about $1.10/gge, with a breakeven crude oil price of about $37/bbl. The CTL-FT-OT plant is simpler, lacking all the recycle equipment, and the plant cost on a fixed coal feed rate basis would be about 10% less. For the 50,000 bpd CTL-FT-OT plant, the total plant cost is estimated at about $5.8 billion. Plant cost per tonne of feed coal is down, but on a stream-day-barrel basis it is up because less liquid fuel is produced per tonne of coal. The liquid fuels produced cost about 25% less than liquid transportation fuels produced in the recycle case. A combination of lower plant cost and improved FT heat utilization, involving integration of steam produced in the FT synthesis into the power island steam cycle leading to higher power generating efficiency, is responsible for this.

For the same CTL OT plant with CCS, the cost of the liquid transportation fuel produced is about $1.40/gge, and the breakeven crude oil price is about $50/bbl. The cost of CO_{2eq} avoided for this case is $18/tonne. This is still lower than that for power generation from coal, which for IGCC is about twice that.

When coal (~63% on an energy basis) and biomass are fed to a smaller CBTL-OT plant (one million tonnes of biomass(dry)/year) producing about 8,100 bbl/day liquid transportation fuel, the total plant cost is about $1.3 billion for a CO_2 venting plant and $1.4 billion for a CCS-based plant. The cost of the fuel produced increases to about $2.10 and $2.48/gge for a CO_2 venting OT plant and for a CCS OT plant respectively when the CO_{2eq} price is zero. The LC GHG emission rate for the CBTL-OT plant is about 80% of the LC GHG emission rate for these products as conventionally produced today, i.e., the GHGI is 0.8. For the CBTL-OT-CCS case, the liquid transportation fuels are essentially zero carbon and the electricity exported is decarbonized on a LC GHG basis (Table 3.6). The CO_{2eq} avoided cost is about \$20/tonne CO_{2eq}. If the MTG synthesis route was used rather than the FT synthesis route to produce liquid transportation fuel, the costs would be less; but the general conclusions would still hold. The product mix would be different[5]. Commercial product demand may be an important driver in process decisions in this situation.

3.4.5   Direct Liquefaction of Coal

A detailed set of engineering design estimates for direct liquefaction of coal has not been reported recently. Cost and performance estimates done here were based on 1993 U.S. DOE design estimates [22, 23]. These estimates were updated to 2007, but they are not considered as definitive or on as consistent a basis as the estimates used for indirect liquefaction. The liquid products of direct liquefaction are typically very aromatic and contain large fraction of sulfur-, nitrogen-, and oxygen-containing molecules. Costs associated with the production of clean transportation fuels that meet U.S. or developed-world fuel specifications have typically not been included in published cost estimates. Direct-liquefaction product quality has more typically been representative of heavy feed to a refinery rather than fuel products from it. For this work, the cost of upgrading all product streams was estimated so that only clean liquid transportation fuels were produced by the process. Total capital cost for a direct liquefaction plant producing 50,000 bpd of liquid transportation fuels, including complete upgrading to finished products specifications was estimated at \$5.5 billion, or about \$115,000 per stream-day-barrel. The best estimates suggest an overall thermal efficiency of about 60%, and a yield of about 2.6–3.0 bbl liquid liquid transport fuels per tonne of coal [24]. This compares with about 2.1 bbl of liquid transportation fuel per tonne of coal, and a thermal efficiency of 50% for indirect liquefaction using FT synthesis. Duddy [25] recently reported a capital cost of \$115,000 per stream-day-barrel and 3.1 bbl of liquid transportation fuels for Axens' direct coal liquefaction process.

Direct liquefaction plant emissions are projected from 5 [25] to 8.5 [17] kg CO_2/ gal of product. This would make direct liquefaction plant CO_2 emissions less than that for the FT plant. The estimated cost of liquid transportation fuels produced is about \$0.20/gal higher than for a comparable indirect CTL plant using FT synthesis. The LC GHG footprint, with CO_2 venting, of a direct liquefaction plant is expected to be slightly better than that for a venting CTL-FT plant. The direct liquefaction plant with CCS will be disadvantaged relative to the indirect liquefaction plant because it will have more flue-gas CO_2 that will have to be recovered, and that

---

[5]It is important to note that although these conclusions are considered valid, there is much less engineering data on MTG and few design calculation relative to FT and more are needed.

is more costly than CO_2 capture in indirect liquefaction. This could be improved through engineering modification to the plant design, but these changes will come at a cost and will not overcome the higher CO_{2eq} avoided cost. To improve the quality of these numbers, more definition to direct liquefaction requires more and better engineering data and up-to-date estimates of capital and operating costs and process importance.

3.5   Other Biomass Options

3.5.1   *Ethanol*

The chapter to this point has focused on converting coal, coal plus biomass, and biomass by thermochemical approaches to liquid fuels that fit directly into the existing fuel infrastructure. Biochemical approaches, such as converting corn grain or sugar cane (starch and sugar) to ethanol by fermentation, or by conversion of lignocelluloses to ethanol by depolymerization to sugars and then fermentation to ethanol are not evaluated in as much detail in this chapter. However, these latter processes may become important for producing ethanol for transportation fuel from biomass and for addressing CO_2 emissions in transportation. Ethanol has a high octane number and contains 80,000 Btu per gallon vs. gasoline, which has 115,000–119,000 Btu per gallon. Thus, about 1.45 gal of ethanol are required to replace 1 gal of gasoline. The cost to produce grain ethanol, which is a mature technology, in 2007 was estimated at about $2.50 per gallon on an energy equivalent basis [26].

Biochemical conversion of starch from corn or wheat involves conversion of starch to six-carbon sugars and its fermentation to ethanol using natural yeast. Sugar from sugar cane is fermented directly. Fermentation technology is mature, commercially robust, and highly optimized. Sugar cane to ethanol is mature and practiced in large volumes in Brazil. Sugar cane-based ethanol has a LC GHG emission that is ~85% lower than petroleum-based fuels. This is in contrast to grain ethanol, as produced in the U.S. from corn, which produces about a 25% reduction in CO_2 emissions over petroleum-derived fuels on an energy-equivalent basis because of all the fossil fuels used in grain ethanol production [27, 28]. Furthermore, ethanol from sugar cane could provide a roughly 90% reduction in petroleum imports vs. petroleum-derived fuels on an energy equivalent basis [27]. However, grain ethanol can reduce petroleum imports by a larger amount also because limited petroleum fuels are used in growing and harvesting corn, as with sugar cane ethanol. However, recent analysis of the full life-cycle of ethanol production from corn grain in the U.S. indicates that it produces only a small net energy gain over fossil fuel energy needed to produce it [27, 28]. Another major issue for grain ethanol is the volume challenge. The U.S. (in 2007) used 25–30% of its corn crop to produce about 3% of its transportation fuels, and this affected world food prices, etc. This is not sustainable.

Production of ethanol from the majority of the plant mass, the lignocellulosic portion, provides a larger feedstock source to convert; and if it is not grown as an energy crop on arable land, it need not have a negative impact on the food chain. Crop wastes can be used without affecting food production, if the not he proper agronomy practices are used [17]. Lignocellulose conversion is more difficult than converting starch in corn or sugar from sugar cane to ethanol, because it first requires that the sugar molecular components must be broken out of the hemicellulose and cellulose structures of the biomass. This produces six-carbon and five-carbon sugars which offer additional challenges to convert to ethanol than starch or sugar cane. This makes the process more challenging and more costly than grain ethanol production. Even at high crude oil prices, grain ethanol still required public subsidies to make it economic because increased demand drove up corn price as crude oil price increased. Cellulosic ethanol has the advantage of cheaper feedstocks and thus potentially, lower production costs. However today, much of this is offset by more process steps and longer process residence time, which drive up cost. Enzyme costs are also high but are expected to be reduced with volume; feedstock costs remain a major fraction of ethanol costs and are not expected to make major reductions. Several cellulosic ethanol commercial demonstration plants are under construction, and the next few years will probably experience much development in the technology.

For cellulosic ethanol, cost remains a major issue. Cost reductions are expected, but the extent achievable will be better known with the first round of demonstration plants. Paustain et al. [29] estimated the cost of cellulosic ethanol at 1.95 ± 0.65 per gallon ethanol for biomass costing $35 per dry tonne. This is about 2.90 ± 1.0 on an energy equivalent basis. At a more realistic biomass cost [17], the ethanol cost would be in the range of $3.0–$4.0 per gallon on an energy equivalent basis. If CO_2 emissions in the transportation sector have a price associated with them, the relative economics would improve.

Other options, driven by the rapid advances that are occurring in the biological sciences and in synthetic biology, could markedly change biochemical production of liquid transportation fuels. Some of this is summarized briefly below; and if one or more are successful, the biofuels picture could change substantially.

3.5.2 Butanol

Biobutanol (butanol or butyl alcohol) is another potential entrant into the liquid transportation fuel (biofuel) market. Butanol is a four-carbon alcohol vs. the two-carbon alcohol, ethanol. Butanol made from biomass is typically referred to as biobutanol. Its longer hydrocarbon chain makes it fairly non-polar and thus more similar to gasoline. Butanol has a number of attractive features as a fuel. Its energy content is closer to that of gasoline (105,000 Btu/gal for butanol, vs. 115,000–119,000 Btu/gal for gasoline); it has a lower vapor pressure; it is not hygroscopic; and thus, it is not sensitive to water, and does not pose the problems of ethanol and water in the distribution system. It is less hazardous to handle and less flammable than gasoline, and it has an octane similar to gasoline. Thus, it can go directly into the existing fuel distribution system. It has been shown to work in gasoline engines without modification [30].

Several technologies to produce butanol are in the R&D phase. The one receiving the most attention is the acetone-butanol-ethanol (ABE) process, which was initially used to produce acetone for making cordite in 1916. The process produced about twice as much butanol as acetone, and also produced acetic, lactic and propionic acids in addition to ethanol and isopropanol. As currently being commercialized, this process involves the bioconversion of sugars or starches using a genetically engineered microorganism, Clostridium beijernickii BA101, which shows greater selectivity to butanol. There are also efforts to develop improved microorganisms that have increased reaction rate and selectivity in the conversion of sugars to butanol. This includes microorganisms or enzymes that can efficiently convert the different sugars that are obtained from cellulose and hemicellulose. Because butanol is toxic to the producing organism, the butanol concentration is limited to about 15–18 g/l even for the native organism that produces it. This problem is similar to the self-inhibition and resulting ethanol concentration limits that occur in grain ethanol and cellulosic ethanol production. Overcoming these limitations and increasing the conversion rate would positively impact these technologies.

Isobutanol is less toxic and is also a good fuel component. It has a higher octane than gasoline but needs to be used in blends because of its high melting point (78°F or 25.5°C). As such, a potentially more-promising approach to improving the process is to engineer organisms that produce mixtures of butanol and isobutanol. Atsumi et al. [31] have recently engineered E. Coli to produce isobutanol in high yield with high specificity from glucose.

An extension of this technology is the conversion of cellulose to butanol. This depends on the development of biotechnologies for the depolymerization of cellulose and hemicelluloses into the basic sugars. These are the same problems faced by cellulosic ethanol. These sugars can then be converted to butanol. The most important development would be microorganisms that could depolymerize the biomass components into sugars and then convert the sugars into butanol, in the same reactor to reduce capital cost. The cellulose approach to butanol is being studied, but the technology is in the research stage and is far from commercial.

Currently, butanol's main drawback is cost. To attack the cost challenge and initiate market entry, Dupont and BP have joined forces to retrofit an existing sugar-based ethanol plant to produce butanol using Dupont-modified biotechnology [32]. Improved next-generation bioengineered organisms could be available within the next few years [32]. According to DuPont, existing ethanol plants can be cost-effectively retrofit to butanol [30].

Because butanol production should markedly reduce some of the highly energy intensive operations associated with grain ethanol production, it should have improved LC GHG performance compared with grain ethanol. If the production route starts with lignocelluloses, the LC GHG reductions should be similar to those of cellulosic ethanol or sugar cane-based ethanol because in these processes the energy-intensive separations involve biomass-generated energy rather than fossil-based energy.

3.5.3   Other Bioconversion Options

Over the last 25 years, significant research efforts have focused on algae for biofuel production and particularly biodiesel production. Algae can be grown in both salt and fresh water environments, in shallow ponds, tubes, or raceways utilizing waste mineral micronutrients. One area of research is the development of algae that have high lipid productivity [33, 34]. The algal oil would be extracted from the collected algae and converted to transportation fuels. Alternately if algae could be developed that excrete the lipids or excrete specific hydrocarbons, the algal collection could be eliminated reducing the cost. Progress had been made in the early 1990s, but due to decreased interest, efforts greatly slowed about a decade ago. Recent reevaluation suggests that current costs are well over $4.00 per gallon of fuel, and much more progress is needed if this technology is to have an impact in the foreseeable future [33]. Many of the challenges are engineering challenges associated with how and where to grow the algae and how to achieve needed productivity. It appears that algal growth needs to be in closed systems to avoid contamination from other strains. Strains that consume large quantities of energy producing lipids tend to multiply slowly; whereas strains that do not produce substantial quantities of lipids multiply rapidly, quickly outnumbering and overwhelming the desirable strains in the "soup". This is a particular problem with open system algal growth. Alternatively, algae could be grown as a source of cellulose for biofuels production.

With the rapid developments in synthetic biology and the increasing ability to engineer new metabolic pathways into organisms to produce specific chemical or fuel products, the areas of synthetic biology and metabolic engineering for renewable fuel production offer significant potential and are receiving intense interest [35]. The approaches being followed include using well-established recombinant DNA techniques to insert genes into microbes to make specific fuel precursors or even direct synthesis of hydrocarbon fuel components. Redesigning genes with computer assistance to accomplish specific synthesis and then synthesizing the desired genes for insertion into microbes hold significant potential. Yeasts can also be engineered to produce larger amounts of lipids which with additional metabolic engineering could be converted to useful products, potentially fuels. This work has not progressed as far as the work on bacteria to date. All of this work is far from the stage of being able to reliably estimate costs for fuels produced, but the intermediate to longer-term offers large potential.

Engineered microbes that produce and excrete specific hydrocarbons minimize the energy-consuming separation costs, although developing organisms that excrete the fuel products are a major challenge since most synthesis products, including hydrocarbons accumulate within the cell. However, properly designed hydrocarbon products in either the diesel range or gasoline range would not require significant refining and could fit directly into the existing infrastructure without building new infrastructure, as is needed for ethanol at larger scale. Although there are no specific processes that are approaching commercial evaluation, the level of activity and the current rate of progress could change that in the future. One example is [36] employing synthetic biology to produce bacteria that make increased amounts of fatty acids and also inserting genes that produce enzymes that convert the fatty acids into hydrocarbons which are then excreted. The bacteria are claimed to make and excrete hydrocarbons of any length and structure that is desired. The hydrocarbon phase-separates from the growth medium, markedly reducing separation costs. The feedstock for the bacteria is renewable sugars which can be obtained from sugar cane or grain or from cellulosic biomass [36]. A number of other attempts [35, 37] are under way.

In another approach, biomass is gasified, producing syngas, and the syngas mixture of carbon monoxide, carbon dioxide, and hydrogen are converted to alcohols using an aqueous bacterial culture [38]. The challenges here are substantial in that this hybrid approach incurs the costs of thermal-chemical conversion and biochemical conversion and must resolve some of the challenges that each has to offer. At the same time, it can also take advantage of the benefits of each.

The array of activities, approaches, and start-up companies that are addressing these challenges offers the potential of evolving successful technology for the production of liquid biofuels that can compete with conventional fossil-based fuels. This potential is increased when (1) a longer-term increase in petroleum prices, and (2) a price or cost associated with CO_2 emissions are considered. The occurrence of these latter two could well correspond with the timing required to develop to commercial readiness any one these bio-based options. Their potential is increased significantly when the on-going advances in synthetic biology and metabolic engineering are considered. Predicting the technology that will and will not succeed commercially is not possible, nor can reliable cost estimates be made at this point. Those technologies in early research and development need to advance further to evaluate their potential for commercial development and economic competitiveness.

3.6   Integration and Interpretation

3.6.1   Fuel-Focused Costs and Emissions

Figure 3.9 shows the range of fuel costs given in $ per gallon of gasoline equivalent (gge) for the thermochemical technologies and process configurations evaluated. The scale economics of CTL and the low cost of coal contribute to the relatively low liquid fuel cost for FT synthesis even thought the process requires a large capital investment. These cost estimates are internally consistent; but as a group, they depend on the capital and operating cost assumptions and the operating performance.

The breakeven crude oil price is about $56/bbl for CTL-FT-RC operated in the CO_2 venting configuration. The fuel cost and breakeven crude oil price are about $1.60/gge and $65/bbl respectively, or about 10% higher, in the case of FT with

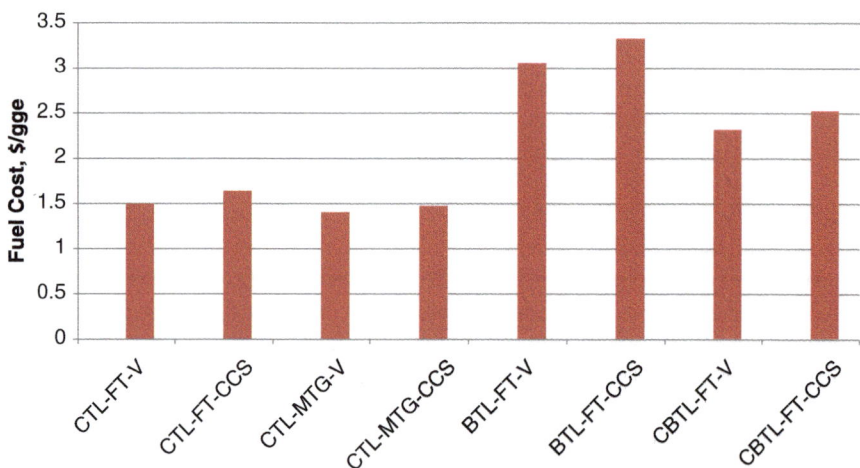

Fig. 3.9 Cost of liquid transportation fuels produced from coal (CTL), biomass (BTL), and coal plus biomass (CBTL) by Fischer-Tropsch (FT) and by methanol synthesis followed by methanol to gasoline (MTG) operated in the recycle mode with either venting the CO_2 (V) or geologic storage of CO_2 (CCS)

CCS than if the CO_2 is vented. The resultant CO_{2eq} avoided cost is \$11 per tonne CO_{2eq}. In the case of methanol synthesis followed by methanol to gasoline (MTG), the capital costs are lower because of higher product selectivity, reduced process complexity, and resultant reduced costs associated with the MTG technology, and the cost of the fuel produced is lower on an energy equivalent basis. The CO_{2eq} avoided cost is also about \$10/tonne. Sale of LPG rather than producing power from it, could further reduce the cost of gasoline produced.

Fuels produced from biomass only are significantly more costly because of the smaller plant size and higher biomass cost (Fig. 3.9). In this case, the breakeven crude oil price is about \$125/bbl (\$3.1/gge fuel) and \$140/bbl (\$3.3/gge fuel) for BTL with venting and with CCS respectively. The CO_{2eq} avoided cost is about \$20/tonne CO_{2eq} in this case, driven by the higher costs. When coal and biomass are combined in the same plant, the gains in economies of scale and in reduced feedstock cost driven by the addition of coal result in reduced liquid transportation fuel cost (\$2.30/gge with CO_2 venting and \$2.50/gge with geologic storage) as shown in Fig. 3.9.

Figure 3.10 shows the full LC GHG emissions for transportation fuels produced by the process configurations discussed above for coal, biomass, and coal plus biomass. LC GHG emissions for petroleum-based gasoline and diesel are included for reference. The figure illustrates that CTL-RC without CCS (CTL-V, FT) produces almost twice the LC CHG emissions as petroleum-derived fuels; and that when CO_2 is geologically stored (CCS), the LC GHG emissions are essentially equivalent to those from the same fuels produced from petroleum. LC GHG emissions from BTL (BTL-V, FT) can be negative without CCS due to unburned

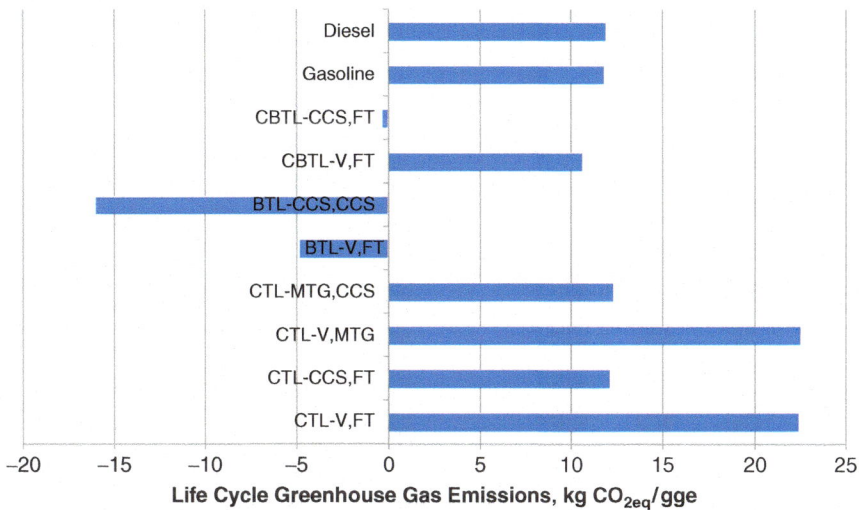

Fig. 3.10 Full life-cycle greenhouse gas emissions for transportation fuels produced from coal, biomass, and coal plus biomass using the indicated technologies, operated in the recycle mode, without and with CO_2 storage

carbon that is permanently sequestered with the char. With CCS, BTL (BTL-CCS, FT) has a large negative LC GHG emissions footprint (~−16 kg CO_{2eq}/gge) because CO_2 is being permanently removed from the atmosphere through geologic storage.

None of these biomass situations involve the potential reductions offered by the storage of carbon in the soil via root mass, which require more controlled agricultural practices. Soil carbon loss can also occur in biomass production, making the LC GHG problem worse. However, this situation is hard to predict, monitor, and verify, and as such, has not been included here; although in the end, soil carbon storage may be important in enhancing CO_2 reductions.

OT configurations have a lower cost of liquid transportation fuels (Table 3.6) than their equivalent process using recycle and optimized for high liquid transportation fuels production (Table 3.5). OT configurations also produce large amounts of electricity to export to the grid (Table 3.6). Although there are many ways to allocate the CO_2 emissions from the process plant, the approach used here was to calculate a greenhouse gas index (GHGI) which is the ratio of the total LC GHG emissions for the synthetic route to the total LC GHG emissions reduced by replacing the products produced by the current petroleum fuel/electricity route. Exported electricity was assumed to replace electricity with LC GHG emission equal to that generated by supercritical PC which vents CO_2. For cost calculations, the exported electricity was valued at 6 ¢/kW$_e$-h [18], which was the average generating cost of electricity in the U.S in 2007 [39].

Figure 3.11 [40] gives the GHGI for CTL and CBTL in several configurations and for CBTL with several biomass to coal feed ratios (energy %). Bars above 1.0

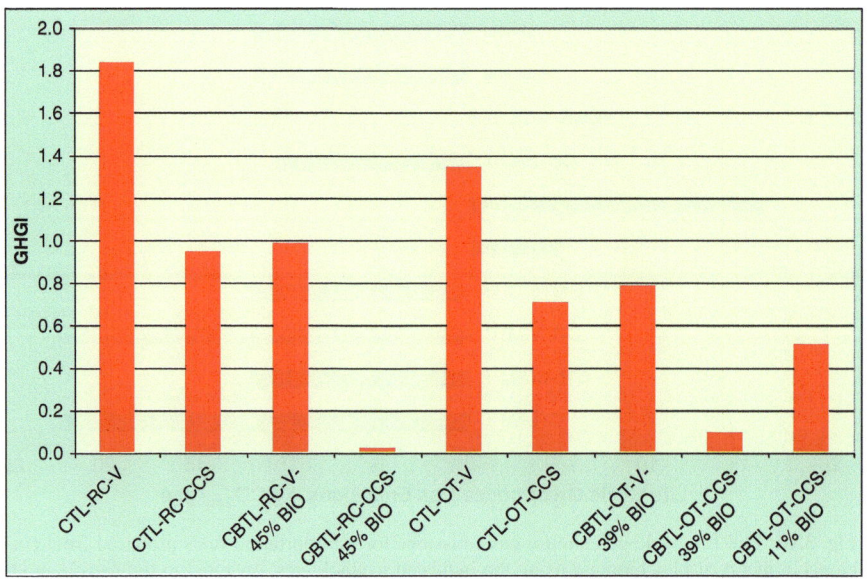

Fig. 3.11 GreenHouse gas index (GHGI) for several coal and coal plus biomass to liquid transportation fuels and electricity (Courtesy Williams and coworkers [20, 40])

result in increased LC GHG emissions vs. business as usual; bars below 1.0 result in reduced LC GHG emissions relative to business as usual. The figure indicates the important role that CBTL with CCS could play in decarbonizing the transportation sector and the electricity sector simultaneously. CBTL-RC-CCS and CBTL-OT-CCS with about 40% (energy basis) biomass in the feed produces essentially zero carbon transport fuels and carbon-free electricity. As indicated in the figure, different configurations provide different levels of LC GHG reduction allowing a country to balance addressing the issues of GHG emissions, energy supply, energy security, and crude oil replacement.

CTL-OT with CCS has a GHGI of 0.7 and thus produces only 70% of the LC GHG emissions of the products (transport fuels and electricity) produced conventionally. It produces lower carbon transportation fuels and partially decarbonized electricity. The extent attributable to each product depends on the allocation used. Furthermore, a high carbon price is not required to induce it, and the technology is effectively ready for commercial implementation today. This approach could address petroleum imports, energy security, and energy diversity while reducing CO_2 emissions from the transportation sector. The lower carbon electricity is competitive with that from a PC or IGCC with venting, and it could be equivalent to IGCC with CCS if the liquid fuels were allocated the CO_{2eq} of petroleum-derived fuels, making it a potential substitute for decarbonized baseload coal plants. A major challenge is the lack of interest shown by power companies in liquid fuels production and the resistance of the integrated oil companies in getting into the power business. This may change in the future.

As discussed in Chap. 2, supercritical CO_2 has been used in enhance oil recovery for over 30 years, and there are three large-scale commercial applications of geologic CO_2 storage that also have many years of successful experience. However, there is more work to do on CCS to answer questions and to gain public and political acceptance. The above analysis shows the importance of CCS in addressing our energy and CO_2 challenges, and it assumes that the technology will be broadly available with careful siting.

For CBTL (with about 40% biomass on an energy basis) with CCS, both liquid transportation fuels and electricity are effectively decarbonized (Table 3.6, Fig. 3.11). CBTL combines the limited biomass resource with coal to produced decarbonized transportation fuels and decarbonized, base-load electricity at a reasonable cost and a low LC GHG emissions price. For biomass alone, the LC GHG emission rate is highly negative, but the cost of the produced fuel is higher. The CO_{2eq} avoided cost for the OT mode is about twice as high as for the high liquid transportation fuels cases ($20/tonne CO_{2eq} for coal once through vs. $11 per tonne CO_{2eq} avoided for coal to liquids via recycle). The once-through cases have a lower cost of CO_{2eq} avoided than the cost of CO_{2eq} avoided for power generation. For IGCC, the CO_{2eq} avoided cost is ~$40/tonne CO_{2eq} (IGCC-venting to IGCC-CCS). Once-through processes with coal, coal plus biomass, and biomass, which produce large fractions of both transportation fuels and electricity, appear to be an important approach for decarbonizing both electricity and liquid transportation fuels and in doing so at an acceptable cost.

Figure 3.12 shows the impact of LC GHG (CO_{2eq}) price on the breakeven crude oil price of the liquids produced by the various feedstock and process combinations. At zero CO_{2eq} price, CTL-OT-V produces the lowest product cost, and CTL-RC-V is next, both venting the CO_2. However, the transportation fuel cost for CTL-OT-CCS is equivalent to the CTL-RC-V option at a zero CO_{2eq} price and decreases with increasing CO_{2eq} price; whereas the transportation fuel cost for CTL-RC-V increases with increasing CO_{2eq} price.

The CBTL options in Fig. 3.12 involve a feed mix of about 60%/40% coal to biomass on an energy basis. CBTL2 is a case that provides liquid transportation fuels that have the same LC-GHG emissions as petroleum-derived liquid fuels and involves roughly 92%/8% coal to biomass on an energy basis. Technologies involving biomass also bring with them the opportunity to pick the option and to adjust the feedstock ratio to optimize cost as the price of CO_{2eq} increases. The most expensive fuel-cost technology at zero CO_{2eq} price, BTL-RC, becomes the cheapest when the price of CO_{2eq} reaches $90/tonne (Fig. 3.12). The point at which lines cross in Fig. 3.12 indicate the LC GHG price at which economics would drive a shift from one technology to the other. The IEA projects a LC GHG emissions price of $90/tonne of CO_{2eq} by 2030 to realize 550 ppm stabilization [41].

It is also important to compare biomass-based hydrocarbon fuels produced by thermochemical routes, such as Fischer-Tropsch, with biomass-based fuels produced by biochemical routes, primarily cellulosic ethanol. With the accuracy that

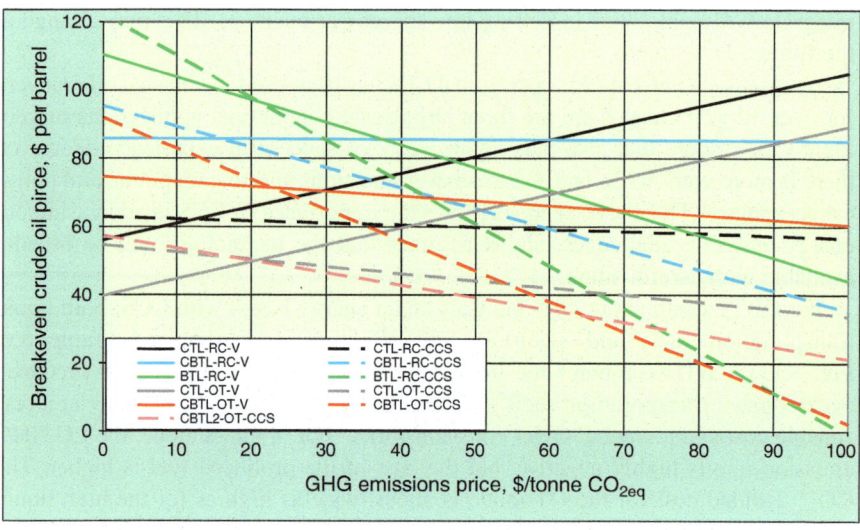

Fig. 3.12 Breakeven crude oil price for Fischer-Tropsch fuels produced by the various combinations of feedstock and process configurations as a function of GHG emissions price (Courtesy Williams and coworkers [18])

the cost numbers are known from the data available, the cost of hydrocarbon fuels produced from biomass by gasification/FT synthesis and ethanol produced by biochemical conversion of cellulose are similar on an energy equivalent basis. Effective commercial demonstration programs of these two different biomass-based technologies would provide the cost, engineering and operating data that are required to move forward with appropriate technology commercialization.

Using available data (for example [17]), the cost of liquid transportation fuels on a gasoline energy equivalent basis is plotted in Fig. 3.13 as a function of LC GHG (CO_{2eq}) price. For the biochemical route the cost range is from $3 to $4 per gge at zero CO_{2eq} price. Starting at the mid-point ($3.55/gge) the cost is almost independent of the GHG emissions price. For BTL-RC-V the transportation fuel cost follows the biochemical-produced fuel cost but decreases slightly. The transportation fuel cost for thermochemical conversion of biomass and biomass plus coal to liquids with CCS decreases more rapidly as shown by BTL-RC-CCS and CBTL-OT-CCS. The cross-over point for gasoline produced from crude oil at $80/bbl is about $30/tonne CO_{2eq} price for CBTL-OT-CCS. Costs favor the thermochemical conversion options, particularly as the GHG emissions price increases because CCS stores more CO_2 and more cheaply. The CBTL-OT-CCS option is the least costly and can provide zero-carbon liquid transportation fuel and decarbonized base-load electricity, each at a competitive cost. If this fuel production were coupled with future highly efficient hybrid electric vehicle (HEV) technology, it could provide a route toward significant reductions in GHG emissions from the light-duty vehicle fleet and in petroleum imports.

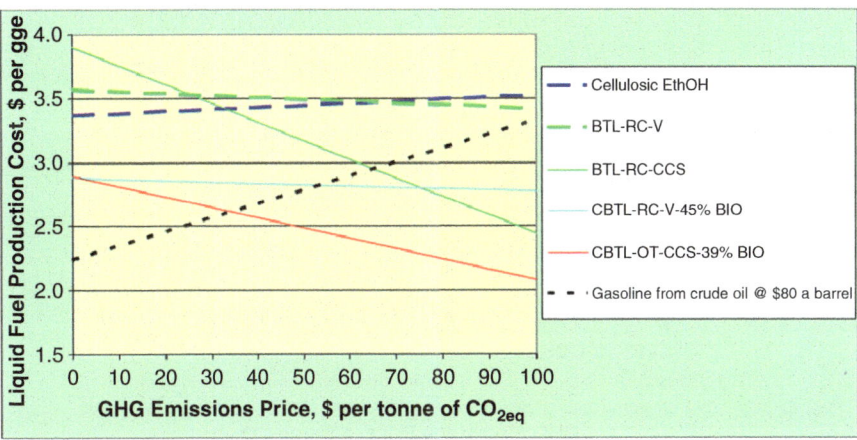

Fig. 3.13 Estimated cost of liquid transportation fuels produced from switchgrass by biochemical conversion and by thermochemical conversion routes as a function of GHG emissions price [14, 17] (Courtesy Williams [40])

3.6.2 Power-Focused Costs and Emissions

To this point, the chapter has focused on the production of liquid transportation fuels, and the net power was sold to the grid at a fixed price; this is a fuel producer view. Another approach is to look at the OT configurations as power producers would, and sell the fuel at the market price. The U.S. has about 314 GW of PC capacity which is mostly mid-size subcritical units that are paid off and reaching the end of their life. They have an average efficiency of 33% and a generating cost of about $26/$MW_e$-h for coal at $1.50/million Btu (Table 2.4). Simbeck and Pooritat [42] estimated the cost of power from a paid-off plant at $33/$MW_e$-h for a coal cost of $2.00/million Btu, consistent with the Chap. 2 estimate. They also estimated the cost of electricity for an amine retrofit to be $90/$MW_e$-h and for an IGCC-CCS rebuild to be $105/$MW_e$-h [42]. Figure 3.14 shows that the CO_2 capture options are economically unattractive without a very high CO_{2eq} price and that a paid-off sub-critical PC would be economically justified to continue to operate and just pay the tax until the price of CO_{2eq} reaches about $80/tonne CO_{2eq}. These are exceedingly high hurdles for coal-based power to overcome, and it suggests that few if any paid-off coal plants would be modified or retired for a long time, even though these are a major contributor to global CO_{2eq} emissions. This is particularly true for developed-world coal-based power generation fleet because few new coal plants will be built to meet demand growth, and CO_2 emissions reduction must be address the existing plants. In the developing world with electricity demand growing rapidly and new coal plants being built, the CO_{2eq} price that would drive the plant technology choice for new-build units from venting technology to capture technology is about half of that or about $40/tonne CO_{2eq} (Table 2.4).

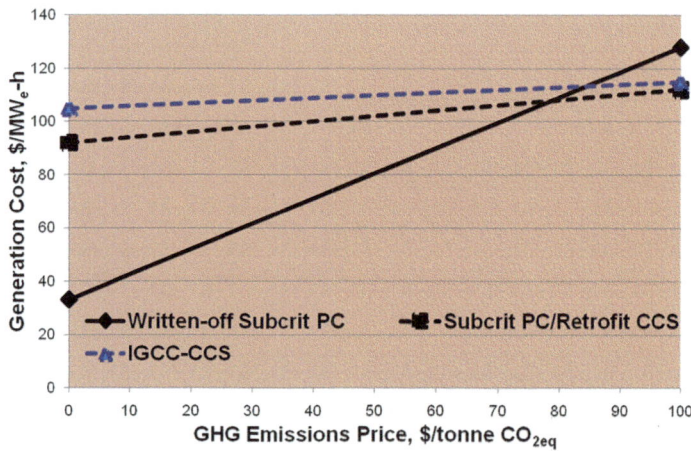

Fig. 3.14 Cost of electricity as a function of GHG price for paid-off subcritical PC and for CO$_2$ capture amine retrofit of a subcritical PC and for a rebuild with an IGCC with CCS

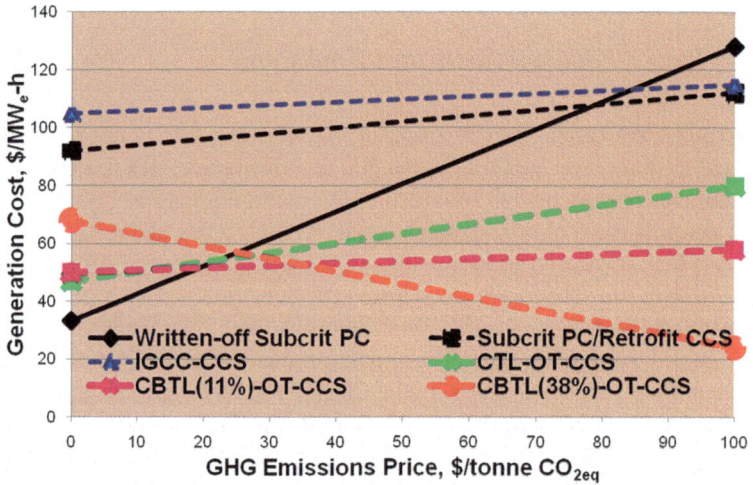

Fig. 3.15 Electricity cost as a function of CO$_{2eq}$ price for coal-based generation and for CTL-OT and CBTL-OT options selling the fuels produced into a fuel market with $100/barrel crude price

Figure 3.15 compares the COE for the three coal-based generating options discussed in Fig. 3.14 with the COE produced by CTL-OT-CCS and two CBTL-OT-CCS options. The transportation fuel produced by these OT options (fuels plus electricity) was sold at market price expected for a crude oil cost of $100/bbl. The figure shows that the electricity produced by these OT options is much cheaper than that for the retrofit or rebuild options at all CO$_{2eq}$ prices. Furthermore, the CO$_{2eq}$ price at which these options become competitive with the paid-off subcritical PC plant is below $30/tonne CO$_{2eq}$. The COE for the OT options is dependent on the crude oil price, and for \$80/bbl the COE produced is ~\$93/MW$_e$-h at zero CO_{2eq} price, i.e. comparable to that for the retrofit and rebuild cases. However, the COE price of the OT options decrease with increasing CO_{2eq} price so that at a low CO_{2eq} price (less than \$20/tonne CO_{2eq}), the OT technologies are economically favored.

Because in most electricity markets, power generators bid into the dispatch market; and in this market, generators will bid down to their minimum dispatch cost which is where their revenue equals their short run cost. Under these conditions OT systems such as CBTL-OT-CCS have two sources of income and can bid sufficiently low that they can defend high capacity factors and force competitors to lower capacity factors. Under these conditions, CBTL-OT-CCS should be able to defend high capacity factors with crude oil prices below \$60/bbl at zero CO_{2eq} price. At low, but expected CO_{2eq} prices and expected crude oil prices, CBTL should be the clear choice on an economic basis [40].

CBTL-OT-CCS offers strategic potential to reduce CO_2 emissions from subcritical PC units. For a biomass availability of one million tonnes (dry)/year and feeding 40% (energy basis) biomass with coal, the CBTL unit would export about 275 MW$_e$ to the grid and produce 8,100 bbl/day of transportation fuel (Table 3.6). This is about the generating capacity of a typical subcritical PC generating unit. With CCS, the CBTL-OT unit would replace the generating capacity of the PC unit with decarbonized (zero carbon) electricity. In addition it would produce 8,100 bbl/day of zero carbon transportation fuels, addressing CO_2 reduction in both the power sector and the transportation sector. This technology option requires careful consideration.

A rough thought experiment suggests that applying CBTL-OT technology to 450 million (dry)tonnes/year of biomass [17] combined with 400 million (AR) tonnes of coal could produce about 3.3 million bbl per day (50 billion gallons/year) of carbon-free gasoline energy-equivalent liquid transportation fuel and provide about 100 GW of decarbonized base-load generating capacity. With increased light duty vehicle (LDV) efficiency this could significantly reduce petroleum imports and significantly reduce LDV CO_2 emissions. This could decarbonize about 30% of the U.S. power generated by coal if it were built as retrofits at existing PC plants or built separately, replacing coal-based electricity. CBTL in the recycle mode would be able to generate about four million bbl/day (60 billion gallons/year) of gasoline equivalent transportation fuels, but less electricity. The cost of CO_2 avoided is much less than projections of the carbon price required to achieve stabilization [41]. If siting and other considerations reduced the amount of biomass that could actually be used by one half, this still clearly represents an important technology option for reducing CO_2 emissions from power generation and from the transportation sector. Furthermore, this also appears to be the lowest cost option to achieving each of these objectives.

3.6.3   Coal and Biomass Availability

A number of countries that typically have limited petroleum reserves have abundant coal reserves. The U.S. has abundant coal reserves. Recently EIA estimated proven U.S. coal reserves to be approximately 270 billion to 275 billion tonnes [43].

A recent NRC study estimates recoverable coal reserves at 227 billion tonnes [44]. These reserve numbers would suggest a 200-year supply of coal at current consumption rates. However, to meet increased demand will require increased mining and opening new mines, and there are environmental impacts associated with these actions. These will have to be dealt with adequately.

Biomass is a scarce resource and also a dispersed resource which raises issues of where and how it can be used best. It is "scarce" in the sense that if the most optimistic estimates of annual biomass availability (to one billion dry tonnes/year) are used to produce transportation fuel, the amount of fuel would be less than 5 million barrels/day, compared to the 12 million barrels/day of transportation fuels that the U. S. consumes [17]. Estimates of the quantity of biomass available on a sustainable basis vary considerably. Walsh et al. [45] estimated total biomass currently available in the U.S. at 460 million tonnes per year at $55/dry tonne in 1995 $ (~$75/dry tonne in 2007 $). Milbrandt [46] estimated the amount of biomass currently technically available (not considering cost) annually in the U.S. at 423 million dry tonnes per year. Perlack et al. [47] estimated over a billion dry tonnes per year of biomass technically available in 35–40 years with technology improvements. The amount of biomass that is sustainably available at an acceptable price is likely to be less then these technically feasible estimates. Estimates of the amount of biomass sustainable globally show that that biomass cannot replace a large fraction of fuel or power used on a global basis.

Furthermore, biomass will be used in a number of applications where it makes economic sense or where it satisfies policy mandates. The major options are power generation and liquid transportation fuels production both of which could be very large; but it will probably also be used in the chemicals and "petrochemicals" areas. Use of biomass in power will be driven by minimum renewables mandates for power generation and generating-unit specific mandates that a given percentage of biomass be fed along with coal. These latter mandates may be a condition imposed when the unit is permitted. For transportation fuels, minimum renewables mandates will continue to drive biomass into fuels applications. Economics should drive the route (process technologies) utilized to get it there, rather than choosing winners. There obviously are other considerations, but energy consumption is so large that economics count.

First, the limited options that exist for liquid transportation fuels production from sources other than petroleum mean that biomass-based fuels will have to be a significant component in transportation. Coal with CCS can provide liquid transportation fuels and diversity the U.S. away from petroleum, but it does not reduce GHG emissions in transportation. However, it also need not increase it; at best, it is neutral. Coal with CCS in power generation reduces GHG emissions associated with power generation. In addition, power generation has a number of options other than biomass to meet CO_2 reduction mandates. More importantly, CBTL-OT has the potential to produce both carbon-free fuels and decarbonized electricity. This suggests that the use of biomass for liquid transportation fuels is an essential and probably preferred component in managing GHG emissions from the transportation sector.

If we use biomass as a component in a CO_2 management strategy, it should be used where it provides the lowest cost per tonne of CO_2 avoided. As shown above, the cost per tonne CO_2 avoided is much lower for liquid transportation fuels production than for power generation. In addition, if we look at the demand side of transportation (vehicle options), the most cost effective option is the hybrid-electric vehicle (HEV) [48]. The HEV is about 20% less costly on an annualized basis vs. a plug-in hybrid electric vehicle (PHEV) and about 20% less costly than a 2030 normally aspirated spark-ignition vehicle [48]. If we now look at the supply-side options, the cost per tonne of CO_2 is lowest for the use of liquid fuel from biomass in an HEV compared to PHEVs or BEVs using electricity from biomass with CCS.

3.6.4   A Way Forward

First, geologic sequestration of CO_2 (carbon capture and storage (CCS)) is an essential technology for coal and coal/biomass to liquid fuels, as it is for coal and coal/biomass to power generation (Chap. 2). The discussion of this subject in Chap. 2 applies here as well. Most of the technologies discussed in this chapter have been demonstrated to be technically ready for commercial deployment.

However, for implementation with coal at the required scale, these technologies are not yet fully robust, cost-reduced, and optimized. The financing hurdle remains very serious, primarily because of the volatility of the energy markets; but deployment is also impacted by uncertainties over climate-change policy and by lack of full-scale commercial demonstration. The energy market uncertainty is illustrated by the price of crude oil between 2005 and 2009, and its over three-fold price collapse in 4 months. These projects have a multi-year time line from planning to operation, and each one requires funding from $1 billion to over $5 billion in capital. In this climate of uncertainty, they are faced with what is often been referred to as a "valley of death" in getting from here to a commercially viable industry. This transition will require certain and durable policies, predictable pricing, and a number of commercial first-mover projects combined with geologic storage of CO_2 that start the technology down the commercial learning curve to robustness, to process optimization, and to significant cost reductions. These commercial first-mover projects would have a major R&D component associated with them to focus on solving issues and problems identified and develop specific improvements. Learning-by-doing at a commercial scale is an important component of reducing cost, increasing operability, and providing engineering data to improve the next generation of processes. This would further develop the technologies, quantify their relative economics, and reduce the risk associated with their commercial deployment if they continue to show economic competitiveness. An aggressive program of commercial first-mover plants demonstrating these technologies integrated with CCS is critical to developing a tool-chest of robust technologies that can be applied when and where needed.

The U.S. consumes about 12 million barrels per day of liquid transportation fuel. To replace 25% of this fuel consumption with liquids from coal would require about 550 million tonnes of coal per year at 2 bbl liquids per tonne conversion efficiency. Current U.S. coal production is about one billion tonnes per year, which would require a 50% increase in U.S. coal production. In addition, we would need about 60 plants of the 50,000 bbl/day scale at a cost in excess of $250 billion.

As for biomass, 450 million tonnes of dry biomass per year could provide about 1.8 million bbl/day of transportation fuels. As noted above, 400 million tonnes (AR)/year of coal plus 450 million tonnes(dry)/year of biomass could produce up to four million bbl/day of liquid fuels (gasoline equivalent) which would be effectively zero carbon transportation fuels along with decarbonized electricity if geologic storage of the CO_2 is used. At the plant scale evaluated here, about 400 plants would be required. This is largely driven by the diverse nature of biomass, limiting plant size. Both of these scope estimates indicate the massive scale of our current transportation fuel consumption, the challenges of replacing a major fraction of it and of reducing CO_2 from it. Achieving major success at this goal will require major gains in vehicle efficiency along with adding zero carbon fuels to the system.

References

1. Marano JJ (2006) Overview of coal to liquids. DOE NETL, Pittsburgh
2. Kung HH (1980) Methanol synthesis. Catal Rev 22:235–259
3. Meisel SL (1981) A new route to liquid fuels from coal. Philos Trans R Soc Lond A300:157–169
4. Yurchak S (1988) Development of mobil's fixed-bed methanol to gasoline (MTG) process. In: Bibby DM (ed) Methane conversion. Elsevier, Amsterdam
5. Tabak S, Zhao X, Branki A (2008) An alternative route from coal to liquid fuels, exxonMobil methanol to gasoline process. In: First world coal-to-liquids conference, Paris
6. Hindman M (2010) Methanol to gasoline (MTG) technology. In: Third world CTL conference, Beijing
7. Zhang Y (2007) Development of Shenhua coal conversion industry in China. In: CTLtech 2007 4th forum on coal conversion technologies and investments, Beijing
8. Bechtel (2003) Gasification plant cost and performance optimization. Final Report, Global Energy, Nexant, San Francisco
9. Gray D, Tomlinson G (2001) Coproduction of ultra clean transportation fuels, hydrogen, and electric power from Coal. Mitretek, Falls Church
10. Gray D (2004) Polygeneration of SNG, hydrogen, power, and carbon dioxide. Mitretek, Falls Church
11. Bechtel (1998) Baseline design/economics for advanced Fischer-Tropsch technology. Final report, Apr 1998
12. Toman M, Curtright AE, Ortiz DS, Darmstadter J, Shannon B (2008) Unconventional fossil-based fuels: economic and environmental tradeoffs. Rand, Santa Monica
13. Tarka T (2009) Affordable, low-carbon diesel fuel from domestic coal and biomass. DOE/NETL, Washington, DC
14. Kreutz TG, Larson ED, Williams RH (2008) Personal communication, Princeton University, PEI

15. Kreutz TG, Larson ED, Liu G, Williams RH (2008) Fischer-Tropsch fuels from coal and biomass. [Paper]. Available from http://www.princeton.edu/pei/energy/publications/texts/Kreutz-et-al-PCC-2008-10-7-08.pdf
16. EIA (2007) Annual energy review. DOE/EIA (ed), Washington, DC
17. NRC (2009) Liquid transportation fuels from coal and biomass. National Academy of Sciences, Washington, DC
18. Kreutz TG, Larson ED, Liu G, Williams RH (2008) Fischer-Tropsch fuels from coal and biomass. In: 25th annual international Pittsburgh coal conference, Pittsburgh
19. Williams, R.H, Larson ED, Liu G, Kreutz TG (2008) Fischer-Tropsch fuels from coal and biomass: strategic advantages of once-through ('Polygeneration') configurations. In: Proceedings of the 9th international conference on greenhouse gas control technologies, Washington, DC
20. Williams R, Kreutz T, Larson E, Liu G (2008) Once-through synfuels production as a strategy for decarbonizing power in a carbon-constrained world, p 1–8
21. Larson ED, Fiorese G, Liu G, Williams RH, Kreutz TG, Consonni S (2008) Co-production of synfuels and electricity from coal+biomass with zero-net carbon emissions: an illinois case study. In: Proceedings of the 9th international conference on greenhouse gas control technologies, Washington, DC
22. U.S. DOE (1993) Direct coal liquefaction baseline design and system analysis. DOE, Pittsburgh
23. Kramer SJ (2008) Design and cost estimates on direct coal liquefaction, p 1–7
24. Tam S (2008) Direct coal liquefaction. In: NRC panel on alternative liquid transportation fuels, Washington
25. Duddy J (2009) Direct coal liquefaction axens H-coal process. In: World CTL 2009 conference, Washington, DC
26. NRC (2008) Transitions to alternative transportation technologies – a focus on hydrogen. National Research Council of the National Academies, Washington, DC, p 55
27. Farrell EA (2006) Ethanol can contribute to energy and environmental goals. Science 311:506–508
28. Hill J, Nelson E, Tilman D, Polasky S, Tiffany D (2006) Environmental, economic, and energetic costs and benefits of biodiesel and ethanol biofuels. Proc Natl Acad Sci USA 103:11206–11210
29. Paustain KJ, Antle J, Sheehan J, Paul E (2006) Agriculture's roll in greenhouse gas mitigation. Pew Center, Washington, DC
30. DuPont Company (2008) DuPont fact sheet on biobutanol, Wilmington, Delaware
31. Atsumi S, Hanai T, Liao JC (2008) Non-fermentative pathways for synthesis of branched-chain higher alcohols as biofuels. Nature 451:86–90
32. Chase R (2006) DuPont, BP join to make butanol: they say it outperforms ethanol as a fuel additive, in USA today. Associated Press, New York
33. Pacheco M (2006) The potential of biofuels to meet commercial and military needs. In: NRC presentation, Washington, DC
34. Briggs M (2004) Widescale biodiesel production from algae. Available from www.unh.edu/p2/biodiesel/article_alge.html
35. Savage N (2007) Better biofuels. Technol Rev 110(4):1
36. LS9 (2008) Renewable petroleum technology. Available from http://www.ls9.com/technology/
37. Amyris (2008) Technology platform. Available from http://www.amyrisbiotech.com/technology.html
38. Bullis K (2008) Ethanol from garbage and old tires. Technol Rev 111(2):96–98
39. EIA (2008) International energy outlook 2008. Available from http://www.eia.doe.gov/oiaf/ieo/electricity.html
40. Williams RH (2010) Senior research scientist, PEI, Princeton University
41. IEA (2008) World energy outlook for 2008. OECD, Paris

42. Simbeck D, Pooritat R (2009) Near-term technologies for retrofit capture and storage of existing coal-fired power plants in the United States. In: Moniz JMDAEJ (ed) Retrofitting of coal-fired power plants for CO_2 emissions reduction. Massachusetts Institute of Technology, Cambridge, pp 87–98
43. IEA (2007) Task force on strategic unconventional fuels. International Energy Agency, Paris
44. NRC (2007) Coal research and development to support national energy policy. National research council, Washington
45. Walsh M et al (2000) Biomass feedstock availability in the United States: 1999 state level analysis. Oak Ridge National Laboratory (ed), Oak Ridge
46. Milbrandt A (2005) A geographic perspective on the current biomass resource availability in the United States. DOE (ed), Washington, DC
47. Perlack R et al (2005) Biomass as feedstock for a bioenergy and bioproducts industry: the technical feasibily of a billion-ton annual supply. DOE-USDA (ed), Washington, DC
48. Kromer MA, Heywood JB (2007) Electric powertrains: opportunities and challenges in the U.S. light-duty fleet. Massachusetts Institute of Technology, Cambridge

Chapter 4
The Role of Nuclear Power in Reducing Greenhouse Gas Emissions*

Anthony Baratta[†]

Abstract As this chapter will point out, nuclear energy is a low greenhouse gas emitter and is capable of providing large amounts of power using proven technology. In the immediate future, it can contribute to greenhouse gas reduction but only on a modest scale, replacing a portion of the electricity produced by coal fired power plants. While it has the potential to do more, there are significant resource issues that must be addressed if nuclear power is to play a larger role in replacing coal or natural gas as a source of electricity.

4.1 Introduction

Currently, nuclear power provides approximately 19% of the total electricity generated in the United States and approximately 15% of the world's electricity. In 2007, 439 nuclear reactors located in 31 countries generated over 2,698 TWh[1] of electricity[2]. Of these, 104 nuclear power plants are located in the United States and generated a total of 806.5 TWh of electricity [1]. Over the next 40 years, the number of reactors in the U.S. and in the world is expected to grow as the world's demand for electricity becomes greater. These increases will be due to both population growth and the expanded use of electricity, with much of the growth associated with increases in electrical use in China, the rest of Asia, and the other third world

---

*The findings included in this chapter do not necessarily reflect the view or policies of the U.S. Nuclear Regulatory Commission or the Environmental Protection Agency. Mention of trade names or commercial products does not constitute Agency endorsement or recommendation for use.

[†] © US Government 2011

[1] TWh = Terawatt-hours = billion kilowatt-hours

[2] http://www.world-nuclear.org/info/reactors.htm

A. Baratta (✉)
Anthony J Baratta, US Nuclear regulatory Commission, Atomic Safety and Licensing Board Panel, MS T3-F23, Washington, DC 20555-0001.
e-mail: ab2@psu.edu

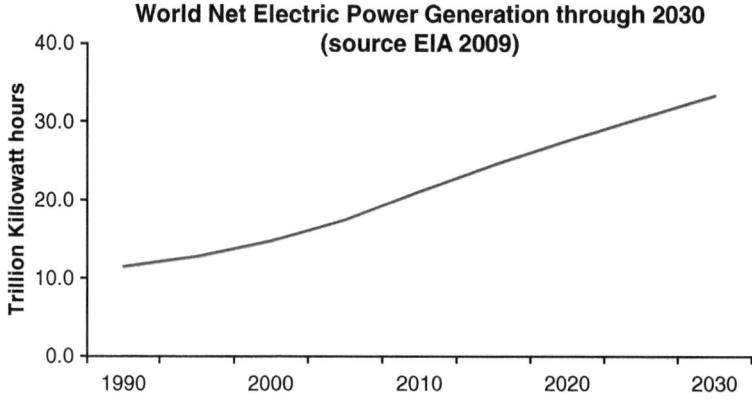

Fig. 4.1 Actual and projected world net electrical energy production in trillion killowatt hours through 2030 (U.S. Energy Information Agency (EIA) [2])

developing economies. Figure 4.1 depicts the expected growth in electricity consumption worldwide.

Another factor influencing the growth of nuclear energy is the concern over carbon emissions. There is growing interest by the public, and by some government officials, in decreasing the emission of greenhouse gasses. The exact way in which these policies will be implemented has yet to be determined. What is clear is that those technologies that emit greenhouse gasses will be penalized in favor of those that are low emitters.

Because nuclear power does not burn fossil fuels to generate electricity, it is often cited as a currently available source of electricity that could contribute to the lowering of greenhouse gases. To do so, nuclear reactors must be built at a rate greater than ever before. Several challenges must be overcome to achieve such a rate and include the availability of funding for construction, the lack of infrastructure for component manufacture, concerns over nuclear safety, the lack of a politically acceptable method to dispose of the spent fuel, and the lack of trained and experienced workers to build and operate these new facilities.

4.2 Nuclear Energy Basics

Nuclear reactors generate electricity by releasing the energy stored in the nuclei of uranium and plutonium atoms. To understand how a reactor works, one must look at the basic structure of an atom and the interaction of its components. An atom consists of a dense central core, called the nucleus, surrounded by electrons that may be thought of as orbiting the nucleus. The nucleus is made up of positively charged protons and neutral particles called neutrons; the strong nuclear forces hold the nucleus together, offsetting the repulsive force between the positively charged protons.

The chemical properties that we associate with an element are determined by the number of protons and electrons. The simplest atom, the hydrogen atom, has one proton in its nucleus and one electron circling the nucleus nearly. All atoms have isotopes which are different forms of that element. Isotopes have different numbers of neutrons in the nucleus, but have the same number of protons. Since it is the positive charge on the nucleus and the negative charges of the electrons that determine the chemical behavior of an element, isotopes of a given element exhibit the same chemical behavior. For hydrogen, there are three isotopes; the simplest form has no neutrons in the nucleus – only a proton. The other two isotopes of hydrogen, deuterium and tritium, contain one and two neutrons, respectively.[3]

Most elements have naturally occurring isotopes. Of particular interest are the isotopes of uranium, uranium 238, and uranium 235. Here, the numbers refer to the total number of neutrons plus protons found in the nucleus. Since the element is determined by the number of protons, the nucleus of an atom of uranium 235 will have the same number of protons as the nucleus of an atom of uranium 238, namely 92. The number of neutrons in the nucleus will be different, however.

As we move up the periodic chart, the neutrons and protons are bound more tightly to one another until we reach the element iron. Once we pass iron, the neutrons and protons are bound more loosely to one another. Hence, in the heavy elements such as uranium, the neutrons and protons are bound less tightly to one another. As a consequence, uranium undergoes a natural form of radioactive transformation called fission. The nucleus literally splits apart into two lighter nuclei in a process called spontaneous fission. Spontaneous fission occurs so very infrequently that it is of little technological interest. However, when struck by a neutron, the nuclei of both uranium 235 and uranium 238 may be induced to undergo fission. When they do, energy is released. If a sufficient number of nuclei fission nearly simultaneously, then a useful amount of energy is released. A nuclear reactor is a device designed to encourage the fissioning of the uranium fuel. It is important to note that the fragments or fission products that result from fission are highly radioactive and must be kept from the environment when produced in large quantities such as in a reactor.

Because of the structure of the uranium 235 nucleus, it is able to fission when struck with a neutron that is moving at any speed, even one moving relatively slowly. Uranium 238, however, requires a neutron that is moving very fast in order for fission to occur. To generate a useful amount of energy, there must be a large number of reactions occurring nearly simultaneously. In nature, the proportion of uranium that is uranium 235 is small, about 0.7%. For most power reactors, this amount is insufficient and the percentage of uranium 235 must be increased or enriched. The enrichment process relies on the slight differences in mass between uranium 235 and uranium 238 and is a very energy intensive one.

Two methods of enrichment are presently in use. The oldest method, the gaseous diffusion method, is based on the fact that a uranium 235 atom bearing molecules of uranium hexafluoride gas is lighter and moves at a higher speed than a uranium

---

[3]The difference in mass between hydrogen and its isotopes does cause the rates of reaction for chemical processes to differ. This difference is used to increase the concentration of deuterium to produce heavy water, which is rich in the isotope deuterium.

238 atom. A gaseous diffusion enrichment plant must have thousands of separation stages since each stage produces only an extremely small amount of enrichment. The gaseous diffusion enrichment method is a very energy intensive enrichment technique and its use is decreasing in favor of the more energy efficient centrifuge enrichment technique.

In the centrifuge method, cylinders containing uranium hexafluoride gas are spun at very high speed. The difference in weight between the uranium 235 bearing molecules and uranium 238 molecules again causes a natural separation to occur. Like the gaseous diffusion method, each stage produces only an extremely small increase in the concentration of uranium 235, requiring many thousands of stages until practical levels of enrichment are obtained. For a typical commercial reactor, the concentration of uranium must be increased to between 3% and 5%. Although still energy intensive, a centrifuge based enrichment facility will use about 1/50 the energy of a gaseous diffusion based enrichment facility.

Once enriched in uranium 235, the uranium hexafluoride gas undergoes a conversion process where the uranium is extracted and converted into uranium dioxide. The uranium dioxide is then formed into pellets about 1 cm in diameter and about 2 cm in length. These pellets are then loaded into cylinders made from an alloy of zirconium which is used since zirconium does not easily absorb neutrons. The cylinders form a cladding around the fuel, preventing the release of fission products into the reactor. Neutron conservation in a reactor is extremely important since it is the neutrons that induce the fissioning of the uranium 235 nuclei. Every effort is made to ensure that neutrons are conserved to encourage their absorption in the uranium fuel pellets.

The cylinders containing the enriched uranium dioxide fuel pellets are hermetically sealed fuel rods. A hundred or more of these rods are then assembled in a square array to form a fuel assembly. Depending on the type of reactor and the number of fuel rods in each assembly, there are between 175 and 400 assemblies needed to form the core of a nuclear reactor. About 100 metric tons of uranium is used to make up a typical reactor core.

The energy released in each fission reaction is far greater than the energy released when a carbon atom is chemically combined with oxygen atoms in a coal fired power plant, each fission releasing about 100 million times as much energy. This energy is in the form of the motion of the split uranium nucleus fragments. The fragments or fission products move off very quickly from the fission site and strike other fuel nuclei, raising the temperature of the fuel. In addition, the fission process produces neutrons that then induce more fissions. If just one neutron from each fission event goes on to induce fission, then the reaction is self-sustaining and the reactor is said to be critical.

The fission products are also highly radioactive and emit various forms of radiation. Some of this radiation is absorbed in the reactor as well, further increasing the temperature of the reactor. The heat generated is then used to heat water in much the same way that the heat from the combustion of coal is used to heat water in a coal-fired power plant. The heated water is then used to generate steam, which in turn generates electricity. The exact process depends on the type of reactor.

4.3   Power Reactor Basics

Two types of reactors are in use today, the pressurized water reactor and the boiling water reactor. In both, water is circulated through the reactor core to extract the heat generated by the fission process. In a pressurized water reactor (Fig. 4.2), the water is kept under very high pressure, about 2,000 psi or 15 Megapascal (MPa).

The pressure is high enough that the water does not boil. Pumps move the heated water to a device called a steam generator that acts as a boiler. In the steam generator, the heated water is passed through tubes where it heats water on the other side of the tube wall. This water is allowed to boil, generating steam. The steam then drives a turbine that is connected to an electrical generator. The advantage of this system is that the water used to generate steam is not exposed to neutrons in the reactor. The water that passes through the reactor core is exposed to neutrons and, as a result, picks up radioactivity.

In a boiling water reactor (see Fig. 4.3), the water that passes through the reactor core is kept under a lower pressure of about 900 psi or 7 MPa.

The water boils, becoming steam which is sent directly to the turbine. Since the water is exposed to the neutrons in the reactor, the steam is radioactive and the areas around the turbine and steam supply system are high radiation areas. The disadvantages of having such levels is somewhat offset by the simplicity of the system and its lower operating pressure. Today, about one-third of the commercial power reactors are boiling water reactors and two-thirds are pressurized water reactors.

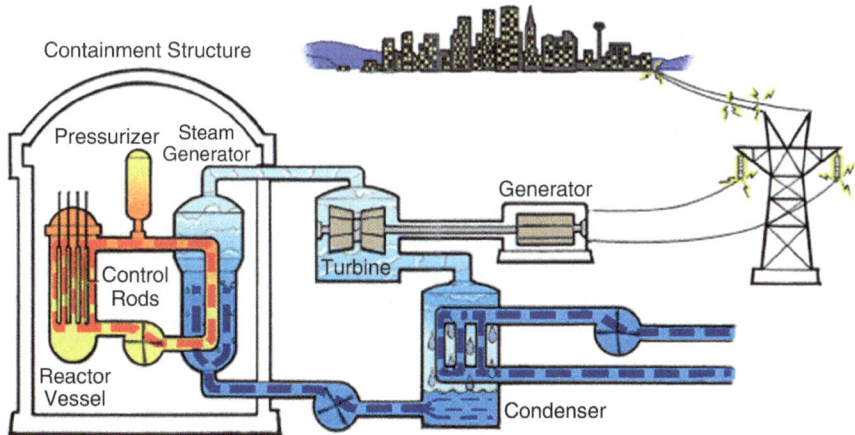

Fig. 4.2 Simplified representation of a pressurized water reactor (Courtesy of US NRC)

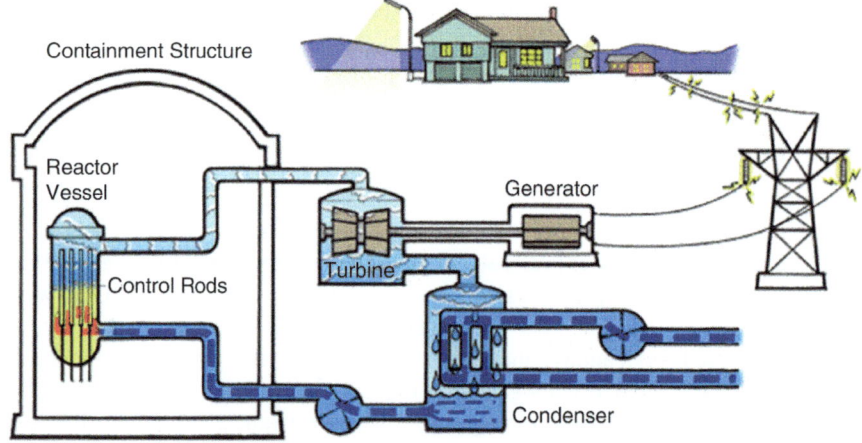

Fig. 4.3 Simplified representation of a boiling water reactor (Courtesy of US NRC)

Other types of reactors are also used, including one that uses natural uranium but has heavy water instead of light water[4], reactors that use gas as a coolant and also reactors that use liquid metals such as sodium to cool the reactor [3].

4.3.1 Spent Fuel Disposal and Reprocessing

As a reactor operates, the uranium 235 is consumed and fission products are built up. These fission products are highly radioactive and emit significant amounts of energy. In a typical reactor, the radioactive fission products account for about 10% of the energy produced during operation. The uranium 238 also fissions since there are neutrons with high enough energy born in the reactor core. However, the uranium 238 is also converted into plutonium 239 through the absorption of a neutron in a non-fission producing reaction. Like uranium 235, plutonium 239 is able to fission with a slow or fast moving neutron. The build up of plutonium is significant, so much so that the fissioning of plutonium eventually produces about one-third of the energy produced in the reactor.

As the amount of uranium 235 in the fuel is used up and drops below about one and one-half percent, the fuel must be replaced. Since the fissioning of the uranium has produced highly radioactive fission products, the spent fuel must be stored and cooled for many years. Initially, the fuel is immersed in a swimming pool like structure called the spent fuel pool where it may remain for many years. After it has

---

[4]The use of heavy water enriched in the deuterium isotope shifts the burden of enrichment from enriching uranium to enriching water.

cooled, the spent fuel is moved to a shielded cask and stored, dry-cooled by air that circulates around the cask. Since the fuel contains material that will remain radioactive for hundreds of thousands of years, the spent fuel must be isolated from the environment throughout that time. As currently planned, the spent fuel is to be disposed of in a deep geological repository.

The spent fuel, however, still contains much of its energy in the form of unburned uranium 235 and plutonium. It is possible to recycle the unburned uranium and plutonium through a process called reprocessing. In reprocessing, the uranium and plutonium are separated from the fission products, formed into new fuel, and then recycled through the reactor. Although the United States currently does not reprocess of spent fuel, other countries that do not have significant domestic supplies of uranium, most notably France and Japan, are currently reprocessing spent fuel and producing new fuel utilizing the unburned uranium and plutonium. Reprocessing has the advantage of removing one of the isotopes with the longest period of radioactivity, plutonium 239.

4.3.2 Breeding

It is also possible to create more fuel than a reactor consumes by converting uranium 238 to plutonium 239 in a process called breeding. As mentioned, uranium 238 is converted into plutonium 239 by the absorption of a neutron and subsequent radioactive decay. In a conventional reactor, the efficiency of the breeding process is not sufficient to make up for the uranium 235 consumed. In certain types of reactors, called breeders, the conversion of uranium 238 to plutonium 239 is enhanced. These reactors typically use fuel containing about 20% plutonium in the fuel and a liquid metal such as sodium as a coolant. Such reactors produce more plutonium than they consume and are, thus, capable of extracting 20 times as much energy from the original uranium fuel as a conventional reactor.

An unusual type of breeder reactor that has recently gained interest is the traveling wave reactor. Initially proposed and studied in 1958 by Saveli Feinberg, who called it a "breed-and-burn" reactor [4], the design generates its own fuel, plutonium 239, by breeding it from uranium 238. Unlike conventional breeders that breed plutonium to be extracted later for fabrication into fuel, the design of the reactor is such that the plutonium that is produced continuously replenishes that which is burned and is sufficient to keep the reactor operating and producing power without the addition of fuel for many years. The traveling wave reactor is so named because the region where the fissions occur and the power is generated moves through the reactor much like a wave moves across the surface of the water. As the fuel in the reactor is used up by generating power, the power producing region moves down the fuel.

The original concept in a modified form is under development by TerraPower. The TerraPower traveling wave reactor [5] is a pool-type breeder reactor cooled by liquid sodium and fueled mostly by depleted uranium. It requires only a small amount of enriched uranium or other fissile fuel such as plutonium to initiate the fission and breeding process. Unlike the original concept, the power producing region doesn't move in the TerraPower design. Instead the fuel is periodically moved in the core to ensure breeding and continued operation. According to analysis performed by TerraPower, the core could operate for 40 years without refueling and could be fueled with depleted uranium or spent fuel from existing light water reactors [6].

The concept has many advantages. It provides a way to dispose of depleted uranium by extracting the large amount of energy still remaining in the uranium. From a proliferation standpoint, there is no need to separate the plutonium from the uranium drastically reducing the likelihood of diversion to weapons production. Still, it remains to be seen if the engineering issues that have limited the deployment of sodium cooled fast breeder reactors can be overcome.

Currently, it is estimated that there are sufficient proven reserves of uranium to fuel the existing reactors for about 80 years [7]. The Energy Information Administration (EIA) estimates that current proven reserves could support the existing reactors as well as those expected to be built between now and 2030. The EIA acknowledges that there is currently insufficient production capacity and that new mines and processing facilities are needed to support this expansion [8]. Long term, the use of recycling and deployment of breeder reactors would extend the fuel supply to 1,000 years at current rates of consumption.

Thorium, a widely abundant element, may also be used as a material for breeding. Through the absorption of a neutron, the naturally occurring isotope of thorium, thorium 232, may be converted to uranium 233, which like uranium 235 is relatively easy to fission. India is embarking on a program to use a thorium-uranium fuel cycle since it lacks sufficient reserves of uranium but has abundant domestic reserves of thorium. Thorium reserves are sufficient to power a breeder-based economy for even longer than a uranium based breeder economy.

4.4   Issues Affecting the Growth of Nuclear Power

There are currently four issues concerning the wider use of nuclear power: safety, waste disposal, cost, and proliferation. Since the fission process produces large amounts of highly radioactive material, the reactor must be designed, built, and operated to prevent the escape of the fission products to the environment. In western design reactors and newer reactors of Russian design, the fission products are prevented from being released to the environment by multiple barriers in the event of an accident. The barriers are the fuel itself, the zirconium cladding, the reactor cooling system and the specially designed concrete and steel building, called the reactor containment building (see Fig. 4.4 below), which houses the reactor and associated systems.

Fig. 4.4 Reactor
containment building
(Courtesy US DOE)

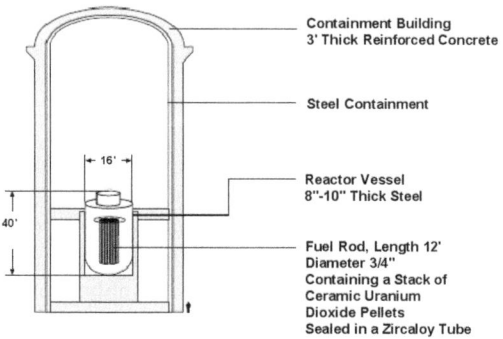

Containment Building
3' Thick Reinforced Concrete

Steel Containment

16'

40'

Reactor Vessel
8"-10" Thick Steel

Fuel Rod, Length 12'
Diameter 3/4"
Containing a Stack of
Ceramic Uranium
Dioxide Pellets
Sealed in a Zircaloy Tube

4.4.1   Reactor Safety

In a reactor accident, an event occurs which causes a mismatch between the cooling of the fuel and the energy generated in the fuel. Even if the fission process is stopped by the insertion of neutron absorbing control rods into the reactor, energy is still released by the radioactive decay of the fission products. These fission products continue to generate large amounts of energy days and even weeks after shutdown of the reactor. If the fuel is not adequately cooled and this energy is not removed, the fuel will overheat, eventually melting and releasing the more volatile fission products from the fuel.

The classic reactor accident occurs when one of the large pipes that carry cooling water to the reactor breaks, causing the cooling water to be lost. The reactor is shutdown by the reactor protective system and large pumps are started to supply cooling water. If for some reason, the cooling is insufficient as happened during the accident at Three Mile Island (TMI) [9, 10]; the fuel will overheat, melt, and release the fission products to the cooling system and eventually outside of the reactor through the break in the piping. The containment building provides the final barrier to release of the radioactive fission products to the environment. The building is a steel reinforced concrete structure designed to withstand the pressures and temperatures that occur during an accident. Although the reactor at Three Mile Island saw significant fuel melting, the containment building prevented any large-scale release of radioactive material to the environment.

Critics are concerned that a more severe event than occurred at Three Mile Island could lead to failure of the containment building and widespread contamination of the surrounding area with fission products. Such an accident did occur at the Russian designed nuclear power plant at Chernobyl in the Ukraine. This reactor did not have a containment building and when the reactor failed catastrophically, it contaminated much of the surrounding area [11].

4.4.2 Waste Disposal

Another concern that is frequently cited is what to do with the spent fuel that is generated. To optimize the burnup of the uranium 235, the reactor is shutdown about every 18–24 months and some of the fuel replaced and the remaining fuel rearranged. A typical fuel assembly stays in the reactor a total of about 6 years before the uranium 235 content becomes too low for its continued use. It is then transferred to the spent fuel storage pool where the faster decaying fission products are allowed to decay and the assembly cools. It may remain there for many years before being transferred to a dry fuel storage cask. Even after removal from the spent fuel pool, the fuel assembly still contains very large amounts of highly radioactive material that must be kept from the environment. Under the National Waste Policy Act, the US Department of Energy is to take custody of the spent fuel and dispose of it through deep geological disposal.

After an extensive scientific assessment, the Department of Energy chose Yucca Mountain, located about 90 miles north of Las Vegas, Nevada, as the site for the repository. In June of 2008, the Department of Energy submitted a construction license application for the repository to the United States Nuclear Regulatory Commission, the licensing authority [12, 13]. The construction and operation of the proposed repository is highly controversial and is strongly opposed by many critics including Nevada's congressional delegation. Even if the licensing and construction of the Yucca Mountain repository continues, it is not likely to be in operation until 2020 or later. Despite assurances that the spent fuel currently at reactor sites and anticipated to be produced in the future may be safely stored there for at least 20 years beyond the life of the reactor, critics are concerned that the nuclear industry has yet to develop and demonstrate a safe long term solution to the issue of spent fuel disposal.

Outside the United States, the approach to the disposal of spent fuel is somewhat different. Through the use of reprocessing, the plutonium and unburned uranium are removed from the waste stream. Since plutonium is one of the materials in the fuel that remains radioactive the longest, its removal reduces the time the waste must be isolated from hundreds of thousands of years to a few thousand years. Although even safe storage of spent fuel for this amount of time has yet to be demonstrated, it is believed to pose much less of a technological challenge and has in fact been implemented at the Waste Isolation Pilot Plant or WIPP [14]. Waste from the U.S.'s nuclear weapons program is presently being disposed of at WIPP in salt domes. The waste consists only of the fission products and not the plutonium or uranium that would be in spent fuel sent directly for disposal without reprocessing.

4.4.3 Cost

The cost associated with the construction of a nuclear power plant is also of concern. Current estimates for the construction costs of the next generation 1,000 MWe nuclear power plant are in the neighborhood of $6 billion. If design and financing costs are included, the cost is estimated to be in the neighborhood of 10–12 dollars. While these estimates do not take into account the recent decline in economic activity and subsequent impact on the commodities market, it is still likely that the investment needed to build a large number of such plants will be very significant.

4.4.4 Proliferation Concerns

The same enrichment technologies used to increase the amount of uranium 235 from 0.7% to 3–5% may also be used to increase the enrichment of uranium 235 to that needed for a nuclear weapon. Because of this, there is great concern an increase in the use of nuclear energy will lead to the spread of nuclear weapons. A classic example currently in the news is that of Iran.

Iran is currently completing the construction of a Russian designed commercial light water reactor. Although fuel for this reactor will initially be provided by Russia, Iran is also building a uranium enrichment facility. Some believe that this facility is not intended for commercial application, but instead is intended for the production of highly enriched, weapons grade uranium. Iran has repeatedly denied this, asserting that the project is intended to provide a secure source of fuel for its civilian nuclear power program. Despite repeated requests for a complete and thorough disclosure of its plans, some believe that Iran has not fulfilled its obligations under the Nuclear Non-Proliferation Treaty to which it is a signatory. Its actions have prompted suspicion about Iran's motivations and led to sanctions being imposed by the United Nations.

The danger here is not the construction of the reactor or its operation. The fear of proliferation from commercial nuclear reactors is based on the assumption that fuel for a commercial nuclear reactor can be used in a nuclear weapon. It cannot. The concentration of uranium 235 atoms in commercial fuel is around 3–5% and this concentration is simply too small to create a nuclear weapon; the concentration of uranium 235 atoms must be increased to levels in excess of 20%. It is true that several countries used the cover of civilian nuclear programs to hide clandestine activities designed to produce weapons grade uranium and plutonium. The existence of the civilian program did not in and of itself lead to proliferation. There are other countries that do not have civilian nuclear power, but are widely acknowledged to possess nuclear weapons.

It is the spread of enrichment technology and the construction of such facilities in countries that may divert the enriched uranium from peaceful uses to weapons. It is the intent, not the construction and operation of commercial nuclear reactors that is the problem. Such a situation is thought to exist in Iran. Efforts are underway by the United Nations, Russia, the United States and other countries to develop a fuel bank to ensure the supply of fuel for any country, eliminating the need for countries to develop their own nuclear enrichment capability.

If other countries were to follow Iran's example and develop their own enrichment capabilities, there is concern that this will eventually lead to large scale proliferation of at least the material for the production of weapons suitable uranium if not the weapons themselves.

Others have also expressed the concern that the plutonium which results from the absorption of a neutron by 238 uranium in the fuel of a commercial reactor could also be used to construct a weapon. The plutonium that is produced in a commercial reactor can be used for a weapon, but it is far less suitable and also very difficult to extract from the fuel due to the tremendous amount of radiation emitted from irradiated reactor fuel. To obtain plutonium from a reactor that is suitable for weapons requires fairly short times in the reactor, something difficult to achieve in most commercial reactors. Those countries that did produce plutonium for a weapon did so using research reactors or specifically designed production reactors.

To obtain commercial nuclear technology, a country must submit to monitoring by the International Atomic Energy Agency (IAEA) as required by the Nuclear Non-Proliferation Treaty. While this does not prevent a country from having a clandestine program, it makes it difficult to do so using commercial reactors. Since it is the enrichment process that is of concern, the fear of proliferation from commercial reactors alone is not well founded.

4.5 Factors Favoring Increased Use of Nuclear Power

Given these concerns, what is driving the recent renewed interest in the construction and operation of new nuclear reactors? The current generation of reactors has demonstrated a very high capacity factor, upwards of 94% [15] and low generating costs compared to nearly all other forms of electricity generation. Typical generating costs for a nuclear station are about $0.018–$0.02 per kWh, the lowest of any source [16]. Further, since fuel costs are relatively stable and low compared to fossil fuels, there is some certainty that the cost of generation will increase much more slowly for nuclear fuel than for gas or coal fired power plants.

The most significant factor favoring nuclear over other forms of electricity generation including renewable is the ability to generate 1,000's of megawatts of power with minimal greenhouse gas emissions. The following section compares the emissions from nuclear with those of other energy sources.

4.5.1 Nuclear Power's Carbon Footprint

Finally, and most importantly, nuclear is cited as extremely low in carbon footprint compared to nearly any other source of electricity. Studies performed in the U.S. and abroad show that while nuclear is not completely carbon free, its carbon emission is lower than any other technology currently available and some studies suggest that it is even comparable to renewables such as wind and solar. A detailed discussion of this topic is beyond the scope of this chapter; hence only a brief summary is presented with details left to the references.

Table 4.1 Estimated greenhouse gas emissions for various technologies (see Refs. [17–20])

Technology	Carbon emissions (g/kWh)
Coal	900–1,000
Combined cycle gas turbine	500
Photovoltaic	50–100
Wind	5–30
Nuclear	6–26
Hydro	3–11

There is always a price for the generation of electricity. Currently, the main source of electricity worldwide is coal. Nuclear and natural gas account for roughly the same percentage. These are followed by hydro, oil, and renewables, mainly wind. To understand the total carbon footprint of an electrical energy source, a complete life cycle analysis must be performed that accounts for all of the sources of energy used throughout the life cycle of the energy source. For an electrical generating source, the life cycle includes the energy and source of that energy for mining the fuel, processing the fuel into a usable form, transporting it, and disposing of the waste. The energy and sources of the energy to construct the facility must also be accounted for as well as the efficiency and capacity factor of the production of energy from that facility. Questions clearly arise as to how far back one must go. For example, the World Nuclear Association questions if the energy used in building the train to transport coal from a coal mine to a coal fired generating station should be counted or if the energy used and carbon emitted to construct a uranium enrichment facility should be included [17].

Most studies limit the life cycle energy costs and carbon emissions to the basic operations of mining the uranium, milling it, converting it to uranium oxide and then to uranium hexafluoride, enrichment, fuel fabrication, construction of the nuclear power station, and ultimate disposal of the spent fuel through either reprocessing and/or burial [17–20]. The carbon contribution is estimated based on various conceivable scenarios. The contributions of the construction emissions are distributed over the assumed life of the power station using an average capacity factor. Collectively, the studies provide a range of carbon emission stated as g CO_2 per kWh of electricity produced. A tabulation (Table 4.1) of these studies shows that nuclear compares favorably with all other forms of generation, even wind and solar, with only hydro having a lower carbon footprint than nuclear.

4.5.2   The Prospects for New Nuclear Power

From these comparisons, nuclear should play a significant role in reducing carbon emissions. What then are the prospects for the future? Can nuclear begin to offset the use of coal and natural gas for electricity generation or at least maintain its share of electric generation for the foreseeable future?

By 2030, electrical consumption worldwide is expected to increase by 70% [21]. To meet this demand, all sources of electricity will need to be increased. Currently in the U.S. and abroad, there is significant interest in new nuclear power plants. As of this writing (March 2009), 17 applications for construction and operating licenses were submitted to the U.S. Nuclear Regulatory Commission for the construction of 26 new reactors. In addition, several reactors whose construction was stopped in the 1980s are now being considered for completion. Construction has already been restarted at one, Watts Bar Unit 2, and at least two other reactors are under review as possible candidates for completion.[5] In Japan, the number of reactors is expected to increase from 55 to 68 over the next few years. A recent study by the Office of Economic Cooperation and Development (OECD) predicts that by the year 2050, there will be between 600 and 1,400 nuclear reactors in operation worldwide [22]. While the actual number will depend as much on economic factors and growth in electrical consumption as anything else, it is reasonable to expect the number of reactors to grow and possibly maintain the percentage of electricity generated worldwide by nuclear power. If such growth occurs, then nuclear may begin to reduce the greenhouse gas emissions contribution from electricity generation by coal and natural gas.

4.5.3   Actions Needed to Expand Nuclear Power's Role

For nuclear to contribute to the reduction of greenhouse gas emissions consistent with the two International Energy Agency (IEA) scenarios considered earlier, 24–32 of the 1,100 MWe nuclear power plants would be needed each year between now and 2050. Since the historical rate of construction for new nuclear power plants is less than this, is it even feasible to achieve this construction rate?

To answer this question, consider the situation in the United States. Figure 4.5 shows actual and projected electricity generation in the United States through 2030 broken down by source [2]. The projections assume that nine new nuclear power plants will be added by 2030, and there will be modest increases in the power produced by existing reactors through a process called power uprates. The EIA projects that electricity generation will increase about 25% by 2030, but that nuclear's share will decline from its current value of 19% to about 17.6%.[6] To maintain nuclear's current percentage, given an overall growth in electricity generation, will require construction of an additional nine nuclear reactors for a total

---

[5]Bellafonte Units 1 and 2 were cancelled in 2006 by the owner Tennessee Valley Authority (TVA). In August of 2008, TVA notified the US Nuclear Regulatory Commission of its interest in resuming construction on these units.

[6]The analysis assumes that all reactors are seeking license renewal to operate for an additional twenty years beyond their original license period.

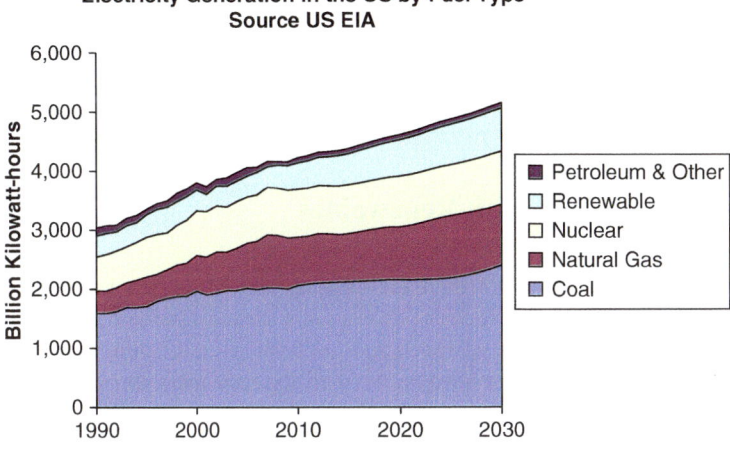

Fig. 4.5   United States electricity generation by source through 2030 (EIA [1])

of 18 new nuclear reactors by 2030.[7] An increase above the EIA's projections would increase nuclear's percentage share, and at the same time decrease the amount of electricity generated by coal, therby reducing the amount of greenhouse gas emitted.

4.6   The Effect of New Nuclear Generation on Greenhouse Gas Emissions

Table 4.2 shows the effect on coal generation and hence greenhouse gas emissions of adding additional nuclear reactors above the nine in the EIA study. The table shows that one can achieve close to a 20% drop in the amount of electricity generated by coal producing a similar drop in greenhouse gas emissions by building a total of 36 additional nuclear reactors above the nine in the original Energy Information Agency (EIA) study. Considering that there are currently a total of 17 applications for 26 reactors under review by the U.S. Nuclear Regulatory Commission, it is reasonable to expect a majority of these will be built. These 26 reactors would correspond closely to the case labeled EIA plus 18 yielding nearly a 10% reduction in the amount of electricity generated by coal and a commensurate reduction in greenhouse gas emissions. If that number is doubled to 36 above the EIA study, then one could achieve a reduction in the percentage of coal fired electricity generation of nearly 20%.

---

[7]Each new reactor is assumed to be capable of producing 1,400 MWe at a capacity factor of 95%. Current designs proposed for constructions are rated from 1,200 MWe for the Westinghouse designed AP1000 to 1,500 MWe for the General Electric designed ESBWR.

Table 4.2 Percentage of electricity generated in the United States in 2030 by coal and nuclear as a function of the number of new nuclear reactors built

	EIA study	EIA plus 9	EIA plus 18	EIA plus 27	EIA plus 36
Coal	46.7	45	43	40	38
Nuclear	17.6	19.6	22	24	26

4.7   Use of Nuclear with Renewables

An additional reduction might be achievable by combining a nuclear generating station with wind and solar. Nuclear power plants are considered base load generators. This means that they are designed and run to provide the minimum amount of electrical power that a utility must provide to its customers throughout the day. Other electrical generation technologies are used to address the peaks in electrical power usage that occur throughout the day as the electrical demand varies. This does not mean that nuclear power units cannot do what is referred to as load following, (which means that it can adjust its power output as demand for electricity fluctuates throughout the day). For example, the French developed advanced reactor called the EPR can be operated efficiently from 50% to 100% power (G. Vanderheydan, 2009, private communication). Thus, reactors could be combined with wind turbines or solar generating technologies so that when wind and sunlight conditions are favorable, the electrical load would be met by a combination of wind, solar, and nuclear. When the wind or the sunshine is not sufficient, then the reactor's power could be increased to meet the electrical demand.

4.8   Factors Affecting Feasibility of Increased Nuclear Generation

Can the number of power reactors be increased in the foreseeable future so as to significantly decrease greenhouse gas emissions? To answer this question successfully, one needs to look at the world situation. Currently, the U.S. and other countries are considering major new nuclear building programs. There are currently 59 countries with nuclear power. An additional 21 others have indicated an interest in the construction of nuclear power plants. These programs feature a new generation of power reactors that are designed to be safer, cheaper to build, and are also expected to be easier to operate and maintain than the original designs now used. While for the most part these designs are based on the current technologies, they do represent some technological and cost risk since most of the design work has yet to be completed and only one of the new generation reactors has actually been built.

As a result of the TMI and Chernobyl accidents, as well as the significant cost overruns associated with the current generation of nuclear power plants, the nuclear industry worked to develop simpler, safer, and easier to build nuclear reactor designs. These designs are referred to as Generation 3+ and are evolutionary in nature resembling the boiling water and pressurized water reactors currently in operation. Modifications to the U.S. licensing process were also made that provide for a one-step license.

The new designs that are currently being considered for construction are Westinghouse's Advanced Passive 1000 (AP 1000), AREVA's Evolutionary Power Reactor (EPR), General Electric-Hitachi's Economic Simplified Boiling Water Reactor (ESBWR), General Electric's Advanced Boiling Water Reactor (ABWR), and Mitsubishi's Advanced Pressurized-Water reactor (APWR) [23]. Of these, only the ABWR has been built and is in operation; four in Japan, and two in Taiwan. AREVA is in the early stages of construction of two EPR's; one in Finland and the other in France.

Each of these designs includes safety features that are intended to reduce the likelihood of an accident and the probability of a catastrophic release of radioactivity should an accident occur. The EPR, for example, has a double walled containment and four redundant electrically powered safety coolant injection systems. Each of these systems is capable of pumping enough water into the reactor to keep the reactor core from melting in the event of a catastrophic pipe break. The AP 1000, in contrast to the EPR, has a specially designed containment system and water tanks that rely on passive phenomena such as gravity and the tendency of hot water to rise, thus requiring no electrical pumps to provide cooling water to the reactor in the event of such an accident. Similar features are designed into the other vendors' reactor designs to increase safety and reduce the risk of the release of radioactivity.

The designs are also intended to be easier to build and less susceptible to construction delay through the use of modular construction techniques. Here, the approach is to construct many of the piping systems, platforms, and structures that make up the reactor systems in a factory remote from the construction site. Construction at the site is then limited to installing these systems and interconnecting them. Modular construction methods have long been in use in the shipbuilding industry and found to minimize schedule risk and improve quality, since much of the work is done in a factory environment not subject to the vagaries of weather. Using modular construction also enables the subassemblies to be fabricated anywhere in the world, making more facilities available. Such an approach is widely used in the aircraft industry for the construction of airliners such as the Boeing 757 and later aircraft.

Another advantage of the use of standardized design is building nearly identical reactors again and again, reducing cost and construction time. Currently, the U.S. has nearly 104 different reactor designs since each of the reactors built during the current generation are essentially one of a kind. By comparison, the French nuclear program has only three different designs in its 58 nuclear reactors. With this next generation of reactors, it is likely that no more than five designs will dominate the market.

Despite the improvements in the designs of the new reactors and the use of modular construction methods, opinion is mixed on whether these reactors can be built on time and within budget constraints. Both of the EPR's currently under construction are considerably behind their original construction schedules. The causes of the delays are not unlike those experienced during the construction of the current generation of nuclear power plants and include poor workmanship requiring costly rework, poor project management, and a lack of understanding of the regulatory requirements. Although these are the first of the new design pressurized water reactors to be built, the types of construction delays are not expected to be design dependent and, thus, could occur on any of the other reactor designs. Only the ABWR's built in Japan experienced minimal construction delays. The ABWR's were built in a carefully defined manner in close cooperation with the regulators and were finished very close to their original schedule. This success points to the need for good project management, attention to detail, and an understanding of the regulatory environment, all of which can be attained with proper planning and personnel.

4.8.1 Cost of New Nuclear Generation

Even using these labor saving construction methods, however, the cost of these new designs is expected be about $6 billion per reactor. At one time, it was thought that the cost of new nuclear power plants would be around $1,200 per kWe for an overall cost of between $1.5 billion and $2 billion. Unfortunately commodity price increases have forced these estimates up considerably to close to $5,000–$6,000 per kWe or $6 billion for a 1,000 MWe unit [24]. If the cost of project management, financing, and detailed design is included, some have suggested that the total cost of a new nuclear power plant would be upwards of $12 billion [25]. Such large costs have led to considerable concern over the ability of a utility to obtain the financing needed for such an expensive project, one which in many cases is equal to 50% of the equity of the company. One possible way to reduce the risk and decrease the financing costs is through government loan guarantees.

4.8.2 Loan Guarantees

Under a loan guarantee program, the cost to the guarantor is only the cost of administration of the program unless a default occurs. The U.S. Department of Energy proposal or the guarantee program planned for the U.S. would pass along all of these administrative costs to the borrower whose loans are guaranteed. If the guarantees work as expected, there is no cost to the guaranteeing government. However, a non-government guarantor would need to put on its financial books a contingent liability in the amount of the guarantee and disclose and discuss the risks of default which might give rise to having to pay out on the guarantee. How the government would determine its liability in advance is not clear.

Previous financing of U.S. nuclear power plants all relied upon leveraged leasing tax advantages (sale-leaseback) as well as the credit worthiness of the fully integrated electric utility which ran the plant. Today, except for those few remaining fully integrated utilities, most applicants for construction and operating licenses in the U.S. do not have the credit to warrant the debt of the size expected to be incurred, and even with respect to those fully integrated utilities, it is doubtful they could afford a new nuclear power plant. The owner would not be able to obtain a reasonable credit rating. The relevance of a low credit rating is that it translates into high interest rates (higher cost of debt service) and requires lowers profit or a larger percentage of equity.

Use of the government guarantee would reduce the interest rate to a small margin above a long-term U.S. Treasury Bill rate (or a small margin over LIBOR[8] plus the cost of an interest rate swap to "fix" the interest rate) which may not be available in today's market. One might rely upon funding from the vendors (AREVA, Toshiba, GE-Hitachi, Westinghouse, or Mitsubishi Heavy Industries) for the components and services they provide, but this would not cover construction costs and probably not the costs of farmed out components. Therefore, that would leave a large portion to be funded by an entity such as the government.

It is likely that guarantees would only be needed for the first few plants of each type. Once experience is gained and there is some certainty as to their cost and construction schedule, it is likely that private investment would be easier to obtain, obviating the need for guarantees. While not inconsequential, the financing problem is only a short term hindrance, one that can be solved by the use of loan guarantees as part of a commitment to the reduction of greenhouse gas emissions through the use of nuclear power as one of the means.

4.8.3   Limited Supplier Base

There are also a very limited number of suppliers for many of the major components that are needed to construct a nuclear power plant. Components such as the pressure vessel that houses the nuclear reactor, the pumps, and valves used in the cooling system, and the steam generators must all be built to extremely high standards. Since very few nuclear power plants were constructed worldwide in the 1980s and 1990s, the number of suppliers of these components has dwindled significantly, particularly in the U.S. In fact, the U.S. currently cannot manufacture these large components and must rely on other countries, mainly Japan and China, for them. Japan has continued building nuclear power plants and even it has only one supplier, Japan Steel Works, which can produce the large forgings needed to make a reactor vessel at a rate of eight to nine per year. Recently Northrop Grumman and AREVA announced

---

[8]London Interbank Offered Rate (or LIBOR); see http://en.wikipedia.org/wiki/LIBOR for a more detailed explanation

a joint venture to build a manufacturing facility to provide another source for the manufacture of large components such as reactor vessels [25]. Their facility would produce two complete sets of the large components per year (Data courtesy of the AREVA Group, 2009, private communication). Despite the addition of a second source, there still would not be enough capacity to produce the material needed for the construction of 24–32 reactors per year. The limited numbers of suppliers simply do not have the capacity to produce these components in the quantities needed to support such a major increase in reactor construction.

To overcome the limited number of suppliers, additional manufacturers are needed. Because of the current uncertainty, it is unlikely that such suppliers will enter the market without incentives. The situation is not unlike that experienced in the 1970s as the U.S. Navy sought to update its nuclear submarine fleet. Then, the competition for components for commercial nuclear power had made it difficult to obtain components needed for the U.S. Navy's ambitious submarine construction program. To increase the number of suppliers, the U.S. Navy developed alternative sources for these large components by paying a premium to bring another supplier into the nuclear component manufacturing business. The premium was paid for the first few sets of equipment. Once the new supplier had produced these components and demonstrated his competence, he was competed against other suppliers, enabling lower costs without subsidies for later sets of components. Such an approach could be adopted to encourage additional suppliers to enter the market, a market that at present has a very large uncertainty. It is interesting to note that the reactor vessel for the very first commercial nuclear power plant in the U.S., the Shippingport Atomic Power Station, was manufactured at the Homestead Works of U.S. Steel which no longer exists. Since these types of facilities employ skilled high paid workers, they have the benefit of increasing employment and making very significant contributions to the local and national economies, easily making it worthwhile for governments to provide initial subsidies to manufacturers as inducements to enter the business.

4.8.4 Uranium Fuel Supply

Once built, each reactor must be provided with fuel in the form of enriched uranium oxide. As discussed earlier, the isotopic content of uranium 235 must be increased from the naturally occurring value of 0.7% to 3–5%. To do so requires a large quantity of raw uranium ore. It takes about 10 metric tons of raw uranium to produce one metric ton of enriched uranium. To fuel all of the 400+ reactors currently operating in the world requires about 65,000 metric tons of ore. The current estimate of proven reserves is 4.7 million tons. In addition it is estimated that about ten million tons have yet to be discovered. Total, this suggests that there is enough fuel to supply the existing reactors for over 200 years. In addition there are a number of non-traditional sources from which one could obtain uranium albeit at a higher cost. These include the extraction of uranium from the ocean as well as from areas of very low concentration such as Chattanooga shale, which contains about 66 ppm of uranium [27]. Thus, it is likely that current reactors, even considering a major expansion of such reactors, will have sufficient fuel for the foreseeable future.

Other technologies are also available and are being considered for commercial exploitation. The current generation of light water reactors such as those discussed earlier mostly employ what is referred to as a once through fuel cycle. The fuel is manufactured, burned in the reactor for about 6 years and then is removed for ultimate disposal. Even when spent, the fuel still contains nearly half the energy it had when it was new. Techniques are available and in use in Europe and Japan to extract the unburned fuel and recycle it. This enables the fuel to be used again for at least two additional cycles. Because of proliferation concerns, the U.S. has not adopted this approach but this policy is now being reevaluated.

4.8.5   Impact of Breeder Reactor on Fuel Supply

As discussed earlier, breeder reactors produce more fuel than they consume, since in these reactors some of the uranium 238 is converted to plutonium 239, which like uranium 235 can be easily fissioned. Such breeder reactors if employed on a wide scale could considerably extend the fuel supply. Several countries including the US, France, and Japan have constructed and operated breeder reactors. Unfortunately to date, these systems have proven to be challenging engineering efforts. Breeder reactors use liquid metal, usually liquid sodium, as a coolant. Sodium when exposed to water reacts violently, oxidizing at a high temperature which produces a sodium fire that is difficult to extinguish. When used in a reactor and exposed to neutrons, sodium becomes radioactive. Because of the high temperatures and the chemical reactivity of the radioactive liquid metal coolant, breeder reactors have been difficult to operate reliably.

Another concern is the large amount of plutonium that is produced in a breeder. Since the plutonium may be separated from the uranium using chemical processes, there is concern that the large quantities of plutonium 239 produced in a breeder might lead to the proliferation of nuclear weapons. The plutonium 239 produced in a breeder although not weapons grade may, like uranium 235, be used to make a crude nuclear weapon.

In spite of these very real challenges, countries that have little or no uranium such as India are willing to accept the difficulties of developing and deploying large numbers of breeders. India is in the process of constructing a 500 MWe breeder that it hopes will become a significant source of fuel for its nuclear power program. Since India lacks significant domestic reserves of uranium, it has to maximize the uranium utilization and plans to do so using breeders to supply its light water reactors with fuel.

4.8.6 Need for Trained Workforce

The last impediment to the large-scale deployment of nuclear power is the availability of skilled, trained, and experienced workers and operators. As the French discovered with the construction of the first two EPR's, one must have trained workers who can produce the materials and components needed for a nuclear power plant and a construction workforce that can build the facilities to exacting standards. Every safety significant component or structure in a nuclear power plant is subject to extensive testing and inspection to ensure that each meets the required standards. The reinforcing bars and concrete used to make the reactor containment must be strong enough in the event of an accident so that the containment building will not fail. Similarly, the welds used to connect the piping of the reactor coolant system must be defect-free to avoid a pipe failure and subsequent loss-of-coolant accident. To achieve these high standards requires a trained, skilled workforce. Training and qualification programs must be developed to create such a workforce. Currently there are not enough workers available to support the construction of 24–32 nuclear power plants per year.

U.S. utilities that have plans to build are working with local technical schools to develop such a workforce. The U.S. Nuclear Regulatory Commission has been tasked by the U.S. Congress to aid in this effort through the use of seed grants to help schools develop the necessary curriculums. Such efforts are needed at all levels if the construction and manufacturing human capital needs are to be met. The current economic slowdown actually provides an opportunity to train this workforce since workers who might otherwise be employed are being let go or are not fully employed and could be retrained for these high paying skilled jobs.

Lastly, operators are needed to run these new facilities. At one time, many of these people in the U.S. received their training from the U.S. Navy. With the downturn in the number of nuclear powered ships and submarines after the end of the cold war, this is no longer a viable source to staff new nuclear power plants. It takes 5–8 years to train a reactor operator. Each person must pass an examination and demonstrate his or her ability to cope with the unusual conditions that might occur should there be an accident. U.S. utilities have recognized this need and are working to establish new training and educational programs to help develop new operators. Duke Power, a large utility located in the southeastern United States, has recently announced the establishment of a comprehensive educational program that would lead high school graduates to not only be qualified reactor operators, but would also provide for advanced college degrees and life long learning. Such approaches can lead to providing the personnel needed to fill these positions.

4.8.7 Conclusions

While these challenges to the widespread deployment of new nuclear plants exist, they are not insurmountable, given the resources and will to succeed. As this chapter has attempted to point out, nuclear is a low greenhouse gas emitter. It is capable of providing large amounts of power using proven technology. In the immediate future, it can contribute to greenhouse gas reduction but only on a modest scale, replacing a portion of the electricity produced by coal fired power plants. While it has the potential to do more, there are significant resources, political and social issues that must be addressed. If these can be overcome, then nuclear power could play a much more significant role in decreasing greenhouse gas emissions.

4.9   Other Uses of Nuclear Energy

Although this chapter has discussed the utilization of nuclear energy for the production of electricity, other ways to use nuclear energy to replace fossil fuels are also being considered. For example, it may be possible to use reactors that operate at very high temperatures to economically extract hydrogen from water for use in the transportation sector. However, such methods will require the development of very high temperature gas cooled reactors and associated chemical systems for the production of hydrogen.

4.10   Very High Temperature Gas Cooled Reactors

In 2007, approximately 40% of the energy consumption in the U.S. was in the form of electrical energy. About 39% of the electrical energy generated is used in the commercial sector, 35% for residential sector, and 26% for industry. The energy for electrical production and comes from a variety of fuels including petroleum, coal, nuclear, natural gas, and renewables. If one looks at the overall energy generation picture, petroleum accounts for 39% of the energy consumed in the U.S. the majority of it being used for transportation. Natural gas generated about 23% of the energy consumed with natural gas being used in roughly equal amounts for industry, electric power generation and for the commercial and residential sector. Coal generated about 22% of the energy used in the U.S. with the majority of the coal consumed being used for electricity generation. Nuclear accounts for about 8% of the energy used with all of it being in the form of electricity. Renewables currently account for about 7% of the energy used in the U.S.

It is difficult to see how nuclear energy in its present form can be used in the industrial sector to replace carbon-based fuels since much of the energy is used in the form of process heat. Process heat applications typically require temperatures well in excess of those that can be achieved with the current generation of light water reactors. Because of physical limitations inherent in light water reactors, it is unlikely that current light water reactor technology can be used to replace fossil fuels in these applications. However, very high temperature gas cooled reactors have the potential to replace carbon based fuels in some applications as these reactors are thought capable of producing temperatures of 900°C and above, sufficient for many industrial processes. A gas cooled reactor, unlike a water-cooled reactor, uses helium as a coolant. Under high pressure the helium is circulated through the reactor core where it removes the heat generated by the fission process. The helium gas can then be used either directly as a source of heat or can be used to generate very high temperature steam that can then be used as a source of heat in an industrial process.

The core of a gas-cooled reactor, like that of a light water reactor, uses enriched uranium. Unlike a light water reactor the uranium is typically enriched to a higher value, typically from 10% to about 19%. Graphite is used extensively in gas-cooled reactors for structural material and also as a means of slowing the fission neutrons down to enhance the likelihood of inducing another fission.

A number of such reactors have operated successfully, including Peach Bottom One in the U.S., a 40 MWe reactor that began commercial operation in 1967 and operated until 1974, the Fort St. Vrain reactor also in the U.S., a 330 MWe reactor that began operation in 1976 and operated until 1989, and the German AVR reactor that operated successfully for over 20. China recently placed a 10 MWt prototype gas cooled reactor into operation and South Africa has embarked on the development of a commercial scale gas cooled reactor based on the German AVR.

Such technology has the potential to efficiently and economically generate hydrogen from water through thermochemical water splitting for use as a replacement for petroleum or for the liquefaction of coal. Such a process involves a set of chemical processes that use heat to decompose water into hydrogen and oxygen at efficiencies of up to 50%. Currently, China, South Africa, and France are actively pursuing the development of these reactors. Unfortunately, the U.S. program is fraught with uncertainty due to political considerations. It is likely that China, and possibly South Africa, will have gas-cooled reactors in operation on a large scale within the next 8–10 years. These will be used for both electricity generation and for process heat.

4.11 Nuclear Fusion

Another possible option is nuclear fusion. Nuclear fusion involves the fusing of nuclei of light elements into heavier ones. In the process, energy is released. Research into the development of nuclear fusion has been ongoing since the 1950s. Despite this long history and considerable support worldwide, harnessing nuclear fusion for peaceful purposes has proven extremely difficult. The big advantage to nuclear fusion is it uses isotopes of hydrogen, one of which is commonly found in seawater, deuterium. As envisioned, a deuterium nucleus would be fused together with another isotope of hydrogen, tritium, releasing energy. The process requires the creation of a plasma, a very high temperature gas-like state of matter. The temperatures needed to create a plasma and to produce fusion are extremely high, comparable to those found on the sun. Technologically, confining a plasma at such high temperatures for a long enough time has proven a daunting task. It is expected though that in the next few years, the ITER[9] project (Iter, Latin meaning "the way") currently under construction in Europe will demonstrate the feasibility of harnessing fusion energy as a commercial energy source. ITER is a full-scale fusion reactor intended to produce more energy from fusion than is consumed in creating the plasma. It will be capable of operating at fusion power levels of up to 500 MW and at conditions prototypical of an electrical generating fusion reactor. The first plasma is expected to be produced in ITER in 2018. Even if successful, full-scale deployment of commercial fusion reactors would likely not occur for several decades.

4.12   The Next Generation of Nuclear Reactors

By the latter part of this century, other types of fission reactors may also find widespread deployment. The U.S. Department of Energy in cooperation with similar organizations in other countries is funding research into a number of alternative reactor concepts called Generation IV. Current reactors are considered to be mostly Generation 3 with the newer reactor designs discussed earlier referred to as Generation 3+. The newer developmental reactors include very high temperature gas cooled reactors, molten salt reactors, liquid metal reactors, and super critical water-cooled reactors [28]. They all employ the uranium fuel cycle, but have distinct advantages over the existing reactor technology including improved proliferation resistance, increased safety, higher efficiencies, and a reduction in the high-level waste produced as well as the radiotoxicity of the waste.

Another novel concept under consideration is the use of an accelerator driven reactor. Current generation reactors require sufficient uranium 235 or plutonium 239 so that at least one neutron from each fission goes on to produce a fission. Accelerator driven reactors do not; instead sufficient neutrons are produced outside the reactor by the accelerator, enabling the reactor to operate. The advantage to this type of system is that it is easily shutdown and is not prone to some of the accidents that can occur in conventional reactors. Because the energy of the neutrons is kept relatively high, the reactor is also able to consume some of the heavier longer-lived radioactive material produced by the fission process. This reduces the amount and radiotoxicity of the waste and can use thorium as a fuel more easily than in a conventional reactor [29].

It is likely that nuclear energy will continue to contribute to greenhouse gas emissions avoidance for the foreseeable future and has the potential to play a significant role in mitigating greenhouse gas emissions from the power generation sector.

---

[9] ITER was originally an acronym for International Thermonuclear Experimental Reactor. But, the name was dropped due to the bad connotations of the "thermonuclear" in conjunction with "experimental".

References

1. Energy Information Administration (2008) Nuclear utility generation by state, 2007. http://www.eia,doe.gov
2. EIA (2009) Annual Energy Outlook 2009 DOE/EIA-0383 (2009), Energy Information Agency, US DOE (March 2009)
3. For a detailed description of these and other reactor types, see Lamarsh JR, Baratta AJ (2001) Introduction to nuclear engineering. Prentice-Hall, Upper Saddle River, pp 136–180. Also, 2008–2009 Information Digest. NUREG 1350 Vol. 20, US NRC, Washington, DC 20555-0001, pp 34–35 (2008)
4. Feinberg SM (1958) Discussion comment. Rec. of Proc. Session B-10, ICPUAE, United Nations, Geneva
5. Michal R, Blake EM (Sept 2009) John Gilleland: on the traveling-wave reactor. Nuclear News, pp 30–32
6. Ellis T et al (2010) Traveling-wave reactors: a truly sustainable and full-scale resource for global energy needs. In: Proceedings of ICAPP '10, San Diego, 13–17 June 2010, Paper 10189
7. World Nuclear Association. Supply of uranium 2010. http://www.world-nuclear.org/info/inf75.html?terms=uranium+resources
8. Uranium Supplies, World Nuclear Association, London, UnitedKingdom (2010).http://www.eia.doe.gov/oiaf/ieo/uranium.html
9. Osif BA, Baratta AJ, Conkling TW (2004) TMI 25 years later. The Pennsylvania State University Press, University Park
10. Samuel WJ (2004) Three Mile Island: a nuclear crisis in historical perspective. University of California Press, Los Angeles
11. IAEA (1991) The International Chernobyl Project: an overview, assessment of radiological consequences and evaluation of protective measures report by an International Advisory Committee. IAEA, Vienna
12. DOE/EIS (2002) Final environmental impact statement for a geologic repository for the disposal of spent nuclear fuel and high-level radioactive waste at Yucca Mountain, Nye County, Nevada, Readers Guide and Summary. US Department of Energy Office of Civilian Radioactive Waste Management, DOE/EIS-0250, February 2002
13. Letter to Michael F. Weber, Director office of NMSS, US NRC, Washington, DC from Edward F. Sproat, Director Office of Civilian Radioactive Waste Management, US DOE, Washington, DC dated June 8, 2008. http:www.ocrwm.doe.gov/factsheets/doeymp0026.shtml
14. Fact Sheet on Licensing Yucca Mountain, US Nuclear Regulatory Commission, Washington, DC 20555-0001. http://www.wipp.energy.gov/
15. The capacity factor is the percent of the total electrical power that could theoretically be produced during a specified period if the plant were operated at full power 100% of the time. Lamarsh JR, Anthony JB (2001) Introduction to nuclear engineering, 3rd edn. Prentice Hall, Upper Saddle River, p 133
16. Nuclear Energy Institute (2008) US electricity production costs and components. Nuclear Energy Institute, Washington, DC
17. World Nuclear Association (2006) Energy analysis of power systems. http://www.world-nuclear.org/info/inf11/html
18. British Energy (2005) Environmental product declaration of electricity from Torness nuclear power station. AEA Technology, Washington, DC
19. Tokimatsu K et al (2006) Evaluation of lifecycle CO_2 emissions from Japanese electric power sector. Energy Policy 34:833–852
20. Krewitt W, Mayerhofer P, Friedrich R, Trukenmüller A, Heck T, Grebmann A, Raptis F, Kaspar F, Sachau J, Rennings K, Diekmann J, Praetorius B (1998) ExternE – Externalities of energy. National implementation in Germany. IER, Stuttgart

21. US DOE (2008) International Energy Outlook 2008, Energy Information Administration, US DOE, DOE/EIA-0484

22. Office of Economic Cooperation and Development (2008) Nuclear Energy Outlook '08 (2008) Office of Economic Cooperation and Development, Paris

23. For a more detailed description of these reactor designs, see http://www.nrc.gov/reactors/new-reactors/design-cert.html

24. Downey J (4 Nov 2008) Duke doubles cost estimate for nuclear plant. The Business Journal of the Greater Triad Area

25. Florida grants early recovery of nuclear costs. Reuters Oct 14 2008, PSC Votes on Nuclear Cost Recovery For Progress Energy," State of Florida Public Service Commission News Release, Tallahassee, FL, October 26, 2010.

26. AREVA, Northrop Grumman Join Forces to Create World-Class Facility in U.S. to Manufacture Heavy Components for American Nuclear Energy Industry, press release, AREVA Inc., Bethesda, MD 20814 (October 23, 2008). http://www.areva.com/servlet/cp_newport_23_10_2008-c-PressRelease-cid-1224679942025-en.html

27. Lamarsh JR, Baratta AJ (2001) Introduction to nuclear engineering. Prentice-Hall, Upper Saddle River

28. GEN IV Nuclear Energy Systems, Office of Nuclear Energy, US   DOE, Washington DC 20585. http://www.ne.doe.gov/genIV/neGenIV1.html

29. C. Rubbia et al., "A High Gain Energy Amplifier Operated with Fast Neutrons", AIP Conference Proceedings 346, International Conference Proceedings on Accelerator Driven Transmutation Technologies and Applications, Las Vegas, July 1994. http://einstein.unh.edu/FWHersman/energy_amplifier.html

Chapter 5
Renewable Energy: Status and Prospects – Status of Electricity Generation from Renewable Energy*

Abstract By 2050, the increased use of renewables such as hydropower, wind, solar, and biomass in power generation is projected to contribute between 9% and 16% of the CO_2 emission reductions. The share of renewables in the generation mix increases from 18% today, to as high as 34% by 2050. Hydropower is already widely deployed and is, in many areas, the cheapest source of power. There is considerable potential for expansion, particularly for small hydro plants. The costs of onshore and offshore wind have declined sharply in recent years through mass deployment, the use of larger blades, and more sophisticated controls. Costs depend on location. The best onshore sites, which can produce power for about USD 0.04 per kWh, are already competitive with other power sources. Offshore installations are more costly, especially in deep water, but are expected to be commercial after 2030. In situations where wind will have a very high share of generation, it will need to be complemented by sophisticated networks, back-up systems, or storage, to accommodate its intermittency. It is projected that power generation from wind turbines is set to increase rapidly. The combustion of biomass for power generation is a well-proven technology. It is commercially attractive where quality fuel is available and affordable. Co-firing a coal-fired power plant with a small portion of biomass requires no major plant modifications, can be highly economic and can also contribute to CO_2 emission reductions. The costs of high-temperature geothermal resources for power generation have dropped substantially since the 1970s. Geothermal's potential is enormous, but it is a site-specific resource that can only

---

*This chapter is an edited version of chapter 4 of Energy Technology Perspectives 2006©IEA/OECD, 2006, reproduced with permission of the International Energy Agency.

The findings included in this chapter do not necessarily reflect the view or policies of the Environmental Protection Agency. Mention of trade names or commercial products does not constitute Agency endorsement or recommendation for use.

be accessed in certain parts of the world for power generation. Lower-temperature geothermal resources for direct uses like district heating and ground-source heat pumps are more widespread. Solar photovoltaic (PV) technology is playing a rapidly growing role in niche applications. Costs have dropped with increased deployment and continuing R&D. Concentrating solar power (CSP) also has promising prospects. By 2050, however, solar's (PV and CSP) share in global power generation is still projected to be below 2%.

5.1   Introduction

In 2003, renewable energy supplied some 18% of global electricity production. Renewable electricity capacity worldwide is estimated at 880 GW (or 160 GW, excluding large hydro). Hydropower supplies the vast majority of renewable energy, generating 16% of world electricity. Biomass supplies an additional 1%. Power generation from geothermal, solar and wind energy combined accounts for 0.7%.

The share of renewable energy in electricity generation is highest in Canada and Latin America, because of the predominant use of hydropower (Table 5.1). The use of geothermal electricity generation explains the rather high non-hydro renewable share in Mexico and other Asia. This share is also relatively high in OECD Europe, at nearly 4%. At the global level, Latin America, China and the OECD countries account for nearly 80% of global hydropower production. The United States and OECD Europe account for nearly 70% of global non-hydro generation.

Growth in hydropower and geothermal electricity production slowed considerably in the 1980s and 1990s (Table 5.2). These more mature renewable technologies did not receive the strong government support that targeted new renewables in the 1990s. Albeit from a low base, the use of solar, wind and biomass energy for electricity generation has grown considerably over the past two decades. Energy production from solar and wind grew by about 22% per year from 1989 to 2000, and the pace has accelerated in the last few years. Hydropower is still the primary source of renewable energy-based generation, supplying 2,645 TWh of generation in 2003. This compares with some 200 TWh for bioenergy, 54 TWh for geothermal and 69 TWh for solar and wind combined.

The largest hydropower producers are Canada, China, Brazil, the United States, and Russia. More than half the world's small hydropower (categorized as from 10 to 30 MW) capacity is in China, where nearly 4 GW were added in 2004. At least 24 countries have geothermal electric capacity, and more than 1 GW of geothermal power was added between 2000 and 2004, mostly in France, Iceland, Indonesia, Kenya, Mexico, the Philippines, and Russia. Most of the current capacity is in Italy, Japan, Mexico, New Zealand, the United States, the Philippines, and Indonesia. The use of renewables other than hydropower and geothermal for power generation has considerable potential, but will require public support and private investment to accelerate commercial use.

Table 5.1 Share of renewable energy in electricity generations, 2003

	Renewable energy share in domestic electricity generation	Non-hydro renewable energy share in domestic electricity generation	Share in global use of hydro for electricity generation	Share in global use of non-hydro renewable energy for electricity generation
OECD				
United States	9.3%	2.4%	10.5%	30.2%
Canada	59.2%	1.7%	12.8%	3.1%
Mexico	13.1%	4.0%	0.8%	2.7%
OECD Europe	17.5%	3.6%	17.6%	37.0%
Japan	11.2%	2.1%	3.6%	6.8%
Korea	2.0%	0.6%	0.2%	0.6%
Australia	8.0%	0.9%	0.6%	0.7%
Transition economies				
Former Soviet Union	16.7%	0.2%	8.4%	0.8%
Non-OECD Europe	24.4%	0.1%	1.7%	0.0%
Developing countries				
Africa	17.1%	0.3%	3.2%	0.5%
China	15.0%	0.1%	10.7%	0.8%
India	12.8%	0.9%	2.8%	1.7%
Other Asia	18.3%	3.2%	5.0%	8.6%
Latin America	70.9%	2.6%	21.4%	6.6%
Middle East	2.9%	0.0%	0.6%	0.0%
World	17.8%	1.9%	100.0%	100.0%

Table 5.2 Global electricity generation from renewables (average annual growth rates) (*Source*: IEA [1]; IEA [2])

	1971–1988	1989[a]–2000	2000–2003
Renewables	3.4%	2.4%	1.0%
Hydro	3.3%	2.2%	0.3%
Geothermal[b]	11.4%	3.9%	1.2%
Biomass	4.0%	3.5%	5.9%
Wind/Solar[c]	4.9%	21.8%	24.8%

[a]There is a break in IEA data for biomass in 1988, necessitating the period breakdown

[b]The IEA Geothermal Implementing Agreement reports growth of 6.1% per year from 1995 to 2000 and 3.2% per year from 2000 to 2004

[c]Wind and solar are not shown separately in IEA statistics

Electricity production from biomass is steadily expanding in Europe, mainly in Austria, Finland, Germany, and the United Kingdom. Cogeneration of wood residues in the pulp and paper industry accounts for the majority of bioelectricity in OECD Europe, followed by generation from the biodegradable portion of municipal solid waste (MSW). The use of sugar cane residues for power production is significant in countries with a large sugar industry, including Brazil, Colombia, Cuba, India, the Philippines, and Thailand. Increasing numbers of small-scale biomass gasifiers are finding applications in rural areas and there are projects demonstrating the use of biomass gasification in high-efficiency combined-cycle power plants in several IEA countries.

Spain, Portugal, Germany, India, the United States, and Italy have led recent growth in wind power. In Denmark, wind turbines supply about 20% of electricity, a portion expected to increase to 25% by 2009. Global wind power capacity was 47 GW at the end of 2004, up from 39 GW in 2003. Wind power from offshore turbines is being developed or is under consideration in the United Kingdom, Denmark, Germany, the Netherlands, and the United States.

Grid-connected solar photovoltaic (PV) installations are concentrated largely in three countries: Japan, Germany, and the United States. The solar thermal power market has remained relatively stagnant since the early 1990s, when 350 MW was constructed in California. Recently, commercial plans in Spain and the United States have led to a resurgence of technology and investment. Projects are also underway in Algeria, Egypt, Israel, Italy, Mexico, and Morocco. Ocean technologies are still in the demonstration stage, with a few projects mainly in Europe and Canada.

The greater use of renewable energy is a key component of government strategies to enhance energy diversity and security, as well as to reduce greenhouse gas emissions. Because biomass absorbs CO_2 as it grows, the full biopower cycle (growing biomass, converting it to electricity and then regrowing it) can result in very low CO_2 emissions.

By using residues, biopower systems can even represent a net sink for GHG emissions by avoiding the methane emissions that would result from the land filling of unused biomass. A typical geothermal power plant emits 1% of the sulphur dioxide, less than 1% of the NO, and 5% of the CO_2 emitted by a coal-fired plant of equal size. A 1-MW hydro plant, producing 6,000 MWh in a typical year, is estimated to avoid the emission of 4,000 tonnes of CO_2 and 275 tonnes of SO_2, compared with a coal-fired power plant.

5.2   Prospects for Electricity Generation from Renewable Energy

First-generation renewable technologies are mostly confined to locations where a particular resource is available. Hydropower, high-temperature geothermal resources, and onshore wind power are site specific, but are competitive in places where the basic resource is plentiful and of good quality. Their future use depends on exploiting the remaining resource potential, which is significant in many countries, and on overcoming challenges related to the environment and public acceptance.

The second-generation of renewables has been commercially deployed, usually with incentives in place intended to ensure further cost reductions through increased scale and market learning. Offshore wind power, advanced biomass, solar PV and concentrating solar power technologies are being deployed now. All have benefited from R&D investments by IEA countries, mainly the 1980s. Markets for these technologies are strong and growing, but only in a few countries. Some of the technologies are already fully competitive in favorable circumstances, but for others, and for more general deployment, further cost reductions are needed. The challenge is to continue to reduce costs and broaden the market base to ensure continued rapid market growth worldwide.

Third-generation renewables, such as advanced biomass gasification, hot dry-rock geothermal power, and ocean energy, are not yet widely demonstrated or commercialized. They are on the horizon and may have estimated high potential comparable to other renewable energy technologies. However, they still depend on attracting sufficient attention and RD&D funding.

Recent IEA analysis suggests that RD&D activities have played a major role in the successful development and commercialization of a range of new renewable energy technologies in recent years [3]. Successful RD&D programs need to be well focused and should be coordinated both with industry efforts to promote commercialization and competitiveness in the market and with international programs. In addition, they must reflect national energy resources, needs, and policies. They also need to have roots in basic science research. Issues related to public acceptability, grid connection, and adaptation and managing intermittency are common to a range of renewable energy technologies and need to be addressed in government RD&D programs.

Each country has its own RD&D priorities based on its own particular resource endowments, technology expertise, industrial strengths, and energy markets. Because of the diverse nature of renewable energy sources, it is important that each country or region promote technologies and options that are well suited to its specific resource availability. RD&D in renewable energy must be strengthened, but priorities must be well selected, in order to address priority policy objectives, especially as they relate to cost-effectiveness. Industry can be expected to play a major role in the development of all technologies, whether or not yet commercially available. It is important to recognize that some renewable technologies will continue to depend to a considerable extent on government RD&D.

A high global share of renewable energy can be only achieved if new renewable technologies are adopted by both developing and developed countries. Governments should consider including, in their renewable technology RD&D programs, an element that specifically concerns the adaptation of renewable technologies to meet the needs of developing countries.

This section looks at prospects for electricity generation using the following renewable energy technologies:

- Bioenergy
- Large and small hydropower
- Geothermal
- Onshore and offshore wind
- Solar photovoltaics
- Concentrating solar power
- Ocean (marine)

5.3   Bioenergy

5.3.1   Technology Description and Status

Biomass encompasses a wide variety of feedstocks, including solid biomass, i.e., forest product wastes, agricultural residues and wastes, and energy crops, biogas, liquid biofuels, and biodegradable component of industrial waste and municipal solid waste. Feedstock quality affects the technology choice, while feedstock costs, including transportation costs determine the process economics. Bioelectric plants are an order of magnitude smaller than coal-fired plants based on similar technology. This roughly doubles investment costs and reduces efficiency relative to coal. Biomass-based electricity generation is a base load technology and, provided that adequate supplies are available, is considered one of the most reliable sources of renewable-based power.

Methods for converting biomass to electricity fall into four main groups:

Combustion.   The burning of biomass can produce steam for electricity generation via a steam-driven turbine. Current plant efficiencies are in the 30% range at capacities of around 20–50 MW. Using uncontaminated wood chips, efficiencies of 33–34% (LHV) can be achieved at 540°C steam temperature in combined heat and power (CHP) plants. Operated in an electricity-only mode this technology would generate at least 40% electricity output. With municipal solid waste, high-temperature corrosion limits the steam temperature that can be generated, thereby holding electric efficiency to around 22% (LHV). Plants with electric efficiencies of 30% (LHV) are in the demonstration phase. Supplying energy for district heating systems from municipal solid waste is expected to generate 28% electricity in CHP mode. Many parts of the world still have large untapped supplies of residues which could be converted into competitively priced electricity using steam turbine power plants. For example, sugar cane residues (bagasse) are often burned in inefficient boilers or left to decay in fields.

Stirling engines have received attention for CHP applications, but such systems are not yet competitive. Small-scale steam cycles also need to see cost reductions.[1]

Co-firing. Fossil fuels can be replaced by biomass in coal power plants, achieving efficiencies on the order of 35–45% in modern plants. Because co-firing with biomass requires no major modifications, this option is economic and plays an important role in several countries' emission reduction strategies (Box 5.1). To raise biomass shares above 10% (in energy terms), technical modifications and investments are necessary. Co-firing systems that use low-cost, locally available biomass can have payback periods as short as two years.

Gasification. At high temperatures, biomass can be gasified. The gas can be used to drive engines, steam or gas turbines. Some of these technologies offer very high conversion efficiencies even at low capacity. The biomass integrated gasifier/gas turbine (BIG/GT) is not commercially employed today, though the overall economics of power generation are expected to be considerably better with an optimized BIG/GT system than with a steam-turbine system. However, the costs are much higher than for co-combustion in coal-fired or fossil-fuelled power plants. Black liquor gasification, (discussed in Chapter 7), is economic for electricity and steam cogeneration. Other technologies being developed include integrated gasification/ fuel cell and bio-refinery concepts.

Anaerobic Digestion. Using a biological process, organic waste can be partly converted into a gas containing primarily methane as an energy carrier. This biogas can be used to generate electricity by means of various engines at capacities of up to 10 MW.[2] While liquid state technologies are currently the most common, recently developed solid-state fermentation technologies are also widely used. Anaerobic digestion technologies are very reliable, but they are site-specific and their capacity for scaling-up is limited; thus, the market attractiveness of this approach is somewhat restricted. The increasing costs of waste disposal, however, are improving the economics of anaerobic digestion processes.

---

[1] A Stirling engine is a highly efficient, combustion-less, quiet engine that harnesses the energy produced when a gas expands and contracts as its temperature changes. Invented by Robert Stirling in 1816, the Stirling engine uses simple gases and natural heat sources, such as sunlight, to regeneratively power the pistons of an engine.

[2] After purification, the gas can be used for production of transport fuels.

Box 5.1 Biomass co-firing potential for CO_2 reduction and economic development

Biomass co-firing has been demonstrated successfully for most combinations of fuels and boiler types in more than 150 installations worldwide. About a hundred of these have been demonstrated in Europe, mainly in Scandinavian countries, the Netherlands, and Germany. There are about 40 plants in the United States and some 10 in Australia. A combination of fuels, such as residues, energy crops, herbaceous and woody biomass, has been co-fired. The proportion of biomass has ranged from 0.5% to 10% in energy terms, with 5% as a typical value.

Co-firing biomass residues with coal in traditional coal-fired boilers for electricity production generally represents the most cost effective and efficient renewable energy and climate change technology, with additional capital costs commonly ranging from USD 100 to USD 300 per kW.

The main reasons for such low capital costs and high efficiencies are (1) optimal use of existing coal infrastructure associated with large coal-based power plants, and (2) high power generation efficiencies generally not achievable in smaller-scale, dedicated biomass facilities. For most regions that have access to both power facilities and biomass, this results in electricity generation costs that are lower than any other available renewable energy option, in addition to a biomass conversion efficiency that is higher than any proven dedicated biomass facility.

Co-firing of woody biomass can result in a modest decrease of boiler efficiency. A typical reduction is 1% point boiler efficiency loss for 10% biomass co-firing, implying combustion efficiency for biomass that is 10% points lower than for the coal that is fired in the same installation. A coal-fired power plant with 40% efficiency would have an efficiency of 30% with co-firing, which is higher than for dedicated biomass-fuelled power plants. Biomass is either injected separately or it is mixed with coal. The challenges for wood co-firing are not so much in the boiler, but in wood-grinding mills. Co-firing of herbaceous biomass is technically possible, but results in a higher chance of slagging and fouling, and its grinding costs and energy use are higher than for other types of biomass.

Worldwide, 40% of electricity is produced using coal. Each percentage point that could be substituted with biomass in all coal-fired power plants results in a biomass capacity of 8 GW, and a reduction of about 60 Mt of CO_2. If 5% of coal energy were displaced by biomass in all coal-fired power plants, this would result in an emission reduction of around 300 Mt CO_2 per year. Furthermore, the biomass used in this process would be approximately twice as effective in reducing CO_2 emissions as it would be in any other process, including dedicated biomass power plants. In the absence of advanced but sensitive flue gas cleaning systems commonly used in industrialized countries, co-firing biomass in traditional coal-based power stations will typically result

(continued)

Box 5.1 (continued)

in lower emissions of dust, NOx, and SO_2 due to the lower concentrations (ash, sulphur, and nitrogen) that causes these emissions. The lower ash content quantities of solid residue from the plant.

Biomass co-firing has additional benefits of particular interest to many developing countries. Co-firing forest product and agricultural residues adds economic values to these industries, which are commonly the backbone of rural economies in developing countries. This economic stimulus addresses a host of societal issues using markets rather than government intervention and involves rural societies with large-scale businesses such as utilities and chemical processing. Co-firing also provides significant environmental relief from field/forest burning of residues that represent the most common processing for residues. All of these benefits exist for both developed and developing countries, but the agriculture and forest product industries commonly represent larger fractions of developing countries' economies, and the incremental value added to the residues from such industries generally represents a more significant marginal increase in income for people in developing countries. Most developing countries are located in climatic regions where biomass yields are high and/or large amounts of residues are available. In countries that primarily import coal, increased use of biomass residues also represents a favorable shift in the trade balance

Co-firing of biomass and waste is now being actively considered. Blending biomass with non-toxic waste materials could regularize fuel supply and could enhance the prospects for co-firing. Certain combinations of biomass and waste have specific advantages for combustor performance, flue gas cleaning or ash behavior.

Source: Implementing Agreement for a Programme of Research, Development, and Demonstration on Bioenergy IEA Bioenergy Implementing Agreement and Implementing Agreement for the IEA Clean Coal Centre [4].

5.3.2   Costs and Potential for Cost Reductions

The additional cost of biomass co-firing with coal is between USD 50 and USD 250 per kW of biomass capacity, depending largely on the cost of biomass feedstock. It is the most attractive near-term option for the large-scale use of biomass for power only electricity generation. Very low generation costs (slightly above USD 0.02 per kWh) can be achieved with co-firing in situations where little additional investment is needed and biomass residues are available for free.

The cost of producing electricity from solid biomass depends on the technology, fuel cost and fuel quality. Solid biomass plants tend to be small, typically 20 MW or less, although there are some CHP plants in Finland and Sweden that are much

Table 5.3 Efficiencies and costs of European biomass plants (in operation and proposed) (*Source*: Novak-Zdravkovic and de Ruyck [5]; IEA data)

	Efficiency (% LHV)	Investment (USD/kW)	Size (MW$_e$)	Typical electricity cost (USD/kWh)
Co-firing	35	1,100–1,300	10–50	0.054
IGCC	30–40	3,000–5,500	10–30	0.112
Gasification + turbine	20–31	2,500–3,000	5–25	0.096
Large steam cycle	30	3,000–5,000	5–25	0.110
Gasification + engine (CHP)	24–31	3,000–4,000	0.25–2	0.107
Small steam cycle (CHP)	10	3,000–5,000	0.5–1	0.130
Stirling engine (CHP)	11–19	5,000–7,000	<0.1	0.132

Note: Based on a biomass price of USD3/GJ. Heat by-product valued at USD5/GJ. Based on a 10% discount rate

larger. In Canada, the capital cost is about USD 2,000–USD 3,000 per kW installed for biomass-based capacity. Plants such as these are connected to district heating systems and are economic in cold climates.

Generation costs are expected to range from USD 0.10 to USD 0.15 per kWh in innovative gasification plants (Table 5.3). Biomass integrated gasifier/gas turbine plants have long-term potential in terms of both efficiency and cost reduction. Larger plants require that biomass is transported greater distances to the generating station, and long transport routes make biomass less attractive in both economic and environmental terms. For even lower capacities, gasification will probably be combined with gas engines or turbines in combined heat and power units, replacing current steam processes.

Table 5.3 provides an overview of European biomass plant efficiencies and cost characteristics. Co-combustion in coal-fired power plants is the least-cost option in the near term – USD 0.054 per kWh. All other systems need to see further cost reductions to be competitive. Costs are typically higher than USD 0.10 per kWh, or more than twice the cost for fossil-fuel power plants. The use of waster biomass will lower costs, but the potential is limited.

5.3.3   Future R&D Efforts

Short-term priorities for bioenergy focus on two primary areas – widening the availability of large quantities of relatively cheap feedstocks and further increasing conversion efficiency of basic processes while reducing their costs [3]. Standards and norms on fuel quality are needed so that a dedicated market emerges to support trade – locally, nationally, and internationally. R&D will also be focused on innovative materials design to reduce cost.

Gasification technologies still need to demonstrate reliable commercial operation. The main barriers are efficient tar removal and economics. The success of the Varnamo plant in Sweden, the first and only plant to demonstrate an integrated gasification combined-cycle based on biomass, and recent advances on tar elimination indicate that the technical problems could be overcome in the short to medium term. Economics, however, may still pose a challenge.

Major R&D efforts are directed at multi-fuel co-firing with biomass and waste, to avoid any negative impact on combustion efficiency, flue gas cleaning requirements or ash behavior. The EU is providing financial incentives for the development of technological environmental sound solutions, both for short and long-term. As an example, the COPOWER European project integrates ten organizations from six countries (United Kingdom, Turkey, Sweden, Portugal, Italy, and Germany) with the aim of developing a comprehensive understanding of process synergy during co-firing of coal with biomass and wastes, in circulating fluidized bed systems. Research is also being carried out on fouling and slagging, and dioxin formation and destruction. Given public sensitivity to waste combustion, this option requires careful consideration.

5.3.4 Challenges to Future Deployment

One of the most significant barriers to accelerated penetration of all biomass conversion technologies is that of adequate resource supply. In the long term, the potential for the sustainable use of biomass in the energy sector will be limited by factors such as competition with food production, the need for biodiversity, and competition between the use of feedstocks as fuels and using them for generating power.[3] The negative effects of intensive farming and long transport distances can reduce the economic and environmental benefits of biopower. In this context, it will be attractive to convert biomass into an energy carrier with higher energy density. This can be achieved with flash-pyrolysis technologies that convert solid biomass into bio-oil, a liquid biofuel that can be transported economically over long distances.

Other challenges to increased market penetration of biomass-fired generation are the high initial cost of replacing boilers with biomass technologies and the higher capital costs for biomass systems compared with conventional technologies. While biomass combustion plants are commercially available at various sizes, their efficiency could still be improved and their costs further reduced. Technology improvements are also needed in areas such as gasification, the development of plants that can use a variety of biomass feedstocks, polygeneration and co-firing using a range of feedstocks.

---

[3]Uncertainty regarding biomass supply potential is discussed in Chap. 5, *Road Transport Technologies and Fuels*.

A significant challenge in developing countries is upgrading the efficiency of cogeneration units run on bagasse.[4] While a handful of countries, including Mauritius and Brazil, use modern generating units, the technology in most countries could be improved considerably. Another issue is storage. For example, power plants in Mauritius only run on bagasse during the harvest season. Techniques for bagasse storage are needed so the plants can operate on biomass year-round.

Market barriers include the limited public awareness of the benefits of biomass technologies, and some unresolved environmental issues, such as emissions from boilers used in urban environments. Because the market for biomass conversion plants is at an early stage of development, there is a perception of high business risk for both suppliers and utilities. Obtaining development and project financing for plants can be lengthy and difficult. Standardisation of feedstocks and technologies could help to overcome these barriers to some extent.

5.4 Large and Small Hydropower

5.4.1 Technology Description and Status

Hydropower is an extremely flexible technology from the perspective of power system operation. Its fast response time enables hydropower to meet sudden fluctuations in demand or to help compensate for the loss of power supply from other sources. Hydro reservoirs provide built-in energy storage, which helps Optimize electricity production across a power grid. The dividing line for Categorization of small-scale and large-scale hydro differs from country to country, but generally it ranges from 10 to 30 MW.

Small-scale hydropower is normally run-of-the-river design and is one of the most environmentally benign energy conversion options available because it does not interfere significantly with river flows. Small hydro is often used in autonomous applications to replace diesel generators or other small-scale power plants or to provide electricity to rural populations.

Large-scale hydropower projects can be controversial because they affect water availability downstream, inundate valuable ecosystems and may require relocation of populations. New less-intrusive low-head turbines are being developed to mitigate these effects. As hydropower usually depends on rainfall in the upstream catchment area, its availability is affected by weather variations. Therefore, backup capacity can be needed to ensure power availability, which adds to hydropower costs.

The IEAs Hydropower Implementing Agreement estimates the world's technically feasible hydro potential at 14,000 TWh per year, of which about 8,000 TWh per year is considered economically feasible for current development. About

---

[4]Bagasse is the fibrous residue remaining after the execution of juice from crushed stalks of sugar cane.

808 GW are in operation or under construction worldwide. Most of the remaining potential for development is in Africa, Asia and Latin America. The technical potential of small hydropower worldwide is estimated at 150–200 GW. Only 5% of global hydropower potential has been exploited through small-scale sites.

At present, OECD countries and the rest of the world produce roughly equal amounts of hydroelectricity. The share in non-OECD countries will likely increase, however, as most large hydro potential that is economically attractive and socially acceptable has already been developed in OECD countries, while untapped potential and pending projects remain in non-OECD countries. China will add some 18.2 GW of capacity by 2009 with the completion of the Three Gorges Dam.

5.4.2   Costs and Potential for Cost Reductions

Existing hydropower is one of the cheapest options on today's energy market because most plants were built many years ago and their initial costs have been fully amortized. For new large plants in OECD countries, capital costs are about USD 2,400 per kW and generating costs are in the USD 0.03–USD 0.04 per kWh range. Small hydropower costs are in the range of USD 0.02–USD 0.06 per kWh. Such systems commonly operate without major replacement costs for 50 years or more.

5.4.3   Future R&D Efforts

The technology challenges facing hydropower include improving efficiency; reducing equipment costs; reducing operating and maintenance costs; improving dependability; integrating with other renewables; developing hybrid systems, including hydrogen; developing innovative technologies to minimize environmental impact; and facilitating education and training of hydropower professionals. Table 5.4 lays out the R&D priorities for large and small hydropower.

Although small-hydro technology is mature and well established in the market, there is a need for further R&D to improve equipment designs, investigate different materials, improve control systems, and optimize generation as part of integrated water management systems. One priority area is increasing the range of head and flow at acceptable costs, particularly in small-capacity and low-head equipment.

Low-head equipment must accommodate considerably more water flow than high head equipment of equivalent capacity. Hence, low-head installations are physically larger and require more extensive engineering.[5] Since output shaft speed is lower as head decreases, low-head schemes generally need speed increasers to drive high-speed generators.

---

[5]The head is the height through which the waterfalls. The flow rate is the amount of water flowing per unit of time.

Table 5.4 Technology needs for hydropower (IEA [3])

Large hydro	Small hydro
Equipment	*Equipment*
• Low-head technologies, including in-stream flow	• Turbines with less impact on fish populations
• Communicate advances in equipment, devices and materials	• Low-head technologies
	• In-stream flow technologies
O&M practices	*O&M practices*
• Increasing use of maintenance-free and remote limited operation technologies	• Develop package plants requiring only O&M
	Hybrid systems
	• Wind-hydro systems

5.4.4   Challenges to Future Deployment

Concerns over undesirable environmental and social effects have been the principal barriers to hydro worldwide. Because most hydroelectric projects depend on dams, a river habitat is often replaced by a reservoir. Conditions for wildlife and aquatic creatures can be radically altered. Proper sitting, design, and operation can mitigate many of these problems, but more difficult challenges arise when human populations are forced to relocate. In some developing countries, the economic well-being and health of affected populations have declined after relocation.

Protection of fisheries is often one of the most contentious environmental issues with hydropower development. Most countries require that a minimum flow be maintained in the river to ensure the life and reproduction of indigenous fish and the free passage of migratory fish. The determination of an acceptable minimum flow is a key issue in the economic viability of any hydro scheme. To date, there is no universally accepted method of determining this flow to the satisfaction of both developers and regulators.

The construction of hydropower systems can have temporary effects on the local environment, particularly on water quality, such as muddying the water downstream from the development. Temporary access for construction vehicles can also cause disturbance, though once established, run-of-river hydro schemes have minimal visual impact. The principal permanent impacts are on the depleted stretch of the watercourse, where mitigating measures need to be taken to sustain river ecology and fisheries. Some types of turbines provide increased oxygenation of the tailrace water, thus improving water quality.

In the last few years, more emphasis has been put on the environmental integration of small hydro plants into river systems in order to minimize environmental damage. The requisite technology can be considered commercially and technically mature, although improvements are possible to make it suitable for export to rapidly expanding non-OECD markets. Innovations in civil engineering design, electromechanical equipment and control are possible, as well as instrumentation and systems which mitigate environmental effects.

Other challenges for small hydro include regulatory delays for sitting and permitting, as well as the burden of lengthy environmental impact reviews and assessments, which are often as rigorous as for large hydropower projects.

5.5 Geothermal[6]

5.5.1 Technology Description and Status

Geothermal power plants can provide an extremely reliable base-load capacity 24 h a day. There are three types of commercial geothermal power plants: dry fields steam, flash steam, and binary cycle. Dry steam sites are rare, with only five fields discovered in the world to date. Reservoirs that contain hot, pressurized water are more common. Flash steam power plants use resources that are hotter than 175°C. Before fluids enter the plant, their pressure is reduced until they begin to boil or flash. The steam is used to drive the turbine and the water is injected back into the reservoir.

Binary-cycle plants use geothermal resources with temperatures as low as 85°C. The plants use heat exchangers to transfer the heat of the water to another working fluid that vaporizes at lower temperatures. This vapor drives a turbine to generate power. This type of geothermal plant has superior environmental characteristics compared to others because the hot water from the reservoir, which tends to contain dissolved salts and minerals, is contained within an entirely closed system before it is injected back into the reservoir. Hence, it has practically no emissions. Binary power plants are the fastest growing geothermal generating technology.

Large-scale geothermal power development is currently limited to regions near tectonic plate boundaries, such as the western United States, Central America, Italy, the Philippines-Indonesia-Japan Pacific area, and East Africa. These areas are likely to be the most promising for large development in the near term. If current R&D efforts are successful, however, geothermal potential will expand to other regions.

5.5.2 Costs and Potential for Cost Reductions

Up-front investments for resource exploration and plant construction make up a large share of overall costs. Drilling costs alone can account for as much as one

---

[6]High temperature geothermal resources can be used in electricity generation, while lower temperature geothermal resources can be tapped for a multitude of direct uses, e.g., district healing and industrial processing. This section only deals with geothermal for electricity generation.

third to one-half of the total cost of a geothermal project. IEAs Geothermal Energy Implementing Agreement, which provides a framework for international collaboration on geothermal issues, is pursuing research into advanced geothermal drilling techniques and investigating aspects of well construction with the aim of reducing costs.

The resource type (steam or hot water) and temperature, as well as reservoir productivity, all influence the number of wells that must be drilled for a given plant capacity. Power-plant size and type (flash or binary), as well as environmental regulations, determine the capital cost of the energy conversion system. Because costs are closely related to the characteristics of the local resource system and reservoir, costs cannot be easily assessed for an average geothermal plant. Capital costs for geothermal plants vary from USD 1,150 per kW installed capacity for large, high-quality resources to USD 5,500 for small, low-quality ones.

Generation costs depend on a number of factors, but particularly on the temperature of the geothermal fluid, which influences the size of the turbine, heat exchangers and cooling system. U.S. sources report current costs of producing power from as low as USD 0.015–USD 0.025 per kWh at The Geysers field in California, to USD 0.02–USD 0.04 for single-flash and USD 0.03–USD 0.05 for binary systems. New construction can deliver power at USD 0.05–USD 0.08 per kWh, depending on the source. The latter figures are similar to those reported in Europe, where generation costs per kWh are USD 0.06–USD 0.11 for traditional power plants (liquid-steam water resources). Projected generation costs per kWh for hot dry rock geothermal systems in Europe are in the USD 0.24–USD 0.36 range.

New approaches are helping to exploit resources that would have been uneconomic in the past. This is the case for both power generation plant and field development. Drawing an experience curve for the geothermal power sector is difficult, not only because of the many site-specific features that affect the technology system, but also because of a lack of good data.

The costs of geothermal energy have dropped substantially from those of systems built in the 1970s and 1980s. Overall costs fell by almost 50% from the mid-1980s to 2000. Large cost reductions, however, were achieved by solving initial problems of science and technology development. Future cost reductions may be more difficult to attain.

5.5.3   Future R&D Efforts

R&D efforts are focused on ways to enhance the productivity of geothermal reservoirs and to use more marginal areas, such as those that have ample heat but are only slightly permeable to water. More complex geothermal systems, including hot dry rock, are in the research phase. To extract energy from hot dry rock, water is injected from the surface through bore holes into hot granite rock underground. The water heats as it flows through cracks in the granite and, when it returns to the surface, the super-heated steam is used to generate electricity. R&D for new approaches, for improving conventional approaches and for producing smaller modular units will allow economies of scale in plant manufacturing.

Several technical issues need further government-funded research and close government collaboration with industry if the exploitation of geothermal resources is to become more attractive to investors. These issues are mainly related to the exploration and enhancement of reservoirs, drilling and power-generation technology, in particular for the exploitation of low-temperature geothermal resources.

5.5.4   Challenges to Future Deployment

Challenges to expanding geothermal energy include long project development times, the risk and cost of exploratory drilling and undesirable environmental effects. Geothermal energy entails higher risks than most other renewable forms of energy because of the geological uncertainties of developing reservoirs which can sustain long-term fluid and heat flow. It is difficult to fully characterize a geothermal reservoir prior to making a major financial commitment. Another potential challenge for hot dry rock is the large quantity of water required in the process. A small 5-MW plant could use 8.5 ML of water per day, while a full-scale commercial plant could use ten times that amount.

Various countries with geothermal resources have devised policies to underwrite risks at both the reservoir assessment and drilling stages. For these countries, it would be impossible to attract private investment without these measures. Some aquifers can produce moderately to highly saline fluids that are corrosive and present a potential pollution hazard, particularly to fresh water drainage systems and groundwater. Re-injection and corrosion management are, therefore, important.

5.6   Onshore and Offshore Wind

5.6.1   Technology Description and Status

The commercial and technological development of wind energy has been closely related to turbine size. From 10 m in the mid-1970s, wind turbines have grown to diameters of 126 m, with multi-MW installed power (Fig. 5.1). Increasing the rotor diameter is an important prerequisite in developing turbines for offshore applications. All new offshore wind farms are expected to have turbines exceeding 1.5 MW.

Modern wind turbines are designed to have a lifetime of 20 years. Other technological developments include variable-pitch (as opposed to fixed-blade) rotors, direct drives, variable-speed conversion systems, power electronics, better materials and improved ratios between the weight of materials and generating capacity.

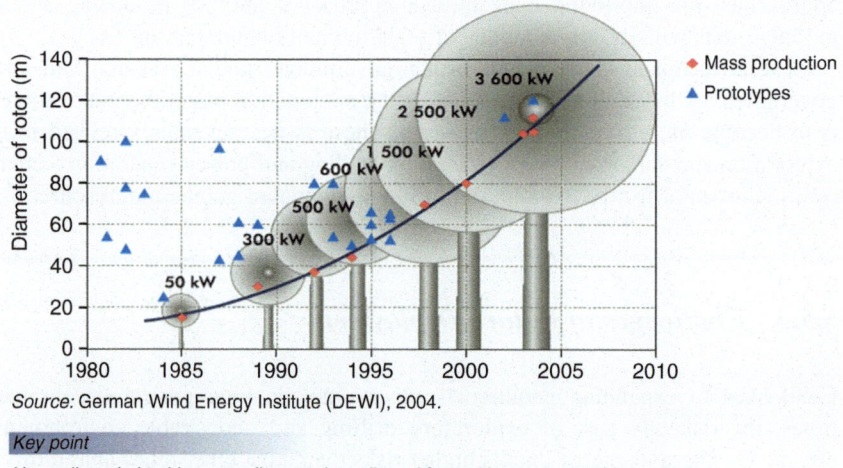

Source: German Wind Energy Institute (DEWI), 2004.

Key point

Up-scaling wind turbine rotor diameter has allowed for multi-megawatt turbine output.

Fig. 5.1 Development of wind turbine size

There are important economies of scale to be achieved in wind turbines. Larger machines can usually deliver electricity at a lower average cost than smaller ones. The reason is that the cost of foundations, road building, maintenance, electrical grid connections and a number of components in the turbine are largely independent of the size of the machine. Large turbines with tall towers use wind resources more efficiently. There is less fluctuation in the electricity output if a number of widely spaced wind parks feed energy into the grid in order to take advantage of the variations in wind regimes at specific sites to offset intermittency.

In 2004, installed global wind capacity exceeded 47 GW, including 578 MW of offshore capacity. Germany has the largest amount of installed capacity, followed by Spain, the United States, and Denmark. India has nearly 3 GW of installed capacity. Offshore wind is currently employed by Denmark, the United Kingdom, Italy, and Sweden.

5.6.2   *Costs and Potential for Cost Reductions*

From a pre-market level of about USD 0.80 per kWh in 1980, wind power costs have declined steadily. Wind power crossed the USD 0.10 per kWh threshold in about 1991, and dropped to about USD 0.05 per kWh in 1998. Since then, costs at the very best sites have dropped to about USD 0.03–USD 0.04 per kWh.[7]

---

[7]These costs are not directly comparable with fossil-fuel-based power generation due to the variable nature of wind electricity and the grid integration costs.

Costs of wind power installations depend on system components and size, as well as on the site. Generating capacity is primarily determined by the rotor-swept area and local wind patterns. Typical turnkey installation costs of onshore wind turbines are USD 850–USD 1,150 per kW. Investment costs differ considerably between onshore and offshore applications. For offshore installations, the foundation accounts for one-third or more of the cost. Turnkey installation costs are now in the range of USD 1,100–USD 2,000 per kW for offshore wind turbines, which is 35–100% higher than for onshore installations.

Operating costs for turbines include servicing, repairs, site rental, insurance, and administration. A study, conducted in Denmark, tracked operating costs for turbines in the size range of 150 kW to 600 kW. It shows nearly contemporary turbines (500–600 kW) having annual operating costs steadily increasing from 1% of the investment cost in the first year to 4.5% after 15 years. These figures are consistent with estimates of 2–4% in Portugal and estimates of 3.4% in the Netherlands for smaller projects. Maintenance and repair costs account for roughly one-third of total operating costs.

5.6.3   Future R&D Efforts

R&D priorities include increasing the accuracy of forecasting power performance, reducing uncertainties related to engineering integrity, improvement, and validation of standards, reducing the cost of new storage options, expanding the cost effective use of existing storage such as hydropower dams, enabling large-scale use and minimizing environmental effects (see Box 5.2). R&D is also focused on new design features, such as tall towers made of lightweight materials and advanced

---

Box 5.2 Priority research and development areas for wind energy

Reducing cost

   Improved site assessment and identifying new locations, especially offshore
   Better models for aerodynamics and aeroelasticity
   New intelligent structures/materials and recycling
   More efficient generators and converters

Increasing value and reducing uncertainties

   Forecasting power performance
   Engineering integrity, improvement of standards
   Storage techniques

(continued)

Box 5.2 (continued)

Enabling large-scale use

Electric load flow control and adaptive loads
Better power quality

Minimizing environmental impacts

Finding suitable locations in terms of wind potential
Compatible use of land and aesthetic integration
Noise studies
Careful consideration of interaction between wind turbines and wildlife

---

aerofoils and on further advances in power electronics. The feasibility of floating wind turbines, individually and in multi-unit formations, has also been the focus of several studies.

Offshore wind development and the role of wind energy within hydrogen-based energy systems are R&D priority areas for the long term. Technology and environmental issues raised by offshore wind energy development is the subject of much research and are likely to form an important part of future activities. In addition to using wind energy for electricity production, the technology could be applied to other energy applications in the long term, particularly hydrogen generation.

5.6.4   Challenges to Future Deployment

The current challenges to increased penetration of wind power are grid integration, forecasting of wind availability, public attitudes and visual impact. For offshore wind energy, a major challenge is cutting costs. The variable nature of wind electricity makes it difficult for wind to fully displace other electricity sources. When wind turbines constitute only a small fraction of generation capacity, their intermittency is hardly noticed by system operators, who are used to adjusting output to sudden changes in demand. At high penetrations, however, the marginal value of wind energy is equal only to the cost of the fuel and other marginal operating costs of power plants that are displaced. But if wind energy could be efficiently stored, wind power could compete economically with other types of electricity generation.

There is a wide range of technologies now available for storing wind energy, but choosing the appropriate one depends critically on the duration of storage required. For small turbines and at durations of only a few seconds to minutes, battery storage is a cheap option, along with flywheels and ultra capacitors. For longer durations, large-scale storage technologies such as pumped hydroelectric storage and compressed air energy storage are available at much lower costs per kWh of stored energy.

Another storage option is to combine back-up generation or integration with existing facilities such as gas turbines or hydro power. The back-up would be used when wind power generation is low. In extreme cases, wind turbines would simply be turned off when wind generation exceeds demand. More typically, the generator would try to maximize wind turbine output and shut off back-up sources of power. At least in the near term, this may be a more cost-effective strategy than large-scale energy storage.

Improved site assessment and identifying new locations, especially offshore, are important challenges. A 10% increase in wind speed will result in an energy gain of 33%. Improved assessment and siting require better models and measurements. Better measures are also needed to predict extreme wind, wave and ice situations. This may eventually make it possible to design site-specific systems that can utilize cheaper, lighter and more reliable turbines.

Several tools have been developed to overcome the aesthetic impacts of wind farms. Mapping of the zone of visual influence is used to show how many turbines will be visible from various locations. Photo-montage and animation techniques are employed to view potential wind parks from various angles. There has also been a great deal of research into the effect of wind turbines on the routes of migratory birds and on sites of special significance to bird populations. Sensitive siting has been found to avoid most of the problems.

Expensive undersea cabling and foundations have, until recently, limited the attractiveness of offshore wind energy. But new approaches in foundation technology, together with multi-megawatt-sized wind turbines, are at the point of making offshore wind energy competitive with onshore wind, at least at shallow water depths up to 15 m. Offshore wind turbines generally yield 50% higher output than turbines on nearby onshore sites because of more favorable and stable wind conditions.

Offshore wind turbines in deep water will become more economic with advances in floating platforms and continuing reductions in undersea cable costs. Recent studies have examined the technical feasibility of using floating platforms that are tethered to the ocean floor at depths of 180 m. Today, these installations are more expensive (USD 0.08 per kWh) relative to shallow installations (USD 0.05–USD 0.06 per kWh). Deep-water wind costs are projected to decrease to nearly the same level as shallow water costs by 2015 and to reach USD 0.04 per kWh by 2025 [7].

Although wind power is already competitive at many locations based on electricity production costs, the additional costs related to grid integration and back-up capacity must be considered as well. With government support for its development, wind power may become generally competitive with conventional technologies between 2015 and 2020. The deep-water offshore share in total wind power will increase, particularly if shallow sites in the United States and in Europe are exploited fairly quickly.

5.7   Solar Photovoltaics

5.7.1   Technology Description and Status

Photovoltaic (PV) cells are based on semiconductors and convert light directly into electricity. They are usually encapsulated within modules with a power up to several hundred watts that can be combined into larger power arrays. These systems are connected to consumers or to the grid via electronics. Solar-photovoltaic technologies include off-grid and on-grid applications. PV systems are made either from crystalline semiconductor modules or from thin films, and PV technologies are characterized by their modularity.

The overall efficiency of systems available on the market varies between 6% and 15%, depending on the type of cell. Crystalline silicon has been the most important PV technology so far. Because of their extremely high cost, other crystalline technologies, such as gallium arsenide (GaAs), are used only in space exploration. The first thin film PV device, an amorphous silicon (a-Si) module, was developed in the 1980s. More recently, other semiconductors, including cadmium telluride (CdTe) and copper indium diselenide (CIS/CIGS), have been used in industrial module production. The potential for thin-film modules is considered very high, but so far their diffusion has been limited by their high cost.

Installed grid-connected capacity is mostly in Japan, Germany and the United States. These three countries account for about 85% of global PV capacity. PV is often perceived as economic only in niche applications, such as traffic lights, weather stations and stand-alone systems for isolated buildings. Stand-alone or off-grid PV systems are particularly well suited for remote areas.

5.7.2   Costs and Potential for Cost Reductions

Costs for PV systems vary widely and depend on the system's size and location, the type of customer, the grid connection and technical specifications. In a standard building-integrated PV system, about two-thirds of the installation cost is for the module. The remainder reflects the cost of components, such as inverters and module support structures. The PV cells account for slightly more than half the total cost of the module itself. Cheaper cells would lower the system cost, but only so long as they deliver good efficiencies. Otherwise, higher balance-of-system costs might outweigh the lower cost of the cells.

Average installation costs are about USD 5–USD 9/W for building-integrated, grid-connected PV systems. Costs vary according to the maturity of the local market and specific conditions. For off-grid systems, investment costs depend on the type of application and the climate. System prices in the off-grid sector up to 1 kW vary considerably from USD 10–USD 18/W. Off-grid systems greater than 1 kW show slightly less variation and lower prices. This wide range is probably due to country and project-specific factors, especially the required storage capacity.

Stand-alone systems cost more, but can be competitive with other autonomous small-scale electricity-supply systems, particularly in remote areas. Solar cells and modules are expensive components of PV systems, so reducing the cost of the cells is vital. A variety of reliable components is available, but the efficiency, lifetime and operation of some components can be further improved, especially those of inverters and batteries.

Investment costs are the most important factor determining the cost of the electricity generated from PV installations. Operation and maintenance costs are relatively low, typically between 1% and 3% of investment costs. The expected lifespan of PV systems is between 20 and 30 years. However, inverters and batteries must be replaced every 5–10 years, and more frequently in hot climates.

Figure 5.2 illustrate electricity generating costs for PV and bulk and peak utility power. In liberalized electricity markets, utilities are likely to charge higher rates in

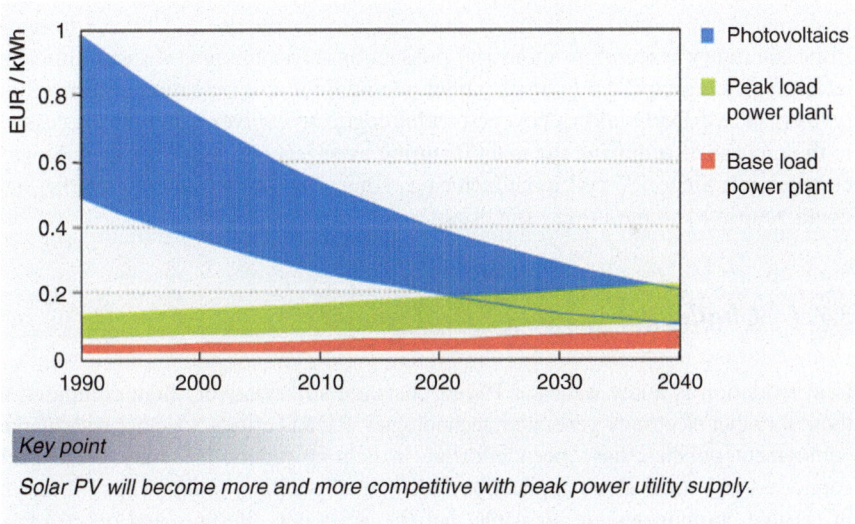

Key point

Solar PV will become more and more competitive with peak power utility supply.

Fig. 5.2 Projected cost reductions for solar PV[8] (Source: Hoffmann [8])

---

[8]The upper and lower boundaries of the PV costs reflect the meteorological situations of Germany and Southern Europe, respectively. The cost increase for conventional electricity is assumed to be 2% per year. External costs are not considered and could lead to a much earlier break-even for PV electricity. It may be estimated that PV electricity can be competitive with peak-load electricity within two decades and to base-load electricity within four decades, depending on the meteorological conditions of the installation site.

periods of peak demand. As a consequence, PV systems will be more competitive with standard peak power utility supply.

About half of potential future cost decreases for PV is expected to result from RD&D directed towards improving materials, processes, conversion efficiency, and design. Substantial cost reductions can also be gained through increased manufacturing volume and economies of scale.

5.7.3   Future R&D Efforts

The following are key drivers for future reductions in PV electricity costs: decreasing cell costs while maintaining or increasing efficiency (based on lower specific material consumption, more efficient production schemes and new cell concepts); increasing module lifespan; reducing specific module costs through higher cell efficiencies and more advanced encapsulation techniques; and reducing specific balance-of-system costs through higher module efficiencies and improved electronics.

RD&D focused on the long term is of high importance for PV. In order to bring new cells and modules to production, new manufacturing techniques and large investments are needed. Such developments typically require 5–10 years to move from laboratory research to industrial production. Over the next decade, thin film technologies are expected to display their potential for cost reduction and improved performance. RD&D allows new cell technologies to evolve, but is also necessary to the process of acquiring the manufacturing experience to make the technologies commercial. Since PV cell manufacturing requires large investments, market and manufacturing volume is very important.

5.7.4   Challenges to Future Deployment

Cost reduction is a key issue for PV, as costs are still relatively high compared to those for other electricity generation technologies. RD&D efforts, together with market deployment policies, have been effective in helping reduce PV costs. Both grid-connected and stand-alone applications need better ancillary components. A variety of reliable components are available, but the efficiency, lifetime, and operation of some components can be further improved, especially for inverters and batteries.

Standardization and quality assurance are crucial, for components as well as for the entire system. Standards exist for testing PV modules, and work has been done on standards for PV systems. To give users and investors more confidence, however; there is a need for standards to be developed for all the main system components, as well as certification or qualifications for designers and installers.

As solar energy is intermittent, storage systems are needed for stand-alone solar systems. However, solar power generation may work well as part of a diversified power supply system. The system compatibility depends on the shape of the electricity load curve. In sunny regions with an electricity demand peak during summer days (often caused by air conditioning), the peak contribution of solar can be high.

Such conditions can be found in California and Japan. However, at higher latitudes with a winter morning peak, solar contribution to peak demand is negligible. This difference affects the need for backup capacity.

Until recent years, supplies of crystalline silicon were abundant. However, as production levels increase, demand from the PV industry versus world market supply of crystalline silicon is becoming a serious issue. In order to resolve this bottleneck, new feedstock production must be developed quickly. This will require significant investment by industry, which previously relied on silicon from the semi-conductor industry. Manufacturing approaches for solar cell technologies are diversifying and many varieties of materials are being investigated.

A number of technologies are in the commercial stage; many others are still in the pilot manufacturing or even laboratory phase. It is likely that different technologies will continue to co-exist for different applications for some time. It would be valuable to undertake an early assessment of production processes, industrial compatibility and costs, including an assessment of generic issues faced by thin-film manufacturing processes.

5.8 Concentrating Solar Power

5.8.1 Technology Description and Status

There are three types of concentrating solar power (CSP) technology: trough, parabolic-dish and power tower.[9] Trough and power tower technologies apply primarily to large, central power generation systems, although trough technology can also be used in smaller systems for heating and cooling and for power generation. The systems use either thermal storage or back-up fuels to offset solar intermittency and thus to increase the commercial value of the energy produced.

The conversion path of concentrating solar power technologies relies on four basic elements: concentrator, receiver, and transport-storage and power conversion. The concentrator captures and concentrates solar radiation, which is then delivered to the receiver. The receiver absorbs the concentrated sunlight, transferring its heat to a working fluid. The transport-storage system passes the fluid from the receiver to the power-conversion system; in some solar-thermal plants a portion of the thermal energy is stored for later use.

The inherent advantage of CSP technologies is their unique capacity for integration into conventional thermal plants. Each technology can be integrated in parallel as "a solar burner" to a fossil burner into conventional thermal cycles. This makes it possible to provide thermal storage or fossil fuel backup firm capacity without the need of separate back-up power plants and without disturbances to the grid. With a small amount of supplementary energy from natural gas or any other fossil

---

[9]CSP is used interchangeably with solar thermal power. D 0.72 per kWh by 2050.

fuel, solar thermal plants can supply electric power on a steady and reliable basis. Thus, solar thermal concepts have the unique capability to internally complement fluctuating solar burner output with thermal storage or a fossil back-up heater.

The efficiency and cost of such combined schemes, however, can be significant. Current costs are about USD 0.10 per kWh and are expected to fall to about USD 0.72 per kWh by 2050. This technology relies on small-scale gas-fired power plants with low efficiency (40–45%), compared to 500-MW centralized plants with efficiencies of 60%. If the efficiency loss is allocated to the hybrid scheme, the economics would be less encouraging.

Fresh impetus was given to solar thermal-power generation by a Spanish law passed in 2004 and revised in 2005. The revised law provides for a feed-in-tariff of approximately EUR 0.22 (USD 0.27) per kWh for 500 MW of solar thermal electricity. In several states in the United States and in other countries, the regulatory framework for such plants is improving. At present, solar plant projects are being developed in Spain (50 MW), in Nevada in the United States (68 MW) and elsewhere. Two U.S. plants will also be constructed in southern California under the state's Renewable Portfolio Standard. A 500 MW solar thermal plant, expected to produce 1,047 GWh, is due for completion in 2012.

There is a current trend toward combining a steam-producing solar collector and a conventional natural gas combined-cycle plant. Projects in Algeria and Egypt, currently at the tendering stage, will combine a solar field with a combined-cycle plant. There are also plans to add a solar field to an existing coal plant in Australia. On a long-term basis, the direct solar production of energy in transportable chemical fuels, such as hydrogen, also holds great promise.

5.8.2 Cost and Potential for Cost Reductions

Since concentrating solar power uses direct sunlight, the best conditions for this technology are in arid or semi-arid climates, including Southern Europe, North and Southern Africa, the Middle East, Western India, Western Australia, the Andean Plateau, Northeastern Brazil, Northern Mexico, and the Southwest United States. The cost of concentrating solar power generated with up-to-date technology at superior locations is between USD 0.10 and USD 0.15 per kWh. CSP technology is still too expensive to compete in domestic markets without subsidies. The goal of ongoing RD&D is to reduce the cost of CSP systems to USD 0.05–USD 0.08 per kWh within 10 years and to below USD 0.05 in the long term. Improved manufacturing technologies are needed to reduce the cost of key components, especially for first plant applications where economies of scale are not yet available. Field demonstration of the performance and reliability of stirling engines are critical.

The European Commission (EC) has undertaken a coordination activity, the European Concentrated Solar Thermal Road-mapping (ECOSTAR), to harmonize the fragmented research methodology previously in place in Europe, which previously led to competing approaches on how to develop and implement CSP technology. Cost-targeted innovation approaches, as well as continuous implementation of this technology, are needed to realize cost-competitiveness in a timely manner.

5.8.3   Future R&D Efforts

Improvements in the concentrator performance and cost will have the most dramatic impact on the penetration of CSP. Because the concentrator is a modular component, it is possible to adopt a straightforward strategy that couples development of prototypes and benchmarks of these innovations in parallel with state-of-the-art technology in real solar-power plant operation conditions. Modular design also makes it possible to focus on specific characteristics of individual components, including reflector materials and supporting structures, both of which would benefit from additional innovation.

Research and development is aimed at producing reflector materials with the following traits [3]:

- Good outdoor durability.
- High solar reflectivity (>92%) for wavelengths within the range of 300–2,500 nm.
- Good mechanical resistance to withstand periodical washing.
- Low soiling co-efficient (<0.15%, similar to that of the back-silvered glass mirrors).

Scaling up to larger power cycles is an essential step for all solar thermal technologies (except for parabolic trough systems using thermal oil, which have already gone through the scaling in the nine solar electric generation stations installations in California, which range from 14 MW to 80 MW). Scaling up reduces unit investment cost, unit operation and maintenance costs and increases performance. The integration into larger cycles, specifically for power tower systems, creates a significant challenge due to their less-modular design. Here the development of low-risk scale-up concepts is still lacking.

Storage systems are another key factor for cost reduction of solar power plants. Development needs are very much linked to the specific system requirements in terms of the heat-transfer medium utilized and the necessary temperature. In general, storage development requires several scale-up steps linked to an extended development time before market acceptance can be achieved. Research and development for storage systems is focused on improving efficiency in terms of energy and energy losses; reducing costs; increasing service life; and lowering parasitic power requirements.

5.8.4   Challenges to Future Deployment

The widespread application of CSP plants is hindered by the heavy investment required for large centralized power plants – the economically viable plant size being on the order of several MW – and by the high ratio of risk to return for investors if long-term power purchase agreements are not in place.

Building a sustainable market for CSP will require taking the lessons learned in the countries where CSP deployment is successful and transferring them to others. This would promote faster market growth, attract larger global companies, and lead to costs that are increasingly competitive with conventional sources.[10]

Most of the technological components of solar thermal-power plant systems still need improvement. For example, higher operating temperatures and efficiencies will become possible in parabolic trough systems by using steam as a heat transfer medium and improved selective absorbers. Advanced storage systems will allow extended daily operation hours and improved plant utilizations. Technical components are already operational in principle, thereby making the development of appropriate support policies the crucial issue for the future of solar thermal power plants.

5.9   Ocean

5.9.1   Technology Description and Status

Ocean (marine) energy technologies for electricity generation are at a relatively early stage of development. Approaches to using ocean energy fall into several categories (Table 5.5): converting potential and kinetic energy associated with ocean waves into electricity; harnessing kinetic energy associated with marine (tidal) currents; converting potential energy associated with tides to electricity by building tidal barrage plants and by using mature hydro-electric turbine/generator technologies; extracting power from temperature differences between the surface and the seabed in deep oceans (ocean thermal energy conversion); using salinity gradients such as the latent heat of dilution at river mouths; and making use of marine biomass.

Wave energy and tidal current energy are the two main areas under development. The IEA Ocean Energy Systems Implementing Agreement is developing programs expected to be operational in 2007.

The technology required to convert tidal energy into electricity is very similar to that used in hydroelectric power plants. Gates and turbines are installed along a dam or "barrage" that goes across a tidal bay or estuary. Electricity can be generated

---

[10]See: CSP Global Market Initiative (GMI; www.solarpaces.org/gmi.htm).

Table 5.5 Status of ocean renewable energy technologies

Sub-sector	Status
Wave	Several demonstration projects up to a capacity of 1 MW and a few large-scale projects are under development. The industry aims to have the first commercial technology by about 2007.
Tidal and marine currents	Three demonstration projects up to a capacity of 300 kW and a few large-scale projects are under development. Industry is aiming for 2007 for the first commercial technology.
Tidal barrage (rise and fall of the tides)	Plants in operation include the 240 MW unit at La Rance in France (built in the 1960s), the 20 MW unit at Annopolis Royal in Canada (built in the 1980s) and a unit in Russia. Tidal barrage projects can be more intrusive to the area surrounding the catch basins than wave or marine current projects.
Ocean thermal energy conversion (OTEC)	There are a few demonstration plants, up to 1 MW, but there is still uncertainly surrounding the commercial viability of OTEC.
Salinity gradient/osmotic energy	A few preliminary laboratory-scale experiments, but limited R&D support.
Marine biomass	Negligible developmental activity or interest.

Note: In addition to the grid-connected electricity generation opportunities, there are potential synergies from the use of ocean renewable energy resources, for example: off-grid electrification in remote coastal areas; aqua-culture; production of compressed air for industrial applications; desalination; integration with other renewables, such as offshore wind and solar PV, for hybrid offshore renewable energy plants; and hydrogen production

by water flowing into and out of a bay (a difference of at least 5 m between high and low tides is required). As there are two high and two low tides each day, electrical generation from tidal power plants is characterized by periods of maximum generation every 6 h. Alternatively, the turbines can be used to pump extra water into the basin behind the barrage during periods of low electricity demand. This water can then be released when demand on the system increases. This allows the tidal plant to function with some of the characteristics of a pumped-storage hydroelectric facility.

Ocean thermal-energy conversion (OTEC) may become important in the long term, after 2030, for certain countries, but it is considered uneconomic in the short or medium term. Salinity gradient and marine biomass systems are currently the object of very limited research activities. Neither seems likely to play a significant role in the short or medium term.

5.9.2   Costs and Potential for Cost Reductions

Opportunities for cost reductions depend on the distance from shore; the choice of maintenance location (onshore or *in situ*); the frequency and duration of maintenance visits and the balance of predicted and unplanned maintenance; and the type and availability of the required vessels and their availability.[11]

5.9.3   *Future R&D Efforts*

R&D efforts are aimed at overcoming technical barriers related to wave and tidal technologies and to salinity gradient. The main focus is on wave behaviour and hydrodynamics of wave absorption; structure and hull design methods; mooring; power take-off systems; and deployment methods. Typical research on tidal stream current systems can be divided into basic research that focuses on areas such as water stream flow pattern and cavitations and into applied science, which would examine supporting structure design, turbines, foundations and deployment methods.

Research efforts on turbines and rotors will need to focus on cost-efficiency, reliability and ease of maintenance. Both components should be manufactured using materials designed to resist marine environments. Special attention should be given to bearings to ensure that they function safely and reliably in the marine environment. Control systems for turbine speed and rotor pitch will also be important to maximize power output. The main challenge for salinity gradient systems is to develop functioning and efficient membranes that can generate sufficient energy to make an energy system competitive.

5.9.4   *Challenges to Future Deployment*

A factor common to all marine technologies is that pilot projects need to be relatively large-scale if they are to withstand offshore conditions. These are costly and carry high perceived risks. These considerations have inhibited early development of these technologies. It is only in recent years that adequate funds have been made available to permit sizeable pilot projects, due largely to government policies to encourage ocean renewables. Once successful pilot projects are completed and confidence in the concepts grows, financing for even larger projects may become easier to obtain. Any major failures would, however, set progress back.

Although the prospects for tidal barrages are good in certain locations, their site-specific environmental effects need careful assessment. The technology reduces the range of the tides inside the barrage. This may affect the mud flats and silt levels in rivers, which would cause changes in the wildlife living in and around the estuary. It could also change the quality of the water retained by the barrage.

---

[11] See http://www.thecarbontrust.co.uk/ctmarine3/Page1.htm

Non-technical challenges include the need for resource assessment, the development of energy-production forecasting and design tools, test and measurement standards, environmental effects, arrays of farms of ocean energy systems and dual-purpose plants that combine energy and other structures.

5.10   Cost Overview

Table 5.6 provides an overview of cost estimates for renewable electricity generation technologies. There is a wide range of costs for each renewable technology due mainly to varying resource quality and to the large number of technologies within each category. Investment includes all installation costs, including those of some demonstration plants in certain categories. Discount rates vary across regions. Because of the wide range in costs, there is no specific year or CO_2 price level for which a renewable energy technology can be expected to become competitive. A gradual increase in the penetration of renewable energy over time is more likely. Energy policies can speed up this process by providing the right market conditions and to accelerate deployment so that costs can be reduced through technology learning.

Technology learning in bioenergy systems has been studied using experiences in Denmark, Finland, and Sweden [9, 10]. In the supply chain, learning rates for wood fuel-chips are 12–15%. For energy conversion in biogas or fluidized bed boiler plants, available data are much more difficult to interpret. An average learning rate of 5% for energy-producing plants appears to be a reasonable average estimate.

Technology learning is a key phenomenon that will determine the future cost of renewable power generation technologies. Unfortunately, the present state-of-the-heart does not allow reliable extrapolations. National data indicate learning rates between 4% and 8% for wind turbines in Denmark and Germany. Learning rates for installation costs are one or two percentage point's higher [11–13]. From 1980 to 1995, the cost of electricity from wind energy in the European Union decreased at a considerably higher rate of 18%. Wind energy is a global technology and experience curves based on deployment in major manufacturing countries like Germany and Denmark may be much lower than learning rates elsewhere analyzed the installation cost of wind farms from a global learning perspective and found learning rates between 15% and 19% [14]. Other recent studies quote learning rates of 5% for recent years.

Technology learning rates are better documented for photovoltaics than for other renewable energy sources. PV modules have shown a steady decrease in price over more than three decades, with a learning rate of about 20% [15, 16]. In 1968, the price of one peak watt of PV module was about USD 100,000 per kW. Today the price is about USD 3,000 per kW. Learning for PV modules is a global phenomenon, but prices for balance-of-system components reflect national or regional conditions. The EU-PHOTEX project found learning rates for balance-of-system in Germany, Italy, and the Netherlands to be from 15% to 18%.

Table 5.6 Key cost and investment assumptions for renewables

	Learning rate (%)	Investment cost 2005 (USD/kW)	2030 (USD/kW)	2050 (USD/kW)	Production cost 2005 (USD/MWh)	2030 (USD/MWh)	2050 (USD/MWh)
Biomass	5	1,000–2,500	950–1,900	900–1,800	31–103	30–96	29–94
Geothermal	5	1,700–5,700	1,500–5,000	1,400–4,900	33–97	30–87	29–84
Large hydro	5	1,500–5,500	1,500–5,500	1,500–5,300	34–117	34–115	33–113
Small hydro	5	2,500	2,200	2,000	56	52	49
Solar PV	18	3,750–3,850	1,400–1,500	1,000–1,100	178–542	70–325	<60–290
Solar thermal	5	2,000–2,300	1,700–1,900	1,600–1,800	106–230	87–190	<60–175
Tidal	5	2,900	2,200	2,100	122	94	90
Wind onshore	5	900–1,100	800–900	750–900	42–221	36–208	35–205
Wind offshore	5	1,500–2,500	1,500–1,900	1,400–1,800	66–217	62–184	60–180

Note: Using 10% discount rate. The actual global range is wider as discount rates, investment cost and fuel prices vary. Wind and solar include grid connection cost learning rate implies

References

1. IEA (2005) Energy balances of non-OECD countries. ©OECD/IEA, Paris
2. IEA (2005) Energy balances of OECD countries. ©OECD/IEA, Paris
3. IEA (2005) Renewable energy: RD&D priorities. ©OECD/IEA, Paris
4. IEA (2005) Implementing agreement for a program of research, development, and demonstration on bioenergy IEA bioenergy implementing agreement and implementing agreement for the IEA Clean Coal Centre
5. Novak-Zdravkovic A, de Ruyck J (2005) Small-scale power generation from biomass. In: Proceedings of the 14th European biomass conference and exhibition, Paris, 17–21 Oct 2005
6. German Wind Energy Institute (DEW) (2004) www.dewi.de/dewi/index.php
7. Greenblatt J (2005) Wind as a source of energy, now and in the future. Inter-Academy Council, Amsterdam
8. Hoffman (2001)
9. Junginger M et al (2005) Technological learning and cost reductions in woodfuel supply chains in Sweden. Biomass Bioenerg, reprinted in: Junginger M, Learning in renewable energy technology development. PhD thesis, Universiteit Utrecht, Faculteit Scheikunde, Utrecht
10. Junginger M et al (2005) Technological learning in bioenergy systems. Energ Policy (submitted), reprinted in: Junginger M, Learning in renewable energy technology development. PhD thesis, Universiteit Utrecht, Faculteit Scheikunde, Utrecht
11. Neij L (1999) Cost dynamics of wind power. Energy 24:375
12. Durstewitz M, Hoppe-Kilpper M (1999) Using information of Germany's '250 MW Wind' – programme for the construction of wind power experience curves. In: Wene CO, Voss A, Fried A (eds) Proceedings of the IEA workshop on experience curves for policy making – the case of energy technologies, Stuggart, Forschungsbericht Band 67, Institut fur Energiewirtschaft und Rationelle Energieanwendung, Universitat Stuttgard, p 129
13. Neij L et al (2004) Experience curves: a tool for energy policy assessment. IMES/EESS Report No. 40. Department of Technology and Society, Environmental and Energy Systems Studies, Lund University
14. Junginger M, Faaij A, Turkenbury WC (2004) Global experience curves for wind farms. Energ Policy 33(2):133–150
15. Harmon C (2000) Experience curves of photovoltaic technology. IIASA Interim Report IR-00-014, March 2000, p 22
16. PHOTEX (Schaeffer GJ et al) (2004) Learning from the sun. ECN Report ECN-C-04-035, Energieonderzoek Centrum Nederland (ECN), Petten

Chapter 6
Mobile Source Mitigation Opportunities[*]

Michael P. Walsh

Abstract The objective of this chapter is to review this history, focusing initially on the historical growth patterns and the resulting environmental consequences; then on the current control efforts around the world; and finally on the emerging efforts to transform vehicles and fuels to accommodate increased vehicle use while minimizing impacts on the environment. Progress in mitigating emissions of criteria air pollutants has been impressive, especially in the developed world. The situation with regard to climate change is particularly challenging. Transportation is already a large contributor to the problem and is a rapidly growing sector. Modest programs to reduce fuel consumption or greenhouse gas emissions from light duty vehicles are being phased in and California and the EU have initiated efforts to reduce the carbon content of vehicle fuels. But much more will need to be done with a likely shift to battery electric vehicles fueled by green electrons or fuel cell vehicles fueled by renewable hydrogen in future decades. As efforts to reduce CO_2 by 70 or 80% by 2050 receive high priority, aggressive short-term actions to reduce short-lived greenhouse pollutants hold promise. The US, Europe and Japan are phasing in high efficiency PM filters which will also reduce black carbon emissions dramatically.

6.1 Introduction

Since the end of the Second World War there has been a strong and steady growth in the world's motor vehicle population. Initially, this growth was focused primarily in North America but over the past six decades it has gradually spread,

---

[*]The findings included in this chapter do not necessarily reflect the view or policies of the Environmental Protection Agency. Mention of trade names or commercial products does not constitute Agency endorsement or recommendation for use.

M.P. Walsh (✉)
International Council on Clean Transportation, Washington, DC, USA
e-mail: mpwalsh@igc.org

F.T. Princiotta (ed.), *Global Climate Change - The Technology Challenge*,
Advances in Global Change Research 38, DOI 10.1007/978-90-481-3153-2_6,
© Springer Science+Business Media B.V. 2011

first to Europe and now Asia and to a lesser extent, Latin America. Vehicles have brought many perceived improvements to the quality of lives – increased mobility, jobs, recreational opportunities – to name but a few. But they have also changed many cities into sprawling conurbations, developed a so far unquenchable thirst for precious and limited oil, become a major if not dominant source of urban air pollution and most recently the most rapidly growing contributor to climate change.

6.2 Trends in World Motor Vehicle Production

Overall growth in the production of motor vehicles, especially since the end of World War II, has been quite dramatic, rising from about 5 million motor vehicles per year to over 60 million. As shown in Fig. 6.1, between 1970 and 2005, approximately 1 million more vehicles have been produced each year compared to the year before with almost 66 million vehicles produced in 2005 [1]. Data regarding motorcycle production is less precise but one major producer [2] estimates that global production exceeded 30 million units in 2003 and is increasing by approximately one million units each year.

Nowhere has the growth been greater than in China which was an insignificant motor vehicle producer two decades ago but is now estimated to be the largest producer of cars, trucks, and buses in the world and is rapidly becoming the major market as well. It is also far and away the largest producer and user of motorcycles.

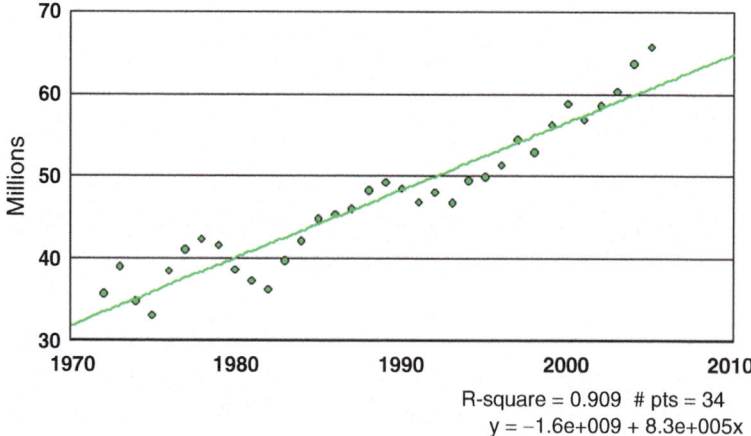

R-square = 0.909 # pts = 34
y = −1.6e+009 + 8.3e+005x

Fig. 6.1 Annual production of cars, trucks, and buses

6.3   Trends in World Motor Vehicle Fleets

Historically, the three primary drivers leading to growth in the world's vehicle fleet have been population growth, urbanization, and economic improvement and all three continue to increase, especially in developing countries. According to the United Nations, the global population increased from approximately 2.5 billion people in 1950 to slightly more than six billion today and it is projected to increase by an additional 50% to approximately nine billion by 2050. Most of this growth will be in urban areas in developing countries.

Annual GDP growth rates over the next two decades are forecast to be highest in China, East Asia, Central and Eastern Europe and the former Soviet Union which will stimulate growth in vehicle populations in these regions.

As a result of these factors, one can anticipate steady and substantial growth in the global vehicle population [3] following the historical trends illustrated in Fig. 6.2. The global vehicle population exceeded one billion units in 2002 and has continued to climb steadily since then.

Since 1990, approximately 27 million additional motor vehicles have been added to the world's roads and highways each year. Newer manufactured vehicles are more durable than in the past which will likely further increase the number of vehicles "on the road."

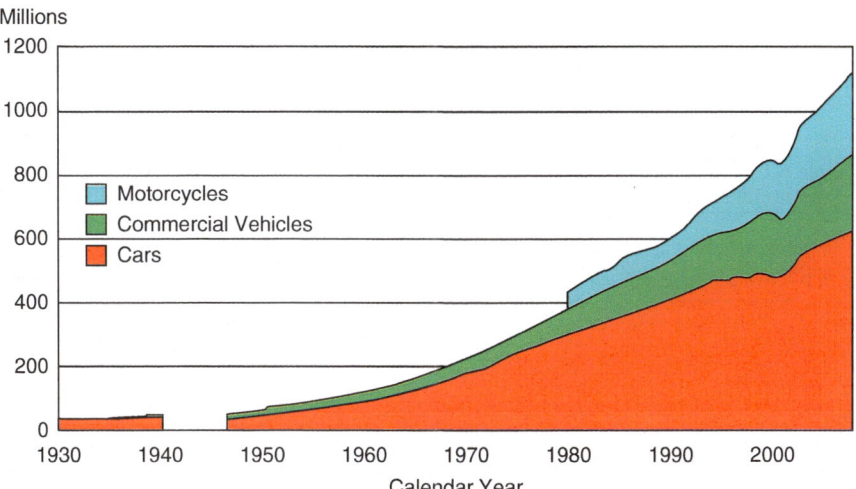

Fig. 6.2 World motor vehicle population

6.4   Motor Vehicle Emissions Trends

Motor vehicles emit large quantities of carbon monoxide (CO), hydrocarbons (HC), nitrogen oxides (NOx), particulate matter (PM), sulfur oxides (SOx) and such toxic substances as benzene, formaldehyde, acetaldehyde, 1,3-butadiene and (where lead is still added to gasoline) lead. Each of these, along with secondary by-products such as ozone and small particles (e.g. nitrates and sulfates), can cause serious adverse effects on health and the environment. Because of growing vehicle populations and resulting emissions, the fraction of health damaging pollution due to motor vehicles remains significant throughout the developed world and is rising in many cities in the developing world.

The greenhouse gases (GHGs) most closely identified with the transportation sector are the main Kyoto gases, CO_2, nitrous oxide and methane as well as the HFC's in vehicle air conditioning systems. The global warming potentials (GWPs)[1] of nitrous oxide and methane, relative to CO_2, are identified in Table 6.1. However, it is important to note that other vehicle-related pollutants also contribute to global warming, although their quantification has been more difficult; these include CO, non-methane hydrocarbons (NMHC) and nitrogen dioxide. It is generally agreed for example that CO emitted from vehicles is eventually converted to carbon dioxide in the atmosphere and in the process consumes hydroxyl radicals which might otherwise reduce methane concentrations. Similarly NMHCs and NOx contribute to global background tropospheric ozone, a very potent greenhouse gas. The GWPs listed in Table 6.1, including those attributed to CO, NMHCs and nitrogen dioxide, are from the original (1990) Intergovernmental Panel on Climate Change (IPCC) report and based on a 100 year timeframe.[2]

The IPCC has changed its GWP estimates with each new assessment report as newer data and information have become available, most recently in 2007 in the fourth assessment report. Even since that report, there has been a growing concern that black carbon (BC) or soot emitted from diesel vehicles and other sources is a

Table 6.1 Global warming potentials of transportation pollutants (Source: Intergovernmental Panel on Climate Change, 1990 and 1996) [4, 5]

GWP	Carbon dioxide (CO_2)	Methane (CH_4)	Nitrous oxide (N_2O)	Carbon monoxide (CO)	Nonmethane hydrocarbons (NMHC)	Nitrogen dioxide (NO_2)
100-Year time horizon	1	21	310	3	11	7

[1]Global warming potential is a measure of how much a given mass of a pollutant will contribute to global warming relative to the same mass of carbon dioxide which by definition is given a value of 1.

[2]Because of difficulty reaching agreement on the appropriate quantification, specific GWPs for these gases were not contained in the most recent Intergovernmental Panel on Climate Change (IPCC) report.

potent greenhouse gas *in part* due to the snow albedo (reflectivity) effect for solar radiation.[3] Hansen et al. for example estimate the GWP for soot is ~2,000 for 20 year, ~500 for 100 years, and ~200 for 500 years [6]. Similarly Jacobsen estimates a BC GWP range from about 800–1,200 [7]. Using a different metric, CO_2 Equivalency Factor, Mark Delucchi of University of California, Davis ascribes a value of 4,684 to black carbon.[4] As recently noted by Ramanathan and Carmichael, "Because of the combination of high absorption, a regional distribution roughly aligned with solar irradiance, and the capacity to form widespread atmospheric brown clouds in a mixture with other aerosols, emissions of black carbon are the second strongest contribution to current global warming, after carbon dioxide emissions. In the Himalayan region, solar heating from black carbon at high elevations may be just as important as carbon dioxide in the melting of snowpack's and glaciers" [8]. Of course, there is not yet a universal consensus on this high ranking of black carbon.

Great progress in reducing emissions of the urban air pollutants and their precursors from gasoline-fueled cars has occurred in the major industrialized countries and stringent requirements for diesel vehicles are starting to be phased in. However, the vehicle population and vehicle kilometers traveled are expected to continue to grow rapidly in the future especially in developing countries which will offset many of the gains to date [3].

6.5   Emissions Reduction Progress to Date

In almost every corner of the world, for every type of road vehicle and fuel, there is a clear trend toward more and more stringent emissions requirements. Over the next decade, this pattern is moving toward similar controls on off road vehicles and fuels. Driving these trends are several factors:

- *Continued growth in the number of vehicles (especially in China and other parts of Asia) and their concentration in urban areas where pollution levels remain unacceptably high,*
- *The growing accumulation of health studies that show adverse impacts at lower and lower levels and in the case of PM at virtually any level, and*
- *Advances in vehicle technology and clean fuels that are making it possible to achieve lower and lower emissions levels at reasonable costs.*

One of the critically important lessons learned to date is that clean vehicles and high quality fuels go hand in hand; they must be treated as a system. The next section

---

[3]Even though many sources emit black carbon, including gasoline fueled vehicles, wildfires, biomass burning, etc. diesel vehicles have been identified by EPA as having the strongest warming impact in the US due to their large BC emissions but small organic carbon (cooling) emissions. While the relative contribution will differ from country to country, diesels are considered an important source in virtually every country.

[4]Personal Communication.

will review the impact of fuel on emissions and the progress to date in improving fuel quality and vehicle technologies.

Over approximately the last 20 years, extensive studies have been carried out to better establish the linkages between fuels, vehicles, and vehicle emissions. One major study, the Auto/Oil Air Quality Improvement Research Program (AQIRP) was established in 1989 in the US and involved 14 oil companies, three domestic automakers, and four associate members [9]. In 1992, the European Commission also initiated a vehicle emissions and air quality program. The motor industry (represented by Association des Constructeurs Européens d'Automobiles (European Automobile Manufacturers Association [ACEA]) and the oil industry (European Petroleum Industry Association [EUROPIA]) were invited to cooperate within a framework program, later known as "the tripartite activity" or European Auto/Oil Program. In June 1993, a contract was signed by the two industries to undertake a common test program, called the European Program on Emissions, Fuels, and Engine Technologies (EPEFE).

The Japan Clean Air Program (JCAP) was conducted by Petroleum Energy Center as a joint research program of the automobile industry (as fuel users) and the petroleum industry (as fuel producers), supported by the Ministry of Economy, Trade and Industry. The program consisted of two stages: the first stage called JCAP I commenced in FY 1997 and terminated in FY 2001; the second called JCAP II commenced in FY 2002 and continued until 2007 to provide a further development of the research activities of JCAP I. In JCAP II, studies focused on future automobile and fuel technologies aimed at realizing Zero Emissions while at the same time improving fuel consumption.

6.5.1 Diesel Vehicles and Fuels

Diesel engines emit more nitrogen oxides (NOx) and particulate matter (PM) than equivalent gasoline engines per mile driven. Reducing PM emissions tends to be the higher priority because ambient PM levels are often above WHO recommended levels and are responsible for hundreds of thousands of premature deaths each year. Diesel particulate (soot) is thought to be particularly hazardous and has been characterized as toxic or potentially toxic by the California Air Resources Board, EPA, the International Agency for Research on Cancer (IARC) the National Institute for Occupational Safety and Health (NIOSH) and others. NOx emissions are also important, however, since they cause or contribute to ambient nitrogen dioxide, ozone, and secondary PM (nitrates).[5]

---

[5]Certain pollutants which are emitted from vehicles as gases undergo transformation in the atmosphere and are converted into particles. For example, some of the gaseous nitrogen oxides (NOx) emitted from vehicles chemically react with other gases and are converted into nitrates which contribute to urban PM air quality levels. Nitrates can account for as much as 20–30% of ambient PM in the US (although that fraction varies regionally).

Modest to significant NO_x control from diesel engines can be achieved by delaying fuel injection timing and adding exhaust gas recirculation (EGR). Very high pressure, computer controlled fuel injection can also be timed to reduce PM emissions. (Modifying engine parameters to simultaneously reduce both NO_x and PM is difficult and limited since the optimal settings for one pollutant frequently increases emissions of the other.) To attain very low levels of NO_x and PM therefore requires exhaust treatment. Lean NO_x catalysts, selective catalytic reduction, NO_x storage traps with periodic reduction, PM filter traps with periodic burn-off, and oxidation catalysts with continuous burn-off are technologies that are being phased in at differing rates in various parts of the world. Japan for example, is tending to lead the world in the widespread use of PM filters on new diesel vehicles whereas Europe is tending to lag.[6] A new type of diesel, the homogeneous charge compression ignition engine, provides another approach to reducing NO_x and particulates that is receiving significant attention and may be introduced on some engines for at least portions of the engine map within a few years.

Diesel fuel is a complex mixture of hydrocarbons with the main groups being paraffins, naphthenes and aromatics. Organic sulfur is also naturally present at varying levels depending on the source of the crude oil. Additives are generally used to influence properties such as the flow, storage, and combustion characteristics of diesel fuel. The actual properties of commercial motor vehicle diesel depend on the refining practices employed and the nature of the crude oils from which the fuel is produced. The quality and composition of diesel fuel can significantly influence emissions from diesel engines.

To reduce PM and NO_x emissions from a diesel engine, the most important fuel characteristic is sulfur because sulfur contributes directly to PM emissions and high sulfur levels precludes the use of or impairs the performance of the most effective PM and NO_x control technologies. For the control of PM, most new vehicles in Japan and the US and a growing fraction in Europe are equipped with filters or traps which reduce over 90% of the particles. NO_x adsorbers and Selective Catalytic Reduction systems are also starting to be introduced; NO_x adsorbers are especially sensitive to sulfur levels in the fuel.

Sulfur occurs naturally in crude oil, and the sulfur content of diesel fuel depends on both the source of the crude oil and the refining process.

The contribution of the sulfur content of diesel fuel to exhaust particulate emissions has been well established with a general linear relationship between fuel sulfur levels and this regulated emission. Shown below (Fig. 6.3) is one estimate of this relationship provided by the U.S. EPA. (This figure shows only the sulfur-related PM and not the total PM emitted from a diesel engine.) An indirect relationship also exists as some emissions of sulfur dioxide will eventually be converted in the atmosphere to sulfate PM.[7] Only a small fraction of the diesel

---

[6] Some European countries are using tax incentives to accelerate the introduction of PM filters beyond the rate required by the Euro new vehicle standards.

[7] Similar to the secondary transformation of NO_x to nitrate discussed earlier.

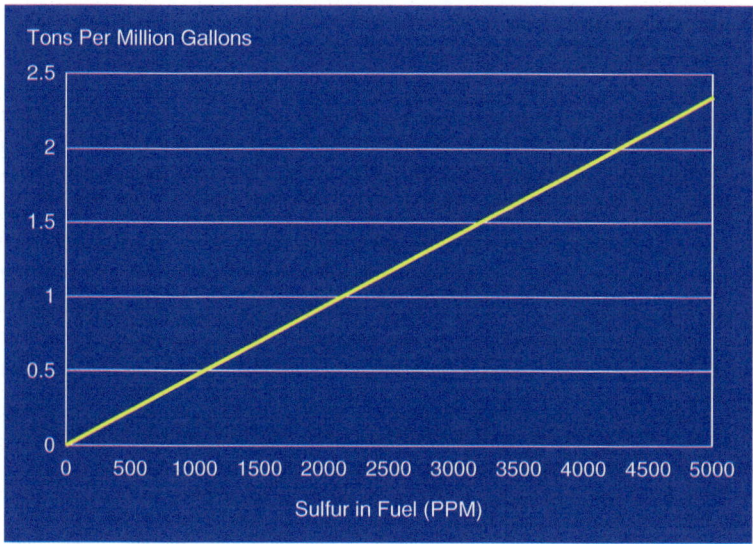

Fig. 6.3 Tons of directly emitted PM from diesel fuels sulfur (Notes: PPM=parts per million. only particulate matter [PM] related to sulfur and not the total PM emitted from a diesel engine are reflected in this figure. Source: United States Environmental Protection Agency [US EPA])

fuel sulfur (1–2%) is converted to sulfate emissions in the exhaust with the remaining 98–99% emitted as gaseous SO2; a substantial fraction of the SO_2 is lost to deposition with the remainder gradually converted in the atmosphere to sulfate PM.

Light duty diesel engines (<3.5 tons gross vehicle weight (GVW)) generally require oxidation catalysts to comply with Euro 2 or more stringent vehicle emission standards. Oxidation catalysts lower hydrocarbons, carbon monoxide, and particle emissions, typically removing around 30% of total particle mass emissions through oxidation of a large proportion of the soluble organic fraction. The conversion of sulfur in the catalyst reduces the availability of active sites on the catalyst surface and therefore reduces catalyst effectiveness. This catalyst deactivation is reversible through high temperature exposure – the sulfur compounds decompose and are released from the catalyst wash coat. However, due to generally low diesel exhaust temperatures, in many diesel engine applications the conditions needed for full catalyst regeneration may rarely be reached. High sulfur content in the fuel can also lead to the formation of sulfates in the converter which are then emitted as additional particles. Therefore, it is important to match fuel sulfur levels to the after-treatment technology present in the vehicle fleet.

To enable compliance with tighter particle emission standards for diesel vehicles, tighter limits on the maximum sulfur content of diesel fuel have been, or are being, introduced in many countries. While substantial reductions in particle emissions can be obtained without reducing sulfur levels, compliance with Euro 2 or tighter vehicle emission standards is generally not possible when fuel sulfur levels are greater than 500 ppm because of the relatively greater proportion of sulfates in the total mass of particle emissions.

In the case of Euro 3 and Euro 4 vehicle emission standards, even lower sulfur levels (350 ppm and 50 ppm, respectively) in diesel fuel will be required to ensure compliance with the standards. Complying with Euro 5 and 6 requirements or US Tier 2 standards will require maximum sulfur levels as low as 10–15 ppm. Apart from contributing to the effective operation of catalysts and reducing particle emissions, these further reductions in sulfur levels will enable tighter emission standards to be met by the use of next generation "de-NOX" catalysts, especially NOx adsorber systems. These are currently extremely sensitive to sulfur. An alternative emission control technology for Euro 5 or cleaner diesel vehicles is Selective Catalytic Reduction (SCR). These systems are not particularly sensitive to sulfur levels in fuel.

Sulfur content is also known to have an effect on engine wear and deposits, particularly under low temperature, intermittent operating conditions. Under these conditions there is more moisture condensation, which combines with sulfur compounds to form acids and results in corrosion and excessive engine wear. Generally lower sulfur levels lessen engine wear. With Euro 4+ or equivalent emission standards, the role of engine oil will be equally critical in ensuring sustained performance of engines/tail pipe devices. Low sulfur levels also allow the use of extended oil-change intervals reducing operating costs.

Diesel fuel has natural lubricity properties from compounds including the heavier hydrocarbons and organo-sulfur. Diesel fuel pumps (especially rotary injection pumps in light duty vehicles), without an external lubrication system, rely on the lubricating properties of the fuel to ensure proper operation. Refining processes to remove sulfur and aromatics from diesel fuel tend to also reduce the components that provide natural lubricity. In addition to excessive pump wear and, in some cases, engine failure, certain modes of deterioration in the injection system could also affect the combustion process, and hence emissions. Additives are available to improve lubricity with very low sulfur fuels and should be used with any fuels with 500-ppm sulfur or less.

A brief summary of the impact of various diesel fuel parameters on diesel vehicle emissions is provided in Tables 6.2 and 6.3.

In summary, from the standpoint of emission control technology, the most important diesel parameter is the sulfur content of the fuel, mainly since it allows for better after-treatment control technologies. Once standards sufficiently stringent to require oxidation catalysts are introduced, the sulfur content should be reduced to a maximum of 500 ppm; for the most advanced NOx and PM controls, the maximum should be 10–15 ppm sulfur. If sulfur levels are higher than these levels, the optimal performance of the pollution control systems will not be achieved and the in-use emissions will likely exceed standards. For cleaner vehicles, depending on the technology selected by the vehicle manufacturer, permanent damage could occur from the use of higher sulfur fuels.

Table 6.2 Impact of diesel fuel characteristics on light duty diesel vehicles

Diesel fuel characteristic	Modest controls	Strong controls	Advanced controls	Comments
Sulfur↑	SO_2, PM↑	If ox cat, SO_3, SO_2, PM↑	If Filter, 50 ppm maximum, 10–15 ppm better	If NO_x adsorber used requires near zero sulfur (<10 ppm) With low S, use lubricity additives
Cetane↑	Lower CO, HC, benzene, 1,3 butadiene, formaldehyde and acetaldehyde			Higher white smoke with low cetane fuels
Density↓	PM, HC, CO, formaldehyde, acetaldehyde and benzene↓, NO_x↑			
Volatility (T95 from 370 to 325 C)	NO_x, HC increase, PM, CO decrease			
Polyaromatics↓	NO_x, PM, formaldehyde and acetaldehyde↓ but HC, benzene and CO↑			Some studies show that total aromatics are important

Table 6.3 Impact of diesel fuel characteristics on heavy duty diesel vehicles

Diesel	Modest controls	Strong controls	Advanced controls	Comments
Sulfur↑	SO_2, PM↑	If ox cat, SO_3, SO_2, PM↑	If Filter, 50 ppm maximum, 10–15 ppm better	If NOx adsorber used requires near zero sulfur (<10 ppm) With low S, use lubricity additives

6.5.2   Gasoline Vehicles and Fuels

The use of catalyst exhaust gas treatment required the elimination of lead from gasoline. Other gasoline properties that can be adjusted to reduce emissions include, roughly in order of effectiveness, sulfur level, vapor pressure, distillation characteristics, light olefin content, and aromatic content. Of these, sulfur is the most important in terms of the impact on advanced pollution control technology.

Gasoline is a complex mixture of volatile hydrocarbons used as a fuel in internal combustion engines. The pollutants of greatest concern from gasoline-fueled vehicles with regard to urban and regional pollution are CO, HC, NOx, lead and certain toxic hydrocarbons such as benzene.[8] Each of these can be influenced by the composition of the gasoline used by the vehicle.

---

[8]PM emissions from gasoline-fueled vehicles have traditionally not been regulated because their emissions are so much lower per mile driven than from diesel vehicles. However, it is now recognized that in many countries and cities where the gasoline vehicle population is much larger than the diesel population, they are a more important source. Also, health studies continue to point to lower and lower levels of ambient PM being acceptable from a public health standpoint. As a result, PM standards from gasoline-fueled vehicles may emerge.

The use of catalyst exhaust gas treatment required the elimination of lead from gasoline. This change, which started in the US and Japan during the 1970s and has now occurred throughout most of the world, has resulted in a dramatic reduction of ambient lead levels. Other gasoline properties that can be adjusted to reduce emissions include, roughly in order of effectiveness, sulfur level, vapor pressure, distillation characteristics, light olefin content, and aromatic content [10].

Modern gasoline engines use computer-controlled intake port fuel injection with feedback control based on an oxygen sensor to meter precisely the quantity and timing of fuel delivered to the engine. Control of in-cylinder mixing and use of high-energy ignition promote nearly complete combustion. The three-way catalyst provides greater than 90% reduction of carbon monoxide, hydrocarbons, and oxides of nitrogen. Designs for rapid warm-up minimize cold-start emissions. On-board diagnostic (OBD) systems sense emissions systems performance and identify component failures. Durability in excess of 160,000 km, with minimal maintenance, is now common in many countries.

6.5.2.1 Lead

Lead additives have been blended with gasoline, primarily to boost octane levels, since the 1920s [11]. Lead is not a natural constituent of gasoline, and is added during the refining process as either tetramethyl lead or tetraethyl lead.

Vehicles using leaded gasoline cannot use a catalytic converter because lead poisons the catalyst, and therefore have much higher levels of CO, HC, and NO_x emissions. In addition, lead itself is toxic. Lead has long been recognized as posing a serious health risk. It is absorbed after being inhaled or ingested, and can result in a wide range of biological effects depending on the level and duration of exposure. Children, especially under the age of 4, are more susceptible to the adverse effects of lead exposure than adults.

Almost every country in the world has eliminated the use of leaded gasoline; the latest estimate is that only 17 countries continue to add lead.

6.5.2.2 Sulfur

Sulfur occurs naturally in crude oil. Its level in refined gasoline depends upon the source of the crude oil used and the extent to which the sulfur is removed during the refining process.

Sulfur in gasoline reduces the efficiency of catalysts designed to limit vehicle emissions and adversely affects heated exhaust-gas oxygen sensors. High sulfur gasoline is a barrier to the introduction of new lean burn technologies using $DeNO_x$ catalysts, while low sulfur gasoline will enable new and future conventional vehicle technologies to realize their full benefits. If sulfur levels are lowered, existing vehicles equipped with catalysts will generally have improved emissions.

Laboratory testing of catalysts has demonstrated reductions in efficiency resulting from higher sulfur levels across a full range of air/fuel ratios. The effect is greater in percentage for low-emission vehicles than for traditional vehicles. Studies have also shown that sulfur adversely affects heated exhaust-gas oxygen sensors; slows the lean-to-rich transition, thereby introducing an unintended rich bias into the emission calibration; and may affect the durability of advanced on-board diagnostic (OBD) systems.

The European Programme on Emissions, Fuels and Engine Technologies (EPEFE) study demonstrated the relationship between reduced gasoline sulfur levels and reductions in vehicle emissions. It found that reducing sulfur reduced exhaust emissions of HC, CO and NO_x (the effects were generally linear at around 8–10% reductions as fuel sulfur is reduced from 382 ppm to 18 ppm)[9]. The study results confirmed that fuel sulfur affects catalyst efficiency with the greatest effect being in the warmed up mode. In the case of air toxins, benzene and C3-12 alkanes were in line with overall hydrocarbon reductions, with larger reductions (around 18%) for methane and ethane.

The combustion of sulfur produces sulfur dioxide (SO_2), an acidic irritant that also leads to acid rain and the formation of sulfate particulate matter.

Certain other additives which are put into gasoline [generally to increase octane] can also affect vehicle emissions. Metallic-based, ash-forming, octane-enhancing additives such as Methylcyclopentadienyl manganese tricarbonyl (MMT) and ferrocene when added to gasoline will increase manganese-oxide and iron oxide emissions respectively from all categories of vehicles. Because of health concerns, participants in a workshop convened by the Scientific Committees on Neurotoxicology and Psychophysiology and Toxicology of Metals of the International Commission on Occupational Health recently published their conclusion that, "The addition of organic manganese compounds to gasoline should be halted immediately in all nations" [12]. The Health Effects Institute noted, "There is a large body of evidence that [13] under certain circumstances, manganese can accumulate in the brain [14–16], chronic exposure can cause irreversible neurotoxic damage over a lifetime of exposure [17], manganese may cause neurobehavioral effects at relatively low doses [5, 18], and these effects follow inhalation of manganese-containing particles."

Vehicle manufacturers have expressed concerns regarding catalyst plugging and oxygen sensor damage with the use of these additives which could lead to higher in-use vehicle emissions especially at higher mileage. The impact seems greatest with vehicles meeting tight emissions standards and using high cell density catalyst substrates.

A brief summary of the impact of various gasoline parameters on vehicle emissions is provided in Table 6.4.

---

[9]The study found that the effects tended to be larger over higher speed driving than in low speed driving.

Table 6.4 Impact of gasoline composition on emissions from light duty vehicles

Gasoline	No catalyst	Early three way catalysts	More advanced catalysts	Comments
Lead↑	Pb, HC↑	CO, HC, NOx all increase dramatically as catalyst destroyed		MIL light may come on incorrectly
Sulfur ↑ (50–450 ppm)	SO₂ ↑	CO, HC, NOx all increase ~15–20% SO₂ and SO₃ increase		Potential deposit buildup
Olefins ↑		Increased 1, 3 butadiene, increased HC reactivity for O3 formation, NOx, small increases in HC for Euro 3 and cleaner		
Aromatics ↑	Increased benzene in exhaust			Deposits on intake valves and combustion chamber tend to increase
	Potential increases in HC, NOx	HC↑, NOx↓, CO↑	HC, NOx, CO ↑	
Benzene ↑		Increased benzene exhaust and evaporative emissions		
Ethanol ↑ up to 3.5% O₂	Lower CO, HC, slight NOx increase(when above 2% oxygen content), higher aldehydes, especially acetaldehyde	Minimal effect with new vehicles equipped with oxygen sensors, adaptive learning systems		Increased evaporative emissions unless RVP adjusted, potential effects on fuel system components, potential deposit issues, small fuel economy penalty
MTBE ↑ up to 2.7% O₂	Lower CO, HC, higher aldehydes, especially formaldehyde	Minimal effect with new vehicles equipped with oxygen sensors, adaptive learning systems		Concerns over water contamination
Distillation characteristics T50, T90↑	Probably HC↑	HC↑		

(continued)

Table 6.4 (continued)

Gasoline	No catalyst	Early three way catalysts	More advanced catalysts	Comments
MMT ↑	Increased manganese emissions	Possible catalyst plugging [PM and other pollutants can also increase, especially with catalyst problems]	Likely catalyst plugging [PM and other pollutants can also increase, especially with catalyst problems]	O$_2$ sensor and OBD may be damaged, MIL light may come on incorrectly
RVP ↑	Increased evaporative HC emissions			Critical parameter for countries with high ambient temperatures
Deposit control additives ↑		Potential HC, NOx emissions benefits		Help to reduce deposits on fuel injectors, carburetors, intake valves, combustion chamber

6.5.3   *Concluding Remarks on Vehicles and Fuels*

One of the most important lessons learned in the approximately 50-year history of vehicle pollution control worldwide is that vehicles and fuels must be treated as a system. Improvements in vehicles and fuels must proceed in parallel if significant improvements in vehicle related air pollution are to occur. A program that focuses on vehicles alone is doomed to failure; conversely, a program designed to improve fuel quality alone also will not be successful.

Reformulated diesel fuels can reduce particulate emissions from all diesel vehicles, as discussed earlier. [Approximately 70–80% of diesel PM is composed of elemental/black carbon. Gasoline PM contains only about 25% elemental/black carbon. Controls on diesel PM, especially catalyzed PM filters, greatly reduce the elemental carbon both in mass and fraction. For example, a 2007 HDD with a cata-lyzed PM trap has lower PM with only ~10% as elemental carbon.]. Especially low sulfur fuels reduce the sulfate contribution. Certain after-treatment technologies are especially sensitive to the sulfur content of the fuel. Therefore if very stringent control of NO_x and PM was needed, sulfur levels will need to be reduced to 50 ppm or less and Euro 4 vehicle standards introduced. Euro 5 or U.S. Tier 2 standards include a fuel sulfur limit of 10–15 ppm. Technologies to achieve these levels already exist and even more advanced technologies are being introduced for new vehicles.

With regard to gasoline-fueled vehicles, the use of catalyst exhaust gas treatment requires the elimination of lead from gasoline. This change, which has occurred throughout most of the world, has resulted in a dramatic reduction of ambient lead levels. Other gasoline properties that can be adjusted to reduce emissions include, roughly in order of effectiveness, sulfur level, vapor pressure, distillation character-istics, light olefin content, and aromatic content [10]. Catalyst technology is emerg-ing for two to three wheeled vehicles and, therefore, lead free and lower sulfur gasoline will be important for these vehicles as well.

6.6   Stringent Vehicle Emissions Standards

The three dominant regulatory programs in the world are the U.S. (including California), the European Union (EU), and Japan. The European and US standards and test procedures or some mixture of them have been adopted by many other coun-tries. For example, China and India have adopted the EU standards for most vehicle categories, although lagging several years behind the EU for implementation. With regard to passenger cars, about 60% of the world's fleet is following the EU regula-tory road map and almost 30% follow the U.S. path. The vast majority, over 90%, of diesel cars are following the EU path. With regard to light trucks, over 60% follow the U.S. standards whereas over 70% of heavy trucks follow the EU emissions stan-dards. No other country outside of Japan requires the Japanese standards.

It is important to emphasize two important points:

1. Standards adopted by the U.S. and the EU will determine the types of technologies and pollution controls used on most light and heavy-duty vehicles around the entire world, so it is important that their standards are sufficiently stringent to address the environmental problems for which they are designed,[10] and
2. While the time gap is narrowing, many developing countries lag the U.S. and the EU by 5 or more years in implementing the standards.

Technologies are now in the market place or rapidly emerging which in combination with the clean fuels discussed above can lower road vehicle emissions of CO, HC, NOx, and PM and other toxins to a very small fraction of those from uncontrolled vehicles per kilometer driven and the major challenge now is to get these technologies adopted around the world.

The most recent light duty vehicle standards for NOx and PM emissions are summarized in Fig. 6.4.[11] While the test procedures used to determine compliance differ somewhat, the control technologies used are very similar and by 2015 when the Euro 6 standards are implemented will be almost identical.

With regard to heavy-duty vehicles and engines, the U.S. and Japan are on track to phase in very stringent NOx and PM requirements before 2010. In the case of Japan, the requirements include a mandatory NO requirement as well as a so-called challenging value for NOx which is only one-third the mandatory requirement. If the challenging value is mandated, the Japanese requirements will be very similar to the US 2010 standards.

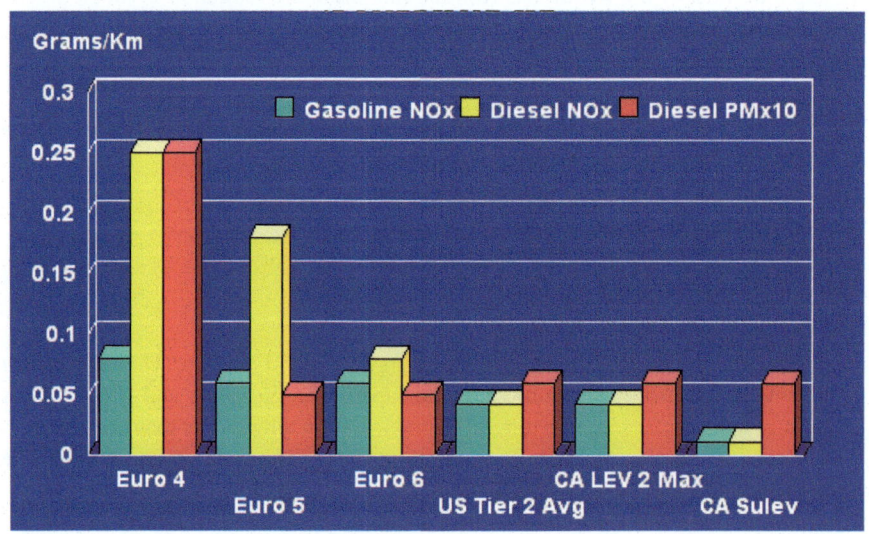

Fig. 6.4 EU and U.S. light duty gasoline and diesel vehicle standards

---

[10] As of early 2010, No other country is following the Japanese vehicle standards roadmap.
[11] The term PMx10 means that PM emissions are multiplied by ten.

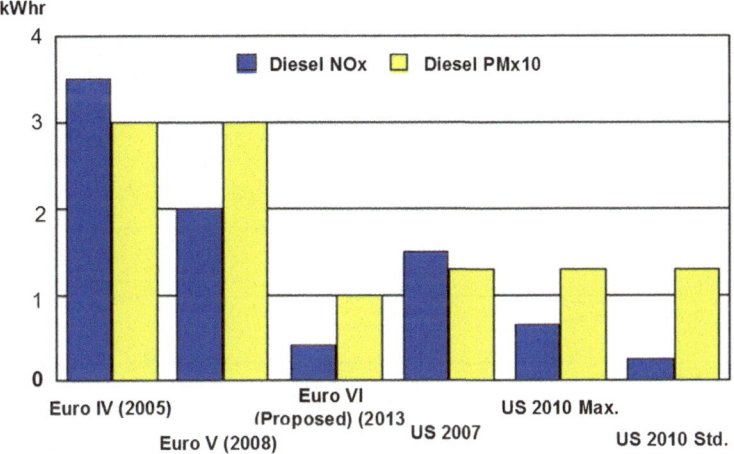

Fig. 6.5 U.S. versus Europe heavy duty transient cycle emissions standards

With regard to Euro VI heavy-duty requirements, the European Commission issued a proposal in December 2007, which it intended to be approximately equivalent to the U.S. 2010 limits (see Fig. 6.5). In December 2008, the European Parliament voted overwhelmingly in favor of the new emissions curbs with 610 votes in favor, 11 against, and 22 abstentions. (MEPs and government negotiators actually arranged the compromise deal in early November.) The new Euro VI regulation will have direct effect and will not require transposition into national law by the 27 EU states.

The agreement backs all the European commission's proposed limit values, including capping emission of nitrogen oxides at 400 mg per kilowatt-hour (mg/kWh) and particulate matter at 10 mg/kWh. All new vehicles of existing models will have to demonstrate compliance with the limits from 1 January 2014 to obtain market approval. This is 9 months earlier than proposed by the European commission. New models will have to meet the standards from January 1, 2013, 3 months earlier than the commission proposed.

In 2006, the European Union introduced Euro III standards for motorcycles (see Table 6.5) which are approximately equivalent to new car standards that applied in the EU in 2000, and these requirements have received a great deal of attention from countries around the world. Both Taiwan (2007) and China (2008) have announced their intention to adopt the EU requirements with only slight variations.

6.6.1 Global Climate Change

With regard to GHGs the prognosis is less promising. As illustrated in Fig. 6.6, CO_2 equivalent emissions from the transportation sector grew significantly in the Developed (Annex 1) countries between 1990 and 2004 [18]. In fact, the growth in the transportation sector, 24%, was by far the largest of any during this period as shown in Fig. 6.7.

Table 6.5 Euro III motorcycle standards (2006)

HC (g/km)	NOx (g/km)	CO (g/km)	Durability (km)
0.8	0.15	2.0	30,000

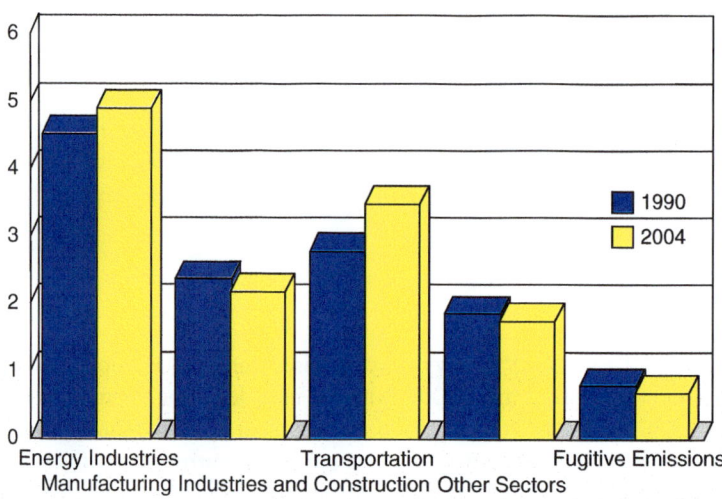

Fig. 6.6 Annex 1 party greenhouse gas emissions in the energy sector

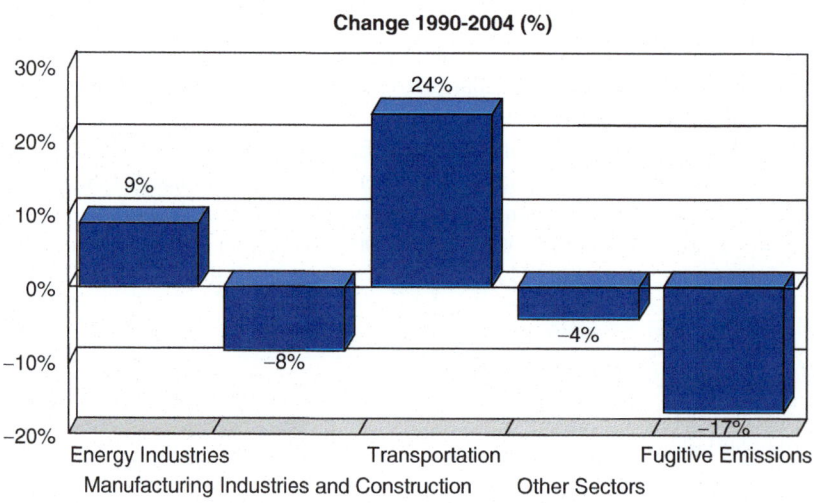

Fig. 6.7 Annex 1 party greenhouse gas emissions growth rates in the energy sector

6.7   Strategies to Reduce Greenhouse Gases and Air Pollution

There are three basic technology-based approaches to reducing GHGs in the transportation sector:

- Setting mandatory or voluntary greenhouse emissions or fuel efficiency standards;
- Shifting to lower-carbon fuels and advanced vehicle technologies; and
- Reducing the use of motorized vehicles.

While the latter approach is critical and will likely be a necessary element of a successful approach it will not be discussed further as it is beyond the scope of this book.

6.7.1   Vehicle Standards

A wide variety of technologies exists or is emerging which can reduce fuel consumption and carbon dioxide emissions from road vehicles. These include variable valve timing, cylinder deactivation, gasoline direct injection, turbo charging and engine downsizing and increased dieselization. The State of California has mandated greenhouse gas emissions standards [19]. At a U.S. national level, mandatory Corporate Average Fuel Economy (CAFÉ) requirements have been in place since the mid 1970s but there had been no significant tightening in over 20 years until Congress mandated further control in 2007; these requirements will result in lower carbon dioxide emissions but do not address the other greenhouse emissions.

In the last three years, American regulators have taken significant steps to improve fuel economy and reduce greenhouse gas emissions from motor vehicles. In April 2007, the U.S. Supreme Court ruled (*Massachusetts v. EPA*), in a 5-4 decision, that GHG emissions are air pollutants potentially subject to federal regulation under the Clean Air Act. In response, the Bush Administration signed an executive order directing the U.S. EPA, in collaboration with the Departments of Transportation and Energy, to develop regulations that could reduce projected oil use by 20 percent within a decade. The Administration suggested that the "Twenty in Ten" goal be achieved by: (1) increasing the use renewable and alternative fuels, which will displace 15 percent of projected annual gasoline use; and (2) by further tightening the CAFE standards for cars and light trucks, which will bring about a further 5 percent reduction in projected gasoline use. The federal Energy Independence and Security Act of 2007 raised the U.S. fuel economy standard for passenger vehicles and light trucks to 35 mpg by the year 2020.

On May 19, 2009, President Barack Obama announced a policy that called for a standard of 35.5 mpg by 2016, essentially requiring light-duty vehicles nationally to meet the same requirements as California. President Obama called for a joint rulemaking by EPA and the National Highway Transportation Safety Administration, which was issued in September 2009.

In a May 21, 2010 memorandum, President Obama directed EPA and DOT to issue a Notice of Intent (NOI) that would lay out a coordinated plan, to propose regulations to extend the national program and to coordinate with the California Air Resources Board (CARB) in developing a technical assessment to inform the NOI and subsequent rulemaking process. NHTSA and EPA, recently announced they will begin the process of developing tougher greenhouse gas and fuel economy standards for passenger cars and trucks built in model years 2017 through 2025, building on the first phase of the national program covering cars from model years 2012–2016. Continuing the national program will help make it possible for manufacturers to build a single national fleet of cars and light trucks that satisfies all federal and California standards.

The U.S. EPA and the U.S. Department of Transportation (DOT) have also announced the first national standards to reduce greenhouse gas (GHG) emissions and improve fuel efficiency of heavy-duty trucks and buses. They are proposing new standards for three categories of heavy trucks: combination tractors, heavy-duty pickups and vans, and vocational vehicles. The categories were established to address specific challenges for manufacturers in each area. For combination tractors, the agencies are proposing engine and vehicle standards that begin in the 2014 model year and achieve up to a 20 percent reduction in carbon dioxide (CO_2) emissions and fuel consumption by 2018 model year. For heavy-duty pickup trucks and vans, the agencies are proposing separate gasoline and diesel truck standards, which phase in starting in the 2014 model year and achieve up to a 10 percent reduction for gasoline vehicles and 15 percent reduction for diesel vehicles by 2018 model year (12 and 17 percent respectively if accounting for air conditioning leakage). Lastly, for vocational vehicles, the agencies are proposing engine and vehicle standards starting in the 2014 model year which would achieve up to a 10 percent reduction in fuel consumption and CO_2 emissions by 2018 model year.

The European Union on the other hand negotiated a voluntary agreement with the European vehicle industry to achieve carbon dioxide targets.[12] This agreement broke down in early 2007 as it became clear that the target of 140 g/km by 2008 will not be met. As a result, the EU will impose a mandatory limit of 130 g/km to be phased in between 2012 and 2015 and will likely further tighten limits to 95 g/km in approximately 2020.

Japan's approach has also focused on fuel consumption using the best in class at a point in time to stimulate industry wide progress. A summary of planned or adopted vehicle requirements is shown in Fig. 6.8 [17].

6.7.2   Low Carbon Fuels

Brazil was the first country to make a significant shift to renewable, lower carbon fuels based on producing ethanol from sugar cane, but many countries around the world are now pursuing similar approaches with mixed success. California recently

---

[12] Similar agreements were also reached with the Japanese and Korean manufacturers.

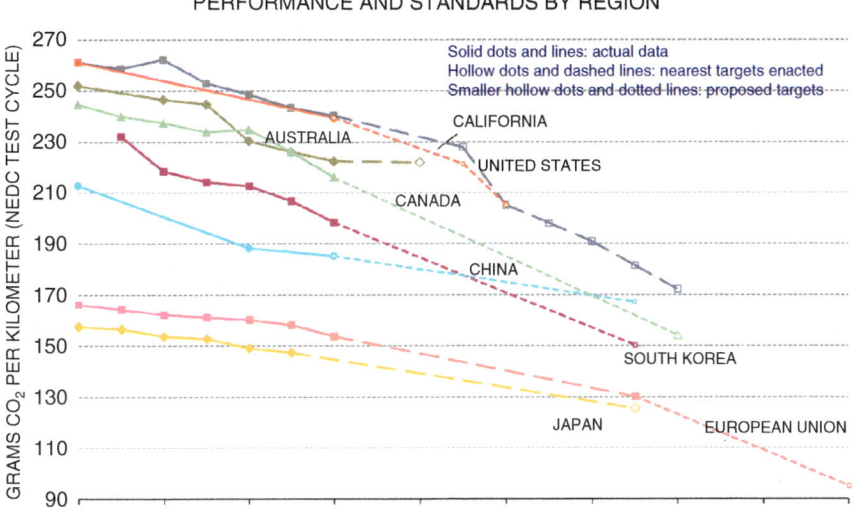

Fig. 6.8 Projected GHG emissions for new passenger vehicles by country/region

proposed carbon based fuels requirements and the EU is pursuing low carbon fuels standards (LCFS) [16]. However, to achieve significant global benefits from low carbon fuels it is increasingly clear that a full life cycle analysis[13] is necessary which includes consideration of indirect land use effects. When such factors are taken into account, it is clear that moving to low carbon fuels that actually achieve significant benefits is a very difficult proposition.

The goal of a LCFS is to promote investment and use of low carbon fuels (e.g., sustainable corn ethanol and biodiesel, CNG, renewable electrons[14] / hydrogen) and dampen demand for high carbon fuels (e.g. Canadian tar sands, Venezuelan shale oil, U.S. coal to liquids).

The current U.S. renewable fuels standard (RFS), mandated by the Energy Independence and Security Act of 2007, takes a step toward a LCFS by requiring life-cycle GHG standards for three categories of biofuels: baseline renewable biofuels 20% below gasoline, advanced biofuels 50% improvement, and cellulosic

---

[13] A full life cycle analysis is an effort to capture all the emissions associated with a given fuel from its extraction or harvest to refinement and transport all the way to the eventual consumption in the vehicle. Only in this way can a fair comparison be made between various fuels and can a fair accounting be made of their impact on climate change.

[14] If a significant portion of the vehicle fleet becomes battery electric or plug in hybrids, it will be important to produce the electricity for these vehicles using clean, renewable fuels. Otherwise the environmental benefits, especially with regard to climate impacts, will be greatly diminished.

biofuels 60% improvement. The RFS however only applies to biofuels and thus does not dampen demand for high carbon fuels (e.g., tar sands, coal to liquids).

California air regulators have adopted a mandate requiring low-carbon fuels, part of the state's wider effort to reduce greenhouse gas emissions. The California Air Resources Board voted 9-1 to approve the standards, which are expected to serve as a template for a national policy. The rules call for reducing the carbon content of fuels sold in the state by 10% by 2020, a plan that includes counting all the emissions required to deliver gasoline and diesel to California consumers' from drilling a new oil well or planting corn to transporting it to gas stations.

The measure also sets the stage for emerging alternative fuels – such as cars that run on compressed natural gas and electric vehicles like plug-in hybrids that run on both gasoline and rechargeable batteries – to compete with second-generation ethanol. That fuel, cellulosic ethanol, is expected to be made in commercial amounts from non-food feedstocks like switch grass and fast-growing trees.

To give fuel producers time to adjust, the bulk of the carbon limits required under the regulation do not go into effect until 2015.

California's regulators ranked 11 different ways of making corn ethanol. They found that traditional distilling methods used in the Midwest, accounting for the bulk of U.S. supplies, emit the most carbon over a lifecycle measured from production to combustion. The state gave much better carbon savings scores to corn ethanol made in California with a distillery fired by a blend of natural gas and crop waste, also known as biomass.

The regulation rates different fuels based on their carbon intensity, measured as the number of grams of carbon dioxide released for every megajoule of energy produced. When the indirect land-use effects of biofuels are included, some types of ethanol rate worse than gasoline.

Fuel Type	Carbon intensity	Carbon intensity (including land use changes)
California gasoline{+1}	95.85	95.85
Midwest ethanol{+2}	75.10	105.10
California ethanol{+3}	50.70	80.70
Brazilian ethanol{+4}	27.40	73.40
Landfill gas(bio-methane){+5}	11.26	11.26

{+1} with 10% ethanol
{+2} with some of the plant's power coming from coal
{+3} with the plant's coming from natural gas
{+4} made from sugarcane and shipped here
{+5} derived from landfills in California
Source: California Air Resources Board

Electric utilities may opt into the program and generate credits if they sell renewable electrons to plug-in hybrids or all-electric vehicles. GHG emissions from direct and indirect land use changes are included in the estimation of fuel lifecycle GHG impacts.

The European Union adopted a LCFS similar to California. The new law will require fuel suppliers to cut life-cycle greenhouse gas emissions from road fuels by 6% over the decade from 2010 to 2020 (intermediate targets: 2% by 31 December 2014 and 4% by 31 December 2017). The cuts will come from production efficiency improvements and switches to cleaner fuels such as biofuels. Biofuels sustainability criteria will be added to the new law once they have been agreed in separate negotiations on a new renewable energy directive.

With this the EU has sent a clear signal that its market is not opened to carbon intensive marginal oils, such as tar sands or coal-to-liquid. Several key technical and economic questions remain with regard to low carbon fuels, including:

- Significant uncertainty around lifecycle GHG emissions from U.S. corn ethanol (range is between average 25% improvement over gasoline to twice the lifecycle GHG emissions of gasoline – thus making corn ethanol potentially worse than tar sands and coal to liquids).
- If biomass for fuels can only be produced sustainably if grown on degraded land or produced from waste products (e.g., corn stover, forest wood waste), then what are the practical limitations of global supplies of sustainable biofuels?
- How much of the recent increase in the world price of corn and other grains is attributable to biofuels mandates in the U.S. and elsewhere?
- What are the prospects of plug-in hybrids and the ability of electric utilities to supply sufficient renewable electrons for this new market?

6.8 Advanced Vehicle Technologies

Advanced vehicle technologies including battery electric cars, hybrids, plug in hybrids and fuel cells are mandated in California and strongly encouraged in Japan with the result that significant advances are occurring [20]; over 300,000 hybrid vehicles were sold around the world in 2006.

There are three main ZEV enabling technologies – energy storage, hydrogen storage, and fuel cells – that are critical to the successful development and deployment of advanced vehicle technologies. The current status and prospects for further advances of these technologies and their integration into vehicles are summarized below.

6.8.1 Vehicle Energy Storage Systems

This section will focus on advanced battery technologies with potential to be fully developed and available for use in Hybrid Electric Vehicles (HEVs), Full Performance Battery Electric Vehicles (FPBEVs) and Plug in Hybrid Electric Vehicles (PHEVs) within the next 5–10 years.

Two of the more important characteristics of batteries are Energy Density (Wh/kg) and Power Density (W/kg). Energy Density is a measure of how much energy a battery can hold. The higher the energy density, the longer the runtime will be. Typical applications are cell phones, laptops, and digital cameras. Power Density indicates how much power a battery can deliver on demand. The focus is on power bursts rather than runtime.

6.8.1.1 Nickel Metal Hydride Batteries (NiMH)

High power NiMH technology for HEVs is now mature and mass manufactured in Japan in plants with capacities up to 500,000 systems annually. High cost remains the greatest challenge for battery and HEV manufacturers, with an estimated cost (price to Original Equipment Manufacturers [OEMs]) of $2,000 for compact and $4,000 for a midsize HEV battery produced at a rate of 100,000 systems per year. These costs appear to account for much of the current price difference between hybrid and conventional vehicles. At a production rate of one million systems, battery costs are projected to drop to $1,300 and $2,500, respectively.

Medium power/medium energy NiMH technology has promise to meet the technical requirements for PHEVs with relatively short (e.g., 10–20 miles) nominal electric range. In mass production, medium power/medium energy NiMH technology's incremental cost over that of HEV batteries, estimated to be about $800–1,200, is probably less than the difference in lifetime fuel costs. However, no substantial efforts to develop or capabilities to fabricate medium power NiMH technology appear to exist.

High-energy NiMH technology was used successfully in FPBEVs manufactured by major automobile manufacturers under the California ZEV program. However, energy density is fundamentally limited and marginal for FPBEV applications, and costs remain as high as or higher than in 2000 and are unlikely to decline. High-energy NiMH technology for possible FPBEV applications does not appear to have advanced in recent years.

6.8.1.2 Lithium Ion Batteries (Li Ion)

Li Ion batteries are making impressive technical progress worldwide especially with regard to calendar and cycle life and safety, the areas of special concern for automotive applications. Promising new materials and chemistries are expanding the capabilities and prospects of all Li Ion technologies.[15]

---

[15]More than half the world's reserves of lithium are located high in the Andes, in a remote corner of Bolivia and there are indications that the country may resist efforts to allow outsiders to control the production. Therefore there has been speculation that shortages may occur as early as 2015 unless other sources are found or an accommodation can be made with Bolivia.

High power Li Ion technology for HEVs appears close to commercialization. A variety of materials, manufacturing techniques and companies are competing to achieve the performance and cost goals for this established battery application which increases the probability of technical and market success. Importantly, for HEV applications Li Ion batteries have potentially lower cost than NiMH because they promise to deliver the required power with smaller capacities and lower specific cost.

Medium energy/power Li Ion technology has sufficient performance for PHEVS and small FPBEVs, and it can be expected to meet the life requirements for FPBEVs. Recent test results indicate good potential to also deliver the very demanding cycle life for PHEVs. The projected costs for shorter-range PHEV Li Ion batteries are about \$3,500–4,000 in mass production; this is generally less than the fuel cost savings expected over the life of the vehicle. Low volume cell production and prototype battery fabrication is underway in Asia and Europe, and limited fleet demonstrations are underway or planned.

High energy Li Ion technology has sufficient performance for small FPBEVs, and good potential to meet all performance requirements also of midsize and larger FPBEVs with batteries of modest weight (e.g., less than 250–300 kg). Cell and battery technology designed for these applications are likely to also meet cycle life goals. However, battery cost remains high even in mass production, well in excess of expected lifetime fuel cost savings. While high energy Li Ion technology probably will benefit from general progress in Li Ion technology, no efforts seem underway to advance technology designed for FPBEV applications.

Batteries assembled from large numbers (typically, 5,000 or more) of small, high energy Li Ion cells mass-manufactured for laptop computers and other electronic applications are now being used in FPBEVs (and PHEVs) fabricated on a small scale. However, such small-cell batteries, although providing early opportunities to demonstrate the technical capabilities of PHEV conversions and modern FPBEVs, have inherently high costs and uncertain calendar and cycle life.

6.8.2   Hydrogen Storage Systems

Storing sufficient hydrogen on a vehicle to power it for adequate distance, safely, and at reasonable cost, without an excessive weight penalty has been and remains a serious challenge for the automobile industry and its suppliers. Hydrogen storage is among the two or three areas of greatest concern, including all of the other cost and technology challenges associated with developing fuel cell systems for consumer vehicles.

Unlike other major technologies being pursued in support of ZEVs, hydrogen storage technologies have advanced relatively little in recent years. However, in the last 3–4 years, as it became apparent that on vehicle fuel reformers for generation of hydrogen from carbon based liquid fuels were not a viable option, many alternative storage concepts have begun to receive significant research attention.

A few concepts (e.g., metal hydrides and carbon nanotubes) that have been investigated at relatively low levels of effort for many years are now receiving increased attention. However, these efforts are fairly young and it is still too early to determine if they will result in technically and economically realistic hydrogen storage system alternatives.

6.8.2.1   Near Term Outlook

In the near term, the dominant form of storing hydrogen onboard light vehicles will continue to be compressed hydrogen gas. Most OEMs preferred 700 bars, which will provide storage of over 50% more fuel in the same space envelope and correspondingly provide almost 50% more range than 350 bars. Using 700 bar storage pressure is not, however, without problems. The volumetric density (kWh/L) will be higher but unit energy cost ($/kWh) is also expected to be higher and the gravimetric energy density (kWh/kg) about the same. It may also require either reduced fill rates or pre-cooling of the hydrogen prior to transferring into the vehicle tank to avoid overheating the tank structural materials.

Liquid hydrogen storage is being demonstrated as workable but with limitations. It provides both higher gravimetric and volumetric density advantages over compressed gas storage but has issues with boil off and dealing with cryogenic liquids. It is not likely to be widely accepted by automobile OEMs.

An important issue with any of the short-term hydrogen storage options is the need for widely accepted codes and standards for permanent storage, onboard storage, and all aspects of transferring and transporting hydrogen.

Cost is another important issue, especially for the short term since none of the storage systems are produced in sufficient volumes to allow significant production economies of scale. Current or near-term costs for the essentially one-of-a-kind hydrogen storage systems are approximately $10,000 or more each for both liquid and compressed gas storage.

6.8.2.2   Longer Term Outlook

For the longer term, some of the alternative storage technologies being researched may prove to be effective. Both solid and liquid carriers are being researched with hydrogen "recharging" being carried out both onboard and off of the vehicle. There don't appear to be any clear winners at the present among these alternatives. It appears to be too early to make reasonably accurate projections.

6.8.2.3   Conclusions

On-board hydrogen storage is a major challenge for hydrogen fuel cell vehicles. At present, the only technology being demonstrated by the OEMs, with the exception of BMW, is compressed hydrogen gas storage which has problems providing sufficient vehicle range without excessive volume, weight, and cost.

The volume issue can be partially resolved by using 700 bar storage (thus a smaller required volume) and by innovative vehicle design or design modification. Such innovations might include utilization of a long, small-diameter tank running longitudinally where the center "tunnel" is located and/or replacing rear coil springs with leaf springs to increase space available for hydrogen tanks. Thus, depending on the type of vehicle and system efficiency, it seems likely that sufficient compressed hydrogen could be stored on a vehicle to provide a range in excess of 200 miles, perhaps reaching 300 miles or more.

Liquid hydrogen storage technology appears to have advanced sufficiently that, within certain constraints, it could be utilized. The advantages of liquid hydrogen, higher storage density and low pressure, suggest that it also could provide an adequate range.

However, it seems unlikely that either compressed or liquid hydrogen storage systems can meet weight or cost targets, especially for 2015. Using the TIAX estimates for mass-manufactured tanks, the system cost would be about $10–$12 per kWh for 350 bar systems and $13–$15 per kWh for 700 bar systems compared to DOE targets of $4 per kWh for 2010 and $2 per kWh for 2015. Assuming that at least 5 kg (165 KWh) of hydrogen will be needed to provide sufficient vehicle range, the cost would be $1,650 even with the lowest TIAX tank cost estimate. For liquid storage, the cost would be even higher. There is little expectation that the cost of either of these systems will go much lower even with higher volumes.

The weight outlook is better than the cost outlook. The TIAX projections for weight fraction are slightly over 6% for both 350 bar and 700 bar systems, compared to the DOE targets of 6% for 2010 and 9% for 2015. The pressure tank manufacturers have also indicated that 6%, and perhaps a bit higher weight fraction is within reach. For a 6% weight fraction system to contain 5 kg of hydrogen, the system would weigh about 83 kg (about 183 lb). Neither TIAX nor the tank manufacturers project that the 2015 target of 9% can be met with pressurized hydrogen tanks.

There are many alternative hydrogen storage systems under investigation. Some of the absorption materials being investigated are relatively inexpensive and have shown, at least in the research phases, the capacity to contain well over 6% hydrogen. However, the remainder of the support system could have a huge effect on both cost and weight fraction.

6.8.3 Automotive Fuel Cell Systems

Automotive fuel cell technology continues to make substantial progress but is not yet proven to be commercially viable. Recent technological and engineering advancements have improved, simplified, and even eliminated components of the fuel cell system. These include major improvements in the membrane electrode assembly (MEA) and fuel cell stack technologies. The Balance of Plant has a reduced number of components and now uses some parts that are of automotive quality and cost. The fuel cell system has a reduced start time and in-vehicle start-up from a frozen condition has been demonstrated. Great strides have been made in the science of materials and operating characteristics of fuel cells. This increase in fundamental understanding shows promise for solving life, abuse, and durability issues for fuel cell systems.

The consensus among the majority of fuel cell system developers is that in order to achieve commercialization there are simultaneous requirements for:

- Higher MEA power per unit area of fuel cell electrodes (goal of 0.8–1.0 W/cm^2)
- Reduced MEA catalyst cost (goal of total MEA catalyst loading <0.1–0.5 mg Pt/cm^2)
- Longer fuel cell system operating life and increased durability (goal of >5,000 h of customer use)
- Proton Exchange Membrane (PEM) materials that are stable and can operate at a higher temperature (above 100°C)
- Engineering advances

An increase in MEA specific power allows a given fuel cell stack to produce more power and thus achieve a lower $/kW. Nearly every stack cost factor, at a given voltage, decreases in inverse proportion to MEA specific power. The MEA catalyst cost is directly related to the price of platinum. The price of this noble metal is rising due to worldwide demand exceeding supply and at current levels it represents a significant barrier to automotive fuel cell commercialization. The life and durability of fuel cells in automotive applications is not yet proven. A life of 5,000 + hours in a light duty vehicle type load cycle has not been demonstrated at the cell or stack level. The development of high temperature membranes can potentially reduce the size and complexity of the Fuel Cell Electric Vehicle (FCEV) thermal system and may possibly eliminate the need for stack humidification. Engineering advances and innovation are focused on materials, stack design, and balance of plant to reduce cost and increase life.

At this time no fuel cell developer has achieved the necessary requirements for automotive fuel cell commercialization. The developers are relying on future technological improvements to meet both cost and life goals. Achieving these goals creates some contradictory requirements for the fuel cell system. These requirements are difficult to achieve separately and because they are interrelated, even more difficult to solve simultaneously. These technological improvements include the development of MEAs that use significantly less catalyst material and that operate at higher specific power and temperature over a longer system life. To simultaneously increase performance, extend life, and reduce cost will likely take ingenuity and invention.

In summary, there is reason to be cautiously optimistic regarding the prospects for fuel cell system commercialization. There are still large technical barriers to be solved but these might well be overcome over the next 5–10 years through the massive efforts underway at the major fuel cell and automobile manufacturers.

However, there are other issues that are beyond the control of any single manufacturer. These include timely availability of adequate and affordable hydrogen refueling, as well as need for a host of sustainable financial incentives to help minimize the capitalization risks of all key stakeholders during the early years of initial commercialization of hydrogen powered FCEVs. Wide spread deployment of FCEVs will require continuous strong support and a long-term commitment from government agencies in resolving these issues.

6.8.4 Vehicle Integration

The status and prospects of vehicle integration of zero emission vehicles (ZEVs), as well as their advanced technology vehicles (ATVs) that could have synergistic benefits supportive to the introduction of ZEVs, are summarized below. In addition to vehicle technical considerations, vehicle business considerations (e.g., manufacturing cost, capital investment, marketability, etc.) also are addressed, in order to forecast the future prospects, introduction timing, and volume milestones of the ZEV and ATV technologies.

6.8.4.1 Full Performance Battery Electric Vehicle (FPBEV)

Full Performance Battery Electric Vehicles are defined in this report as BEVs fully capable of high speed U.S. urban/suburban freeway driving.

Despite substantial technology progress, prior efforts to introduce FPBEVs were unsuccessful. Specifically, the large batteries required to provide the necessary driving range, as well as an acceptable "cushion," remain very expensive.

Higher fuel prices and less demanding driving conditions in Japan and Europe provide lower barriers to success and as a result a few OEMs are developing small FPBEVs with Li Ion batteries for these markets, and they may bring them to the U.S. as niche vehicles.

FPBEVs are not likely to become mass market ZEVs in the near future due to the high cost for the battery not being recoverable with fuel cost savings and limited customer acceptance due to range and recharge time issues.

6.8.4.2 City Electric Vehicle (CEV)

City Electric Vehicles are defined in this report as BEVs with limited acceleration and top speed (e.g. 50/60 mph) and thus not suitable for high speed U.S. urban/suburban freeway driving, although at present they must meet all Federal Motor Vehicle Safety Standards (FMVSS) requirements. These performance limitations allow a smaller size battery and lower power electric drive system, so that the vehicle can have a lower manufacturing cost and thus be made more affordable to the customer.

Prior efforts to produce CEVs were unsuccessful due to high cost and limited mass-market customer acceptance. A special CEV FMVSS similar in concept to FMVSS 500 (e.g., restrict CEVs from freeway driving, etc.) may help stimulate development in the U.S.

CEVs are more likely to become future mass market ZEVs in Japan and Europe than in the U.S. due to performance limitations.

6.8.4.3 Neighborhood Electric Vehicle (NEV)

Neighborhood Electric Vehicles are defined in this report as BEVs capable of top speeds between 20 and 25 mph that meet FMVSS 500 and are limited to roads with posted speeds of 35 mph or less.

NEV technology appears to be commercially successful but has low volume potential due to limited applicability. Also, because they use very simple technology, NEVs have very little synergy with larger BEVs.

NEVs provide no significant technical benefits to future mass market ZEVs due to their simple technology and performance limitations.

6.8.4.4 Hybrid Electric Vehicle (HEV)

HEVs have no customer compromises and therefore appeal to mass-market customers willing to pay a premium. While producers are driving down the costs of electric drive components and systems, high manufacturing cost is still an issue. However, OEMs are introducing many new entries, despite the cost issue, mostly for competitive reasons. Overall, HEV sales volume rises and falls with the price of gasoline – making future growth forecasts uncertain.

HEVs, due to their success, are providing major support to future mass market ZEVs by continuing to stimulate advances in electric drive systems, electric accessories, and battery technologies. In addition, they are increasing customer awareness of electric drive technology and the associated benefits.

6.8.4.5 Plug-in Hybrid Electric Vehicle (PHEV)

PHEVs have no expected customer compromises while promising several benefits to customers and society. The relatively small battery capacity can be fully used daily for maximum customer fuel savings payback of the initial vehicle premium.

Recently, some OEMs have become interested in PHEVs, and GM and Ford have shown concept PHEVs at recent auto shows and other events – which is attracting major media attention and establishing high consumer expectations.

All Electric Range (AER) is the distance the vehicle can travel without utilizing the internal combustion engine. The longer the AER, the higher the impact on manufacturing costs, as well as capital investment requirements if unique and more powerful electric drive systems are necessary. Therefore AER could have a significant impact on the early success of the technology.

PHEVs are likely to become available in the near future. They may foster future mass market BEVs by stimulating energy battery development and conditioning mass market customers to accept plugging in.

6.8.4.6   Fuel Cell Electric Vehicle (FCEV)

FCEVs are considered the ultimate solution by several OEMs with massive R&D efforts underway. However, simultaneously achieving performance, durability, and cost objectives with FCEVs continues to be very difficult.

The cost, weight, and volume of adequate vehicle hydrogen storage and availability of a hydrogen infrastructure are major issues.

Plug-in series hybrid FCEVs operating "steady state" have potential to simultaneously achieve performance, durability and cost objectives.

With the past rate of success and the massive intellectual and financial resources being devoted to this technology, FCEVs continue to be a promising candidate for a future mass market true ZEV.

6.8.5   Overall Conclusions: The Prospects of ZEVs

PHEVs with modest energy storage capacity will be derived from HEVs and will likely proliferate rapidly, stimulating further development and cost reduction of energy batteries and leading to commercially viable PHEVs and, in the longer term, FPBEVs. While PHEVs will continue to grow rapidly, as they have no functional limitations, FPBEVs will grow more slowly due to customer acceptance of limited range and long recharge time. NEVs are commercially viable now and will continue to grow, but will grow slowly due to limited functionality. CEVs will become commercially viable in Japan and Europe in the not too distant future due to lower hurdles for BEVs to overcome. CEVs may be offered in the U.S. as energy batteries continue to mature, but growth will be slow due to functional limitations of BEVs in general, and the specific limitations of CEVs, especially urban freeway driving. The intense effort on FCEVs will result in technically capable vehicles by the 2015–2020 time frame, but successful commercialization is dependent on meeting challenging cost goals and the availability of an adequate hydrogen infrastructure. If these challenges are met, FCEVs will likely grow rapidly.

6.9   Conclusions

The number of motor vehicles and their miles driven has literally exploded over the past 60 years and the likelihood is that this growth will continue for the foreseeable future. While vehicle populations are roughly stabilizing in the highly industrialized OECD countries, they are accelerating in rapidly industrializing highly populous countries. In response to strong and aggressive regulatory programs, especially in the United States, new vehicles being sold in many countries today are much cleaner than in the past. For example, over 95% of all gasoline sold in the world today is lead free and over 95% of all new gasoline fueled cars are equipped with a three way catalyst which dramatically lowers CO, HC and NOx emissions per mile driven. As a result, air quality in urban areas in developed countries has generally improved.

However, even in the developed world, air pollution levels in major cities continue to exceed levels necessary to protect public health. And in the rapidly industrializing countries, pollution in many cities is worsening. Therefore, with regard to urban and regional pollution, two major challenges remain:

1. To accelerate the introduction of the state of the art technologies for clean vehicles and fuels in rapidly industrializing countries such as China, India and Brazil and get at least modest controls in places where none currently exist such as in much of Africa and the Middle East, and
2. To phase out or clean up the so-called legacy fleet of existing high polluting vehicles. (California, for example, has embarked on an effort to eliminate every diesel vehicle in the State not equipped with a diesel PM filter either by mandatory retrofit or scrappage.)

With regard to climate change the picture is much more bleak and the challenge more daunting. Transportation is already a large contributor to the problem and is a rapidly growing sector. Modest programs to reduce fuel consumption or greenhouse gas emissions from light duty vehicles are being phased in and California and the EU have initiated efforts to reduce the carbon content of vehicle fuels. Much more will need to be done with a likely shift to battery electric vehicles fueled by green electrons or fuel cell vehicles fueled by renewable hydrogen in future decades.

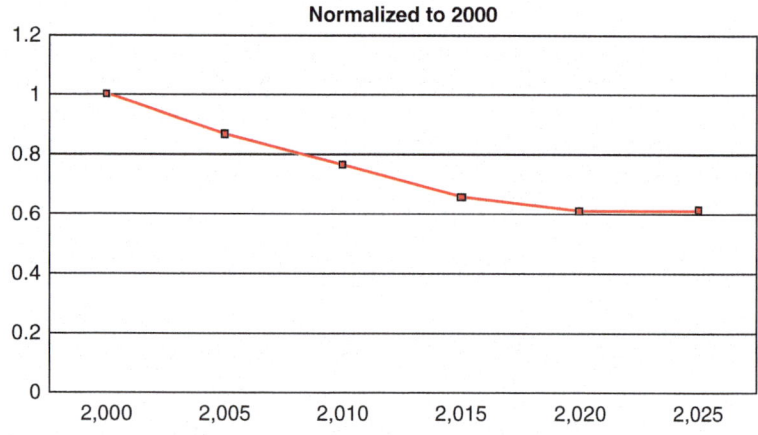

Fig. 6.9 Global trends in on road vehicle black carbon emissions

As efforts to reduce CO2 by 70 or 80% by 2050 receive high priority, aggressive short-term actions to reduce short-lived greenhouse pollutants hold promise. The US, Europe and Japan are phasing in high efficiency PM filters which will also reduce black carbon emissions dramatically. On a global basis, driven by these efforts, road vehicle emissions of black carbon are declining rapidly as shown in Fig. 6.9.

However, after 2025 they will start to increase unless the rapidly industrializing countries also move to require PM filters on new diesel vehicles soon. If they do, the downward trend in black carbon can continue while CO_2 reduction measures are developed and implemented.

References

1. Ward's World Motor Vehicle Data (2007) Ward's Atuomative Group, 3000 Town Center, Suite 2750, Southfield MI 48075
2. Honda Motor Company (2004) World motorcycle facts and figures Corporate Communications Division, 2-1-1, Minami Aoyama, Minatu-ku, Tokyo, Japan
3. World Business Council on Sustainable Development (2004) Mobility 2030: meeting the challenges to sustainability. The Sustainable Mobility Project, Full Report, WBCD, Geneva
4. IPCC First Assessment Report (FAR), Climate Change: The IPCC Scientific Assessment (1990) Report prepared for Intergovernmental Panel on Climate Change by Working Group I, J.T. Houghton, G.J. Jenkins and J.J. Ephraums (eds.)., Cambridge University Press, Cambridge, Great Britain, New York, NY, USA and Melbourne, Australia, 410 pp.
5. Climate Change 1995, The Science of Climate Change, Edited by J.T. Houghton, L.G. Meira Filho, B.A. Callander, N. Harris, A. Kattenberg and K. Maskell, Production Editor: J.A. Lakeman, Contribution of WGI to the Second Assessment Report of the Intergovernmental Panel on Climate Change, *Published for the Intergovernmental Panel on Climate Change*, Cambridge University Press, Published by the Press Syndicate of the University of Cambridge, The Pitt Building, Trumpington Street, Cambridge CB2 IRP, 40 West 20th Street, New York, NY 10011-4211, USA, 10 Stamford Road, Oakleigh, Melbourne 3166, Australia, First published 1996, Printed in Great Britain at the University Press, Cambridge
6. Hansen J, Sato M, Kharecha P, Russell G, Lea D, Siddall M (2007) Climate change and trace gases. Royal Transactions – Non CO_2 GWP
7. Jacobson M (2005) Correction to 'Control of fossil fuel particulate black carbon and organic matter, possibly the most effective method of slowing global warming. J Geophys Res 23 July 2005
8. Ramanathan V, Carmicheal G (2008) Global and regional climate changes due to black carbon. Nat Geosci 1:221–227. www.nature.com/naturegeoscience
9. Auto/Oil Air Quality Improvement Research Program (1997) Final report, Jan 1997
10. Sawyer RF (1992) Reformulated gasoline for automotive emissions reduction. In: Twenty-fourth symposium (International) on combustion, Sydney. The Combustion Institute, Pittsburgh, pp 1423–1432
11. Rosner D, Markowitz G (1985) A 'Gift of God?: the public health controversy over leaded gasoline during the 1920s. Am J Public Health 75(4):344–352
12. Landrigan P, Nordberg M, Lucchini R, Nordberg G, Grandjean P, Iregren A, Alessio L (2006) The declaration of Brescia on prevention of the neurotoxicity of metals. Am J Ind Med 50:709–711
13. Bellagio memorandum on motor vehicle policy: principles for vehicles and fuels in response to global environmental and health imperatives (2001) Consensus document, Bellagio, 19–21 June 2001

14. Hauser RA, Zesiewicz TA, Martinez C, Rosemurgy AS, Olanow CW (1996) Blood manganese correlates with brain magnetic resonance imaging changes in patients with liver disease. Can J Neurol Sci 23(2):95–98
15. Lucchini R, Albini E, Placidi D, Gasparotti R, Pigozzi MG, Montani G, Alessio L (2000) Brain magnetic resonance imaging and manganese exposure. Neurotoxicology 21(5): 769–75
16. Farrell A, Sperling D, Arons S, Brandt A, Delucchi M, Eggert A, Farrell A, Haya B, Hughes J, Jenkins B, Jones A, Kammen D, Kaffka S, Knittel C, Lemoine D, Martin E, Melaina M, Ogden J, Plevin R, Sperling D, Turner B, Williams R, Yang C (2007) A low carbon fuel standard for California, part 1: technical analysis. University of California Berkeley, Berkeley, 29 May 2007
17. The International Council on Clean Transportation (2008) Passenger vehicle CO_2 and fuel economy standards: a global update. ICCT, Washington, DC
18. Roels HA, Ghyselen P, Buchet JP, Ceulemans E, Lauwerys RR (1992) Assessment of the permissible exposure level to manganese in workers exposed to manganese dioxide dust. Br J Ind Med 49(1):25–34
19. United Nations Framework Convention on Climate Change (UNFCCC) (2006) National greenhouse gas inventory data for the period 1990–2004, and status of reporting. UNFCC, Bonn, 19 Oct 2006
20. Walsh MP, Kalhammer FR, Kopf BM, Swan DH, Roan VP (2007) Status and prospects for zero emissions vehicle technology. Report of the ARB Independent Expert Panel 2007, Prepared for State of California Air Resources Board, Sacramento, 13 Apr 2007

Chapter 7
Buildings: Mitigation Opportunities with a Focus on Health Implications*

Robert Thompson[†], James Jetter, David Marr, and Clyde Owens

Abstract Addressing building energy use is the critical first step in any strategic plan for mitigating climate change. Buildings have a direct impact on estimated global climate change due to their large carbon footprint. Energy use in the building sector is the largest man-made contributor to climate change, and coincidentally a key sector to start mitigating climate change. To avoid revisiting problems such as sick building syndrome arising from aggressive building weatherization programs in the 1970s, it is critical that policy makers, regulators, and strategic planners remember that the primary function of buildings is not saving energy. The bottom line of why we build buildings is for safety and comfort in our homes, to enhance productivity in the workplace, and to ensure an optimal learning environment in our schools. The fundamental services of improving human health, comfort, productivity, and performance should not be compromised as we strive to minimize energy use in buildings. A one-dimensional focus on energy could result in unsustainable policies and practices. Much is understood about technologies, materials, and design techniques that can reduce energy use in buildings. However, much attention must be paid to recognizing how these approaches can enhance or damage human health and productivity as well as the environment. The focus of this chapter is not existing energy sectors and conservation technologies that have been extensively understood and considered in the literature, but on underutilized mitigation techniques that both increase the sustainability of our buildings while maintaining a focus on human health and the environment. A key intersection between climate change, buildings, and human health is building materials and products, and an effective testing and information transfer program is urgently

---

*The findings included in this chapter do not necessarily reflect the view or policies of the Environmental Protection Agency. Mention of trade names or commercial products does not constitute Agency endorsement or recommendation for use.

[†] © US Government 2011

R. Thompson (✉), J. Jetter, D. Marr, and C. Owens
U.S. Environmental Protection Agency, Office of Research and Development,
National Risk Management Research Laboratory, Air Pollution Prevention
and Control Division, Cincinnati, OH, USA
email: bob.thompson@epa.gov

F.T. Princiotta (ed.), *Global Climate Change - The Technology Challenge*,
Advances in Global Change Research 38, DOI 10.1007/978-90-481-3153-2_7,
© Springer Science+Business Media B.V. 2011

needed so that building stakeholders have the information and tools they need to make good decisions during the design, construction, operation, and renovation phases of buildings.

7.1 Introduction

Energy use in the building sector is the largest man-made global contributor to climate change, and coincidentally a key sector to start mitigating climate change [1]. Most major reports and studies on climate change address the impacts of energy use in buildings [2–10]. To avoid revisiting problems such as sick building syndrome arising from aggressive building weatherization programs in the 1970s, it is critical that policy makers, regulators, and strategic planners remember that the primary function of buildings is not saving energy.

The bottom line of why we build buildings is for safety and comfort in our homes, to enhance productivity in the workplace, and to ensure an optimal learning environment in our schools. The fundamental services of improving human health, comfort, productivity, and performance should not be compromised as we strive to minimize energy use in buildings. A simple example of the relative value of energy as compared to just productivity alone is that the typical energy cost for a commercial building is less than \$2 per ft^2 per year, while the workers occupying the building cost approximately \$200 per ft^2 per year – a hundredfold greater value. A one-dimensional focus on energy could result in unsustainable policies and practices. That being said, this chapter is focused on the specific topic of underutilized mitigation techniques that both increase the sustainability of our buildings while maintaining a focus on human health and the environment. This point is emphasized by the contrasting knowledge base between well-known single attribute mitigation programs (e.g. Energy Star [11], WaterSense [12]) and more holistic, multi-attribute programs that focus on sustainability and health within the building/energy sector. This chapter acknowledges, but does not delve into, the importance of each topic which must considered for such an approach, including other energy sectors, water use, etc.

A major report on climate change, the IPCC report [2], recognizes the following potential co-benefits of greenhouse gas mitigation in the residential and commercial building sectors: (1) Reduction in local/regional air pollution; (2) Improved health, quality of life and comfort; (3) Improved productivity; (4) Employment creation and new business opportunities; (5) Improved social welfare and poverty alleviation; and (6) Energy security.

7.2 Why Building are Important to Climate Change

Buildings have a direct impact on estimated global climate change due to their large carbon footprint. From an energy use perspective, buildings in the United States are responsible for about 40% of the national total carbon dioxide emissions [13] and approximately 9.8% of the global carbon dioxide emissions, which is greater than the carbon emissions of the complete economies of Japan, France, and the United Kingdom combined [3]. This direct production of greenhouse gasses (GHGs) becomes even more significant when one considers the additional GHGs that are produced through the manufacture of the materials and products used in building construction. Combined with the transport of both construction and waste products, the carbon footprint of the built environment is substantial. It should be noted here that building energy usage and CO_2 emissions are influenced by the type of materials and construction practices used in their construction. This is exemplified by the extraction of natural resources and the production, transportation, and disposal of non-food goods being estimated to be 35% of U.S. CO_2 emissions [14].

Residential and commercial buildings account for 35% of the total *global* energy consumption. In the United States, the energy usage by the built environment outpaces both transportation (28%) and industrial energy use (32%) (See Fig. 7.1) [3]. Studies indicate that buildings are responsible for 71% of the total electricity consumption [3]. There has been a prolonged and substantial increase in electrical demand since 1973. Data in 2003 indicated that electricity accounted for 38% of energy consumption in the commercial and residential sectors in the Organization of Economic Cooperation and Development (OECD) countries [17]. Over the last 30 years there has been a shift in the fuel that buildings use. Oil, although it has

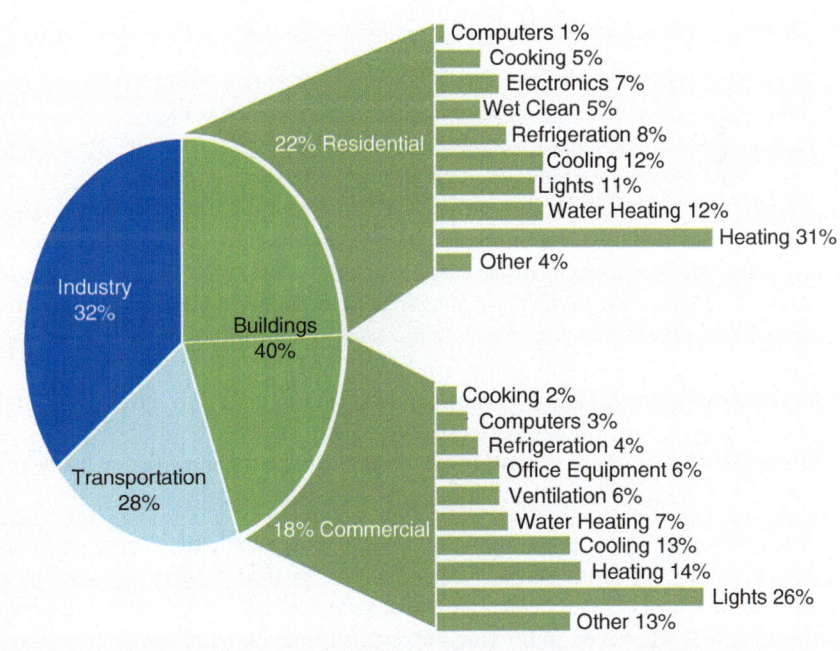

Fig. 7.1 Energy consumption in the United States. 2007 DOE buildings energy book (Tables 1.1.3, 1.2.3, 1.3.3). Note: "industry" includes energy used by facilities and equipment

continued to increase in absolute terms has lost ground to natural gas and electricity and currently accounts for 18% of the energy supply. Natural gas has now become the dominant fuel in residential buildings, while electricity is the dominant fuel in commercial buildings [4].

New products and technologies are being developed to make buildings more environmentally acceptable. There are many R&D efforts aimed at lowering the consumption of energy, material resources, water, and the associated volatile organic compound (VOC) emissions and pollutant impacts on the building occupants. These activities are changing the way that buildings are designed and maintained, engineered, renovated, and demolished. It has been reported in several sources that current high performance building practices could reduce energy consumption by an average of 30–50% [16].

7.3 Why Buildings Should be First

In December of 2007, the consulting firm McKinsey & Company developed a comprehensive and detailed cost base analysis to understand the costs and potential options for reducing Greenhouse Gas (GHG) emissions in the United States [5]. Shown in the Fig. 7.2 below, the curve illustrates that given aggressive policies to overcome market barriers; emissions can be cut up to 30% from current levels by 2030.

The cost curve suggests two major conclusions:

- **Significant reductions in greenhouse gas emissions are achievable.** The analysis implies that implementing all measures below $50 per ton of CO_2e (equivalent carbon dioxide) would reduce emissions to 40% below 1990 levels in 2030, assuming full realization of the identified opportunities in the given timeframe [5].
- **Significant quantities of 'negative-cost' opportunities are available**. These opportunities would allow the U.S. to reduce emissions at no net cost to the economy. *Many of these negative-cost opportunities relate to building heating and cooling, lighting and building appliances.* These reductions would involve pursuing abatement options available at marginal costs less than $50 per ton, with the average net cost to the economy being far lower if the nation can capture sizable gains in energy efficiency. The options on the left, falling below the zero-line, actually save money over their lifecycle; those on the right above the zero-line have a positive lifecycle cost. Forty percent of the abatement options identified actually save money – they return more than they cost over their lifecycle.

Replacing carbon-emitting fossil fuels with most alternative energy sources, including clean-coal (with carbon sequestration), nuclear power, biomass, wind, solar photovoltaics (PV) and concentrated solar power (CSP) will have substantial abatement costs. However, improving energy efficiency in buildings has the potential to save the economy money. Achieving these reductions at the lowest cost to the financial market, however, will require strong, coordinated, economy-wide action that begins in the near future.

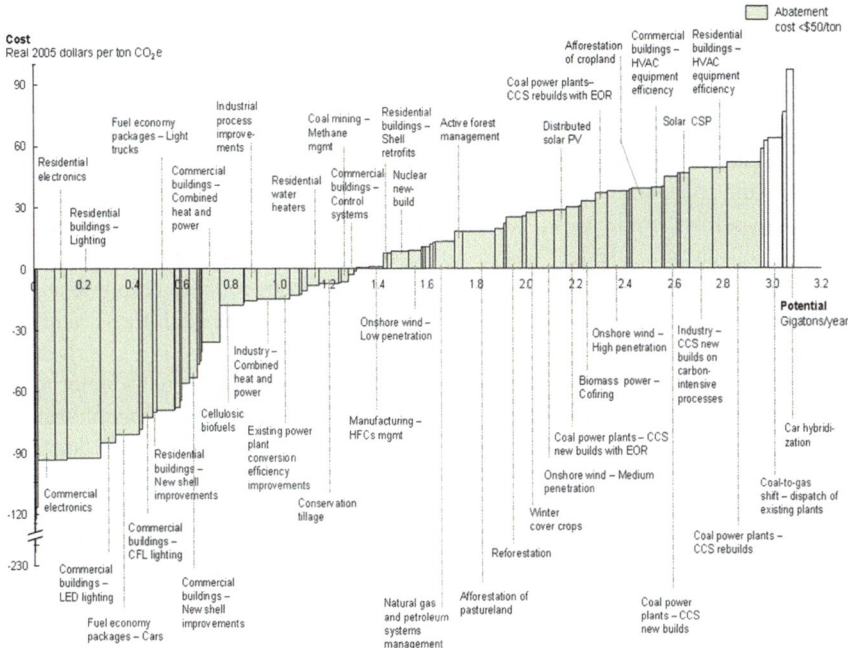

Fig. 7.2 The figure, published by the consulting firm Mckinsey & Company, presents emissions reducing options that represent in the best return on investment by plotting costs against potential for CO_2 emissions reductions. The curve illustrates that given aggressive policies to overcome market barriers; emissions can be cut up to 30% from current levels by 2030 [5]

7.4   The Impact of Emissions from Building Materials on Energy Use

Minimum ventilation rates for buildings are specified in building codes that typically reference ANSI/ASHRAE Standard 62.1 [17]. A portion, estimated to be around 40%, of the minimum ventilation rate is based upon the emissions from building materials and products into the indoor air. When emissions are reduced, indoor concentrations of contaminants are reduced, and human health, productivity, and performance are improved. Therefore, when emissions are reduced, outdoor air ventilation rates may be decreased with potentially significant energy savings. Since energy is required to move, heat, cool, and dehumidify the outdoor air that is used to ventilate buildings, a substantial amount of the total energy used nationally could be saved if low emission materials and products were adopted into building design and use.

The following is an overview on the relationship between building ventilation and the emissions from building materials. Standard 62.1 provides two procedures for determining outdoor air intake rates, described as follows:

Ventilation Rate Procedure. "This is a prescriptive procedure in which outdoor air intake rates are determined based on space type/application, occupancy level, and floor area. Note: The Ventilation Rate Procedure minimum rates are based on contaminant sources and source strengths that are typical for the listed space types" (Sect. 7.6.1.1).

IAQ Procedure. "This is a design procedure in which outdoor air intake rates and other system design parameters are based on an analysis of contaminant sources, contaminant concentration targets, and perceived acceptability targets. The IAQ Procedure allows credit to be taken for controls that remove contaminants (for example, air cleaning devices) or for other design techniques (for example, selection of materials with lower source strengths) that can be reliably demonstrated to result in indoor contaminant concentrations equal to or lower than those achieved using the Ventilation Rate Procedure..." (Sect. 7.6.1.2).

Outdoor air intake rates for new commercial buildings are usually determined using the Ventilation Rate Procedure. Outdoor air rates based on people-related pollutant sources and area-related sources are listed in ANSI/ASHRAE Standard 62.1. Outdoor air intake rates for area-related sources (the building and its materials) are converted to units of cubic feet per minute (cfm)/person using default occupant density values from the Standard. "People" and "area" outdoor air rates are added to determine the total outdoor air intake rate. The Ventilation Rate Procedure is a relatively simple method for determining outdoor air intake rates based on assumed pollutant emissions from people and materials in buildings, and is the chosen procedure for nearly 100% of building covered by the standard.

Outdoor air intake rates can be reduced through the use of the IAQ Procedure with the selection of building materials with lower emissions (contaminant source strengths), and energy savings can result. Potential nationwide energy savings are difficult to quantify due to a lack of data. Research is needed to quantify the potential savings, because limited existing data indicate that energy savings may be substantial. In applications where gas-phase filtering is an option, such as in typical offices or school classrooms, ventilation can typically be reduced from 20 cfm/person to 5 cfm/person or 5% of supply air (whichever is greater). A case study [15] of a multiplex movie theater showed that the ventilation rate could be reduced from 20 cfm/person to 5 cfm/person through application of the IAQ Procedure by cleaning and recirculating air. Low-emission building materials could have a similar impact by removing indoor air pollutant sources. McDowell et al. [18] conducted a modeling study to estimate energy usage of 25 buildings representing the U.S. core office building stock. Decreasing ventilation from 20 cfm/person to 5 cfm/person decreased heating load by 17% and decreased cooling load by 6%. Decreasing ventilation from 20 cfm/person to 10 cfm/person decreased heating load by 9% and decreased cooling load by 4.5%. For all energy used in commercial buildings, space heating accounts for approximately 34% and cooling accounts for about 15% [6]. Assuming that, for commercial buildings, the average outdoor air intake rate could be reduced from 20 cfm/person to 5 cfm/person through low-emission building materials, total commercial building energy savings is calculated as follows

$$(0.34 \times 0.17) + (0.15 \times 0.06) = 0.067 = 6.7\%$$

Commercial buildings consume approximately 18% of all U.S. energy, so savings of total U.S. energy could be as much as 1.2%, although there is much uncertainty in this estimate. Turnover time of the building stock is relatively long, so it would take many years to realize the full potential savings, but the average time for renovation of buildings is relatively short, so potential savings can start immediately.

Most residential buildings in the U.S. are ventilated with air that leaks through the building shell by design. However, new residential construction practices, especially in colder climates, are changing to provide tighter building shells and, in some cases, controlled mechanical ventilation. As new residential buildings become more like commercial buildings with controlled ventilation, low-emission building materials can provide potential long-term energy savings similar to the projected savings for commercial buildings.

In addition to building materials, human activities and many products used inside buildings affect indoor air quality. Human activities include cleaning, cooking, personal care, hobbies, crafts, etc. Products include cleaners, personal care chemicals, office equipment, solvents, adhesives, coatings, etc. When indoor air contamination is reduced by use of lower emitting products, less ventilation air is required, and energy savings results.

7.5 Building Materials: Their Embodied Energy and Greenhouse Gas Impact

Embodied energy is defined in the U.S. DOE's *Buildings Energy Data Book* as the energy used during the entire life cycle of a product including the energy used for manufacturing, transporting, and disposing of the product. For example, the embodied energy in dimensional lumber includes the energy used to grow, harvest and process the trees into boards, transport the lumber to its final destination, and ultimately dispose of the wood at the end of its useful life. Embodied energy, also called life cycle assessment (LCA), is a useful tool for evaluating the relative environmental impact of various building materials because it takes production, transportation, and disposal into account, all things that can have a pronounced environmental impact but are not necessarily reflected in the price [7].

Embodied energy is associated with LCA, but it should be noted that LCA includes other environmental impacts in addition to embodied energy.

Embodied energy is sometimes defined as the *non-renewable* energy used during the entire life cycle of a product. *Initial* embodied energy in buildings is the energy used in extracting raw materials, processing materials, manufacturing, transporting, and constructing buildings. *Recurring* embodied energy is the energy used to maintain, repair, or replace materials during the life of a building. Cole and Kernan [19] analyzed a typical Canadian office building and found that recurring embodied energy becomes important over the life of a building. Cole and Kernan showed that when an office building is 50 years old, recurring embodied energy is about 144% of the initial embodied energy.

Lippke et al. [20] found that for a typical U.S. home, embodied energy is small but significant compared to energy used to operate the home over its lifetime. Embodied energy in materials is about 15%, and operating energy is about 85% of the total energy used over the lifetime of a typical U.S. home. Buildings with high-energy efficiency have a greater percentage of total energy embodied in materials. For example, Thormark [21] found that for energy-efficient apartment housing in Sweden, embodied energy is about 45% of total energy over a life span of 50 years. Thormark found that about 40% of embodied energy can be recovered through recycling at the end of the building's lifecycle, if buildings are designed with recycling potential in mind.

Embodied energy for a building material may vary depending on extraction processes, production methods, transportation distances, and other variables. Nevertheless, estimates of embodied energy are available for many building products, so architects, builders, purchasers, and others can make informed decisions. The U.S. Energy Research and Development Administration published an extensive report on embodied energy in 1976, and many subsequent publications have referred to the report. Research is needed to update estimates of embodied energy for new materials, building construction practices, recycling processes, etc. Researchers use various methodologies for calculating embodied energy, so comparing embodied energy between studies is difficult. Standards for determining embodied energy are needed.

Embodied energy can usually be reduced by using recycled materials. For example, embodied energy in recycled aluminum is approximately 5% of embodied energy in aluminum produced from raw material. Reusing materials, as well as reusing whole buildings, can reduce embodied energy even more than recycling in some cases. A recent EPA report [8] describes opportunities for reducing greenhouse gas emissions, thereby reducing embodied energy, during construction of buildings. Embodied energy in a building product can be reduced by minimizing energy used at any stage of production. For example, cement produced by a dry-process kiln can use about 50% less energy than cement produced in a wet-process kiln. Using durable, long-lasting materials in construction can reduce embodied energy over the life of a building. Using local materials can reduce transportation costs and thereby reduce embodied energy.

The International Energy Agency [4] estimated that worldwide efforts to improve materials and product efficiency could annually save 5 EJ (Exajoule: 10^{18} J) of energy and could reduce carbon emissions by 0.3 Gt CO_2. Life-cycle evaluation is needed to realize potential savings, but most companies do not consider the entire life cycle. Government measures are needed to encourage life-cycle considerations. The IEA [4] also estimated that feedstock substitution could annually save 5–10 EJ of energy and could reduce carbon emissions by 0.4 Gt CO_2.

Incentive options for reducing embodied energy in buildings could include "green" building rating systems such as LEED, tax incentives, carbon-trading systems, and voluntary measures. Economic incentives for reducing embodied energy grow as the cost of energy increases. The 2020 Vision Workgroup, a collaborative effort between EPA and the States, is developing recommendations for a comprehensive materials management strategy that would significantly reduce embodied energy in materials [22]. Effective policies, supported by scientific research, can encourage innovative ways to reduce embodied energy in all buildings.

7.6 Increasing Energy Efficiency and CO_2 Mitigation in Buildings

For each of the two categories, the technologies are listed in the order of their potential impact in 2,050 according to IEA [4] for two global emission reduction scenarios: the ACT and Blue scenarios (Table 7.1). The Blue scenario is the most aggressive in that it calls for 2,050 global CO_2 emissions to be reduced to 50% of actual 2005 emissions. The ACT scenario would constrain 2,050 emissions at 2005 levels. The technologies are either aimed at enhancing end use efficiency or are new alternative building heating/cooling technologies. It is important to note that those high-efficiency appliances and heating and cooling technologies are currently commercial, although there is potential for even higher efficiencies assuming a focused, successful research program. Lack of incentive and higher initial costs are the primary reasons for the slow rate of utilization. This is in contrast to the power generation sector, which is constrained both by unavailable or undemonstrated technology and underutilized technology.

Due to slow building turnover, increasing the energy efficiency of existing building stock and maximizing energy efficiency of new buildings is essential in reducing total building energy use. This includes a concerted effort both in building systems, design and in selection/replacement of mechanical systems and appliances. The process is most efficient when comparing all building system components against each other, as commonly done in a building energy audit. Example tools for determining a building's energy use and potential energy savings can be found on the Department of Energy web page [23]. To be truly effective, an integrated design approach is needed both during initial building construction and retrofitting to determine the most effective method of maximizing individual building efficiency. For example, a relatively simple aspect such as building siting (based on reference angle with the sun instead of the road) can greatly affect the available natural lighting and building cooling load requirement. This can allow for a reduction in both cooling needs and electrical lighting, which is a significant component of electricity use in buildings worldwide.

Immediate, off-the-shelf options exist for increasing the energy efficiency of residential and commercial buildings regardless of building age or type. The method resulting in maximum energy use reduction will vary by region, building type, etc., making a nationwide or global drive complex, likely relying on regional incentives and techniques. The IPCC [2] presents just such a regionally based technology recommendation to increase energy efficiency in buildings, including parameters such as cost effectiveness, technology stage, and appropriateness. An example of this regional dependence is the use of structural insulation panels in building design for

Table 7.1 Ref. [1] summarizes major building technologies capable of achieving significant global reductions in Gt CO_2 generation in the 2050 time frame. The technologies are divided into two categories: (1) heating and cooling and (2) appliances, which include lighting. The Blue column shows Gt of global CO_2 emission reductions in 2050. The Blue scenario calls for 2050 global CO_2 emissions from all sources to be reduced to 50% of actual 2005 emissions

	Technology	Current state of the art	Blue 2050 impact	Issues	Technology RD&D priority and needs	Potential environmental impacts
Heating & cooling	Enhanced energy mgt. and high efficiency building envelope: insulation, sealants, windows, etc.	Commercial	2.5	Lack of incentive, high initial costs, long building lifetime	Low/medium priority, incremental improvements to lower cost and enhance performance	Less fossil fuel and nuclear power generation, and less on site fossil fuel combustion, yield reductions in coal & natural gas emissions, and nuclear wastes
	High efficiency building heating and cooling, including heat pumps	Commercial	0.8	Lack of incentive, high initial costs	Low/medium priority, incremental improvements to lower cost and enhance performance	Same as above
	Solar heating and cooling	First generation commercial	0.5	High initial costs, availability of low cost efficient biomass heating systems	Medium, focus on development of advanced biomass stoves and solar heating technology in developing countries	Same as above

Appliances	More efficient electric appliances	Commercial	4.5	Higher initial costs and lack of information to the consumer	Low/medium priority, incremental improvements to lower cost and enhance performance	Less fossil fuel and nuclear power generation, yields reduction in coal & natural gas emissions, and nuclear wastes
	More efficient lighting systems	Commercial-compact fluorescent		Lack of incentive given higher initial costs	Medium, LED and OLED technology needs further development with aim of lowering initial cost	Same as above; however, mercury content of fluorescent bulbs could cause health and env. problems
	Reduce stand-by losses from appliances, computer peripherals, etc.	Commercial		Lack of incentive from vendors and lack of knowledge form end-users	Low	Less fossil fuel and nuclear power generation, yields reduction in coal & natural gas emissions, and nuclear wastes

cold climates that are not as appropriate for use in warm climates. In fact, Levy et al. [24] projected a potential savings of 800 trillion British thermal units (TBTU) per year due to insulation retrofits in existing housing. In warmer climates however, it is projected that technology such as solar thermal water heaters are a much more appropriate way to conserve energy, as solar energy is more readily available. Technologies which use direct sunlight to reduce energy demand for heating water exist and are especially economical in temperate regions.

In new construction, primary barriers for use of energy efficient design may include concern over increased initial building cost (including a lack of incentives), designer lack of knowledge, owner specifications, owner lack of interest, and an unknown potential for energy savings via a detailed energy analysis. Building appliances are rarely replaced when a newer version is available due to initial cost of new equipment, and the long operating life of many existing household appliances. Often, building occupants are accustomed to a consistent energy bill and are put off by the high initial cost of appliance upgrade considering the effort required to quantify the effect it will have on the energy bill. Although initial costs can often be recouped in reduced energy bills in a reasonable time frame, this is not immediately obvious or important for consumers who are not responsible for the building energy bill, a frequent scenario in rental properties. Newly purchased appliances are not always used as a replacement for aging equipment either. A new refrigerator may provide a more efficient means of cooling food but the older system, if operable, may continue to be used for owner convenience [25]. In fact, even when older appliances are discarded, increased energy efficiency may not be fully realized due to increase in new appliance use (termed the 'rebound effect') [26]. Truly an integrated approach that results in a well informed consumer AND user is necessary to increase the energy efficiency of our household products.

It is important that every government incentive be thoroughly analyzed to ensure that the intended effects are realized. Young [27] found that individual incentives had to be tailored to individual population groups in order to be most effective. The success of incentives to increase the turnover of household appliances can range greatly depending on socioeconomic factors such as income. This drive towards increased building energy efficiency happens on multiple fronts. In the effort to reduce final building price, it is not uncommon for builders to begin the design process with a focus on minimizing initial design and construction costs. Standards and codes exist by regions that define the minimum requirements with which a building can be constructed. Changes in these codes are a means of leveling the field in regards to building construction and operation. If all buildings are kept to the same energy efficiency standard, there can be no one builder undercutting the competition by reducing the building efficiency or quality. This has also driven the development of high performance building standards and codes which allow perspective homeowners to compare similar buildings during the purchase process. For example the Energy Star for Homes label [28] and ASHRAE Standard 189.1 are straightforward metrics the normally uninformed buyer can use to compare all perspective properties against one another. More recently, federal dollars have been allocated to help U.S. homeowners and builders reduce the cost of building energy efficiency upgrades and new building construction [29]. Incentives to buy Energy Star appliances allow consumers to upgrade their current systems for a reduced initial price. Similar programs provide homeowners incentives to install photovoltaic panels or passive water heating technology. Regardless of increased building efficiency, building energy use will continue to impact the energy needs of our nation.

The IPCC [2] discusses a number of co-benefits as potential outcomes of a reduced building energy use scenario including; reduction in local/regional air pollution, improved health (discussed in the following section), quality of life and comfort, improved productivity, employment creation, improved social welfare and poverty alleviation and energy security. Such benefits are useful incentives in regions where direct benefits may not be realized. Many of these benefits may be just as important as the anticipated side effects of global climate change.

A myriad of options exist for increasing energy efficiency in buildings across the United States. While the optimal mitigation technique is dependant on region and is often building specific, it is believed that a nation-wide concerted effort is needed to significantly reduce energy use in buildings.

7.7 Focus on Building Materials and Product Labels

Buildings use a massive amount of materials and products that have direct impacts on humans as well as the environment during all phases in the life of a building. Materials and products include structural components, equipment and systems, finishes and furnishings, and products used for operation and maintenance. The following highlights how buildings and building materials are at the intersection of climate change and human health, and how addressing them is an important early step in mitigation of climate change.

The current trend of a rapidly growing number of 'green' testing and labeling efforts is causing confusion in the marketplace and limiting the potential growth in the sale and use of products that are better for climate change and human health. Currently, the labels are either not recognized by users, are not trusted by users, or do not address the broad attributes of green, which is a careful integration of the energy, indoor air quality, water, and recycling implications, and the impacts of operating and maintaining products over their lifetime.

Just as the Energy Star and WaterSense labels have helped level the playing field and lead to transformation of the energy aspects of buildings, materials, and products; an equivalent program that integrates the multi-attributes that comprise 'green' is needed to help ensure a level playing field and lead to the transformation of the green marketplace by providing incentive for manufacturers and retailers to supply greener products, and by providing the necessary information so that consumers can identify, demand, and use greener products. An expanded R&D effort for testing and certification is needed to ensure that a level playing field is created.

7.8 Health Impacts

Americans spend approximately 90% of their time indoors where the air quality can be many times more polluted than ambient air due to product emissions, mold growth, and a myriad of other sources [30]. Therefore there is a concern for any future increase in time spent indoors due to increasing outdoor temperatures and the effects this increased time has on human health. The possibility of emissions from materials combined with increased building tightness increases the potential exposure to harmful air pollutants if not properly addressed. The level of exposure is difficult to quantify as it will vary by region, building age and a myriad of other parameters. While current U.S. EPA policy is to avoid regulation of materials brought into the American home, policy has been shifting towards increasing public knowledge to create a well-informed consumer through federally funded programs such as Energy Star, Water Sense, and Design for the Environment. Certainly, global climate effects on the indoor environment and the occupants therein have been a topic of interest both in the academic and government communities [9, 10, 31, 32]. Unfortunately, limited research exists for climate change impacts on the indoor environment, as discussed by Nazaroff [33] "It is noteworthy that studies of energy use in buildings and related climate impacts often fail to make evident the clear linkages of these issues with indoor environmental quality and health."

7.9 Conclusion

Addressing building energy use is the critical first step in any strategic plan for mitigating climate change. Much is understood about technologies, materials, and design techniques that can reduce energy use in buildings. However, much attention must be paid to recognizing how these approaches can enhance or damage human health and productivity as well as the environment. This point is emphasized by the contrasting knowledge base between well-known single attribute mitigation programs and more holistic, multi-attribute programs that focus on sustainability and health within the building/energy sector. A key intersection between climate change, buildings, and human health is building materials and products, and an effective testing and information transfer program is urgently needed so that building stakeholders have the information and tools they need to make good decisions during the design, construction, operation, and renovation phases of buildings.

References

1. Princiotta F (October 2009) Global climate change and the mitigation challenge. J Air Waste MA, US EPA, RTP, NC
2. Levine M, Ürge-Vorsatz D, Blok K, Geng L, Harvey D, Lang S, Levermore G, Mongameli Mehlwana G, Mirasgedis S, Novikova A, Rilling J, Yoshino H (2007) Residential and

commercial buildings. In: Metz B, Davidson OR, Bosch PR, Dave R, Meyer LA (eds) In climate change 2007: mitigation contribution of working group III to the fourth assessment report of the intergovernmental panel on climate change. Cambridge University Press, Cambridge/New York

3. U.S. Department of Energy (2009) Buildings energy data book. http://buildingsdatabook.eren. doe.gov/docs/DataBooks/2007_BEDB.pdf. Accessed 22 Mar 2009
4. International Energy Agency (2009) Energy technology perspectives: in support of the G8 plan of action. http://www.iea.org/Textbase/techno/etp/. Accessed 22 Mar 2009
5. McKinsey & Company (2009) Reducing U.S. Greenhouse gas emissions: how much at what cost? http://www.mckinsey.com/clientservice/ccsi/pdf/US_ghg_final_report.pdf. Accessed 22 Mar 2009
6. U.S. Department of Energy, Energy Information Administration (2009) Commercial buildings energy consumption survey. http://www.eia.doe.gov/emeu/cbecs/contents.html. Accessed 22 Mar 2009
7. U.S. Department of Energy (2009) Buildings energy data book.http://buildingsdatabook.eere. energy.gov/TableView.aspx?table=Notes. Accessed 22 Mar 2009
8. U.S. Environmental Protection Agency (2009) Potential for reducing greenhouse gas emissions in the construction sector. Sector strategies.http://www.epa.gov/ispd/pdf/construction-sector-report.pdf. Accessed 7 May 2009
9. U.S. Environmental Protection Agency (1989) The potential effects of global climate change on the United States, EPA-230-05-89-050, Office of Policy, Planning and Evaluation, Office of Research and Development: Research Triangle Park
10. U.S. Department of Energy (2007) Effects of climate change on energy production and use in the United States. A Report by the U.S. Climate change science program and the subcommittee on global change research; climate change science program. Office of Biological and Environmental Research, Washington, pp 1–160
11. Energy Star (2010) http://www.energystar.gov/. Accessed 28 May 2010
12. Water Sense (2010) http://www.epa.gov/watersense/. Accessed 28 May 2010
13. U.S. EPA Green Building Workgroup (2009) Buildings and the environment: a statistical summary. http://www.epa.gov/greenbuilding/pubs/gbstats.pdf. Accessed 22 Apr 2009
14. U.S. EPA (2009) Design for the Environment (DfE). http://www.epa.gov/dfe/. Accessed 22 Mar 2009
15. Stanley WBM, Muller CO (2002) Practical application of energy conservation with ASHRAE Standard 62. In: IA2002 Proceedings, Proceedings from the 9th international conference on indoor air quality and climate, Monterey
16. Long CM, Suh HH, Kobizik L, Catalano PJ, Ning YY, Loutrakis P (2001) A pilot investigation of the relative toxicity of indoor and outdoor fine particles: in vitro effects of endotoxin and other particulate properties. Environ Health Perspect 109(10):1019–26
17. ANSI/ASHRAE (2007) Standard 62.1-2007, ventilation for acceptable indoor air quality, American society of heating, refrigerating and air-conditioning engineers, Inc. Atlanta
18. McDowell TP, Emmerich S, Thornton JW, Walton G (2003) Integration of airflow and energy simulation using CONTAM and TRNSYS. ASHRAE Trans 109(Part 2):757–770
19. Cole RJ, Kernan PC (1996) Life-cycle energy use in office buildings. Build Environ 31(4):307–317
20. Lippke B, Wilson J, Perez-Garcia J, Bowyer J, Meil J (2004) CORRIM: life-cycle environmental performance of renewable building materials. For Prod J 54(6):1–1
21. Thormark C (2002) A low energy building in a life cycle – its embodied energy, energy need for operation and recycling potential. Build Environ 37:429–435
22. U.S. Environmental Protection Agency (2003) Beyond RCRA: prospects for waste and materials management in the Year 2020, final white paper, EPA530-R-02-009, office of solid waste, Washington
23. U.S. Department of Energy (2009) Building energy software tools directory. http://apps1.eere. energy.gov/buildings/tools_directory/. Accessed 23 Mar 2009
24. Levy J, Nishioka Y, Spengler J (2003) The public health benefits of insulation retrofits in existing housing in the United States. Environ Health Glob Access Sci Source 2:4

25. Young D (2008) Who pays for the 'beer fridge'? Evidence from Canada. Energy Policy 36:553–560
26. Sorrell S (2007) The Rebound Effect: an assessment of the evidence for economy-wide energy savings from improved energy efficiency; 1-903144-0-35. Energy Research Centre, London
27. Young D (2008) When do energy-efficient appliances generate energy savings? Some evidence from Canada. Energy Policy 36:34–46
28. Energy Star (2009) Energy star qualified new homes. http://www.energystar.gov/index.cfm?c=new_homes.hm_index. Accessed 23 Mar 2009
29. Energy Star (2009) Federal tax credits for energy efficiency. http://www.energystar.gov/index.cfm?c=products.pr_tax_credits. Accessed 23 Mar 2009
30. Fisk WJ (2000) Health and productivity gains from better indoor environments and their implications for the U.S. Department of energy. Annu Rev Energy Env 25:537–566
31. Girman J (2008) Impacts of climate change on indoor environments. Presented at indoor air 2008, the 11th international conference on indoor air quality and climate, Copenhagen, 18 Aug 2008
32. Levin H (2008) Indoor climate and global climate change: exploring connections. In: IA 2008 Proceedings, Proceedings from the 11th international conference on indoor air quality and climate, Copenhagen
33. Nazaroff WW (2008) Climate change, building energy use, and indoor environmental quality. Indoor Air 18(4):259–60

Chapter 8
Reduction of Multi-pollutant Emissions from Industrial Sectors: The U.S. Cement Industry – A Case Study[*]

Ravi K. Srivastava[†], Samudra Vijay, and Elineth Torres

Abstract Carbon dioxide (CO_2) accounts for more than 90% of worldwide CO_2-eq greenhouse gas (GHG) emissions from industrial sectors other than power generation. Amongst these sectors, the cement industry is one of the larger industrial sources of CO_2 emissions. In 2005, this industry accounted for about 6% of the global anthropogenic CO_2 emissions. Further, global production of cement has been growing steadily, with the main growth being in Asia. Considering these trends, the worldwide cement industry is a key industrial sector relative to CO_2 emissions.

The development of policy options for managing emissions and air quality can be made more effective and efficient through sophisticated analyses of relevant technical and economic factors. Such analyses are greatly enhanced by the use of an appropriate modeling framework. Accordingly, the Industrial Sectors Integrated Solutions (*ISIS*) model for industrial sectors is under development at the U.S. Environmental Protection Agency (U.S. EPA). Currently, this model is populated with data on the U.S. cement-manufacturing sector and efforts are underway to build representations of the U.S. pulp and paper and iron and steel sectors.

[*]The findings included in this chapter do not necessarily reflect the view or policies of the Environmental Protection Agency. Mention of trade names or commercial products does not constitute Agency endorsement or recommendation for use.

[†]© US Government 2011

R.K. Srivastava (✉)
U.S. Environmental Protection Agency, Office of Research and Development, National Risk Management Research Laboratory, Air Pollution Prevention and Control Division, Cincinnati, OH 45268, USA

S. Vijay
ORISE Research Fellow, U.S. Environmental Protection Agency, Office of Research and Development, National Risk Management Research Laboratory, Air Pollution Prevention and Control Division, Cincinnati, OH 45268, USA
currently at
Sam Analytic Solutions, LLC, 614 Willingham Road, Morrisville, NC 27560, USA

E. Torres
U.S. Environmental Protection Agency, Office of Air Quality Planning and Standards, Sector Policy and Programs Division, Research Triangle Park, NC 27711, USA

F.T. Princiotta (ed.), *Global Climate Change - The Technology Challenge*,
Advances in Global Change Research 38, DOI 10.1007/978-90-481-3153-2_8,
© Springer Science+Business Media B.V. 2011

In this chapter, ISIS was used to conduct an example analysis of the U.S. cement sector to gain some insights relative to two broad questions: (1) what range of CO_2 reductions may be practicable in the near-term, and (2) for that range, what may be the market characteristics for the U.S. cement industry. These questions are relevant because in the absence of carbon capture and sequestration (CCS) technology, the path forward for reducing CO_2 emissions in the near-term (e.g., decade ending 2020) will need to depend on the currently available energy efficiency measures and raw material and product substitution approaches.

8.1   Introduction

Globally, carbon dioxide (CO_2) accounts for more than 90% of CO_2-eq greenhouse gas (GHG) emissions from industrial sectors [1]. Also, industrial sector CO_2 emissions from energy consumption and industrial processes account for about one-fourth of global CO_2 emissions [2]. Among the industries, steel manufacturing (~30%), non-metallic minerals (~27%), and chemical and petrochemical (~16%) were the three largest CO_2 emitters in 2005 (see Table 8.1).

The non-metallic minerals sector includes cement manufacturing, brick kilns, glass and ceramic manufacturing, and building materials. Although this sector accounts for only about 10% of industrial energy consumption, its share of CO_2 emissions is significantly larger due to large process related emissions from cement manufacturing. Of the total CO_2 emissions from non-metallic minerals sector, cement manufacturing accounted for 83% of the total energy use and 94% of CO_2 emissions. In 2005, cement manufacturing was responsible for about 6% of the global CO_2 emissions [2].

Cement is a key building and construction material. Global production of cement has been growing steadily, with the main growth being in Asia. China, in particular, now accounts for almost half of the global cement production [3]. Notably, cement industry appears to have large CO_2 abatement potential [1].

Table 8.1 Global industrial sector energy consumption and CO_2 emissions (2005) Source: IEA 2008 [2]

Sector	Energy consumption (Mtoe)[a]	Share (%)	CO_2 emissions (Mt)[b]	Share (%)
Chemical and petro-chemicals	809	29.3	1,086	16.3
Iron and steel	560	20.3	1,992	29.9
Non-metallic minerals	263	9.5	1,770	26.6
Pulb and paper	154	5.6	189	2.8
Other	977	35.4	1,623	24.4
Total	2,763	100	6,660	100

[a]Million metric tons oil equivalent
[b]Million metric tons

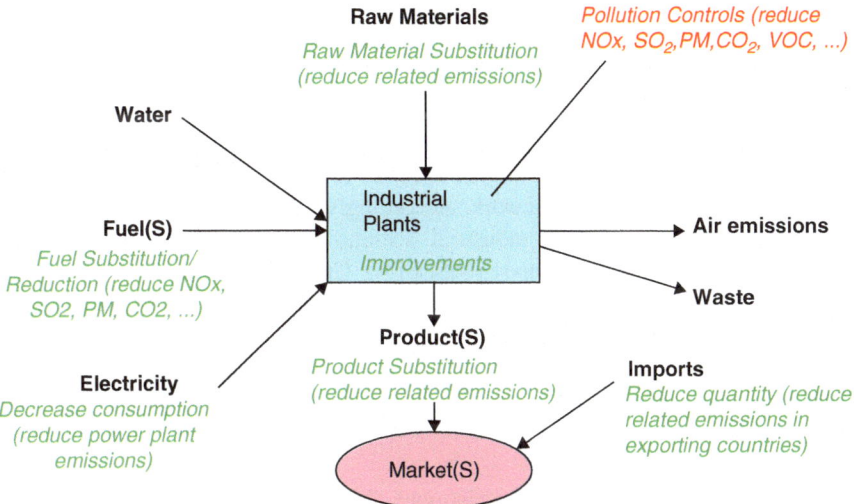

Fig. 8.1 An integrated view of pollution generation pathways, emissions abatement approaches, and multimedia impacts for an industrial sector

Within an industrial sector, generally CO_2 and other emissions arise from four pathways: (1) on-site emissions due to combustion of fossil fuels for energy at plants, (2) on-site emissions due to processing of certain raw materials (e.g., limestone calcination in cement plants, non-energy uses of fossil fuels in chemical processing and metal smelting), (3) off-site emissions due to combustion of fossil fuels at power plants to generate the electricity needed by the industrial sector, and (4) emissions associated with imports. These pathways are depicted in Fig. 8.1.

Also, shown in Fig. 8.1 are the potential options for mitigating CO_2 and other emissions from industrial sectors. Those in green are pollution prevention measures and the ones in red are mitigation measures. Clearly, the integrated picture presented in the figure makes a compelling case for considering commodity production/supply activities along with emissions, while developing holistic emission reduction strategies.

Mckinsey & Company [4] estimate that use of energy efficiency and other measures, including carbon capture and sequestration (CCS), can provide almost 25% reduction in global CO_2 emissions from worldwide cement industry, relative to the business-as-usual case, by 2030. The study also indicates that 80% of this abatement potential may be achieved using measures other than CCS. However broad deployment of cost-effective abatement options (e.g., energy efficiency measures) will only be possible if appropriate policy drivers are in effect and barriers such as lack of availability of substitute materials (e.g., blast furnace slag, fly ash, biomass) are addressed.

In another study, the World Business Council for Sustainable Development (WBCSD) used an economic model to analyze the global cement and carbon flows under a number of CO_2 abatement options [5]. The study indicates that there is relatively little potential for reducing CO_2 emissions via energy efficiency gains at cement plants because older plants are being retired and new plants are already quite efficient. The study also reflects that a sector based policy option could be quite effective in abating CO_2 emissions.

The development of policy options for managing emissions and air quality can be made more effective and efficient through sophisticated analyses of relevant technical and economic factors. Such analyses are greatly enhanced by the use of an appropriate modeling framework. Accordingly, the Industrial Sector Integrated Solutions (*ISIS*) model for industrial sectors is under development at U.S. Environmental Protection Agency (U.S. EPA). Currently, the ISIS model is populated with data on the U.S. cement-manufacturing sector, and efforts are underway to build representations of the U.S. pulp and paper and iron and steel sectors.

The ensuing sections describe the U.S. cement industry, CO_2 abatement approaches for this industry, the ISIS model framework, and the U.S. cement industry-related data included in ISIS. Subsequently, an example analysis of the U.S. cement industry, investigating the potential for near-term reductions in CO_2 and other pollutants, associated costs, and industry operation, is presented. Two broad questions were investigated in our example analysis: (1) what range of CO_2 reductions may be practicable in the near-term (i.e., by the decade ending 2020 for this study), and (2) for that range, what may be the market characteristics for the U.S. cement industry. Finally, this chapter concludes with a summary and thoughts on future directions.

8.2 The U.S. Cement Industry

8.2.1 Cement Types

Cement is a finely ground powder, which, when mixed with water, forms a hardening paste of calcium silicate hydrates and calcium aluminum hydrates. Cement is used in mortars (to bind together bricks or stones) and concrete (bulk rock-like building material made from cement, aggregate, sand, and water). Most of the cement produced is used for making concrete.

Of the types of cement produced, Portland cement is most commonly used for concrete production. By modifying the raw material mix and, to some degree, the temperatures in the manufacturing process, slight compositional variations can be achieved to produce Portland cements with slightly different properties. In the U.S., the different varieties of Portland cement are described in the American Society for Testing and Materials (ASTM) Specification C-150-07 [6].

Portland cements are usually gray in color, but a more expensive white Portland cement can be obtained by processing raw materials with low contents of iron and transition elements. In addition, small volumes of specialty cements are also manufactured including blended cement (Portland cement mixed together with blast furnace slag or other pozzolan materials), pozzolan-lime cement, masonry cement, and aluminous cement [7].

Since Portland cement accounts for approximately 95% of the U.S. cement industry's total production [8], the costs and trends of this industrial sector can be adequately captured by describing the market processes associated with production, distribution, and use of Portland cement. In the remainder of this chapter, "cement" is synonymous with "Portland cement."

8.2.2   Overview of the Cement Manufacturing Process

Cement is produced from raw materials such as limestone, chalk, shale, clay, and sand. These materials are quarried, crushed, finely ground, and blended to the correct chemical composition. Small quantities of iron ore, alumina, and other minerals may be added to adjust the raw material composition. The finely ground raw material is fed into a large kiln, where it is heated to high temperatures (about 1,500°C), which causes the raw material to react and form a hard nodular material called "clinker." Clinker is cooled and then ground with small amounts of gypsum and other minor additives to produce cement. The main steps in the cement manufacturing process are illustrated in Fig. 8.2.

The heart of the clinker production process is the kiln, which can be rotary or vertical shaft designs. Rotary kilns are commonly used in the U.S. and elsewhere.

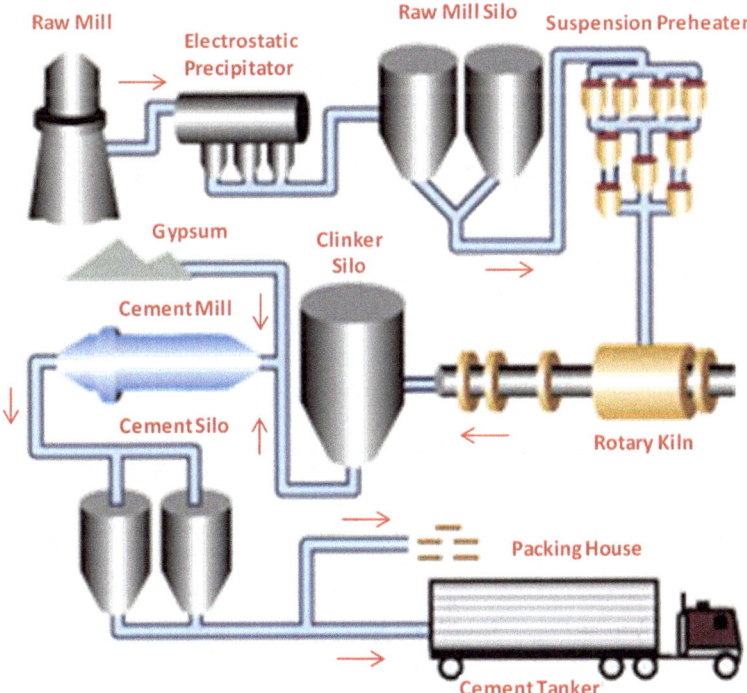

Fig. 8.2 A schematic of the cement manufacturing process

These kilns are 6–8 m in diameter and 60 m to well over 100 m long. They are set at a slight incline and rotate at 1–3 rpm. The kiln is fired at the lower end and the feed materials move toward the flame as the kiln rotates. The materials reach temperatures between 1,400°C and 1,500°C in the kiln. Three steps occur with the raw material mixture during pyro-processing. First, all moisture is driven off from the materials. Then the calcium carbonate in limestone dissociates into CO_2 and calcium oxide (free lime) during calcination. Finally, the lime and other minerals in the raw materials react to form calcium silicates and calcium aluminates, the main components of clinker.

8.2.3   Rotary Kiln Types and Their Use in the U.S.

Rotary kilns are broadly categorized as dry- and wet-process kilns, depending on how the raw materials are prepared. Wet-process kilns are fed raw material slurry with moisture content ranging between 30% and 40%. A wet-process kiln needs additional length to evaporate the water contained in the raw material feed. Nearly 33% additional kiln energy is consumed in evaporating the water in the slurry.

Dry-process kilns are fed dry powdered raw materials. Three designs of dry-process kiln systems are in operation: long dry (hereafter referred to as "dry"), preheater, and precalciner. In preheater and precalciner kilns, the early stages of pyro-processing occur before the materials enter the rotary kiln. These kiln systems have greater fuel efficiency compared to other types of cement kilns. Table 8.2 shows the specific heat input requirements for various types of rotary kilns [9]. As these data reflect, preheater and precalciner kilns operate with greater fuel efficiency. As such, replacement of wet and certain dry process kiln capacity with modern kiln processes can yield substantial reductions in fuel use.

Over the recent years, the trend in the cement sector has been replacement of smaller, inefficient, wet, and long-dry kilns with larger more efficient kilns. This trend is expected to continue.

Between 1995 and 2004, the number of kilns in the U.S. decreased by 11%, but total clinker production capacity increased by 18.6%. The Portland Cement Association (PCA) data show that average kiln capacity also increased by 27%, from 367,000 to 504,000 metric tons per year, in the same period [10]. The trend in kiln designs and average kiln capacity is shown in Fig. 8.3. Note that "dry" in this figure includes dry, preheater, and precalciner kilns.

Table 8.2 Heat input requirements of cement kiln types

Kiln type	Heat input (10^6 Btu per short ton of clinker)
Wet	5.309
Dry	4.319
Preheater	2.969
Precalciner	2.825

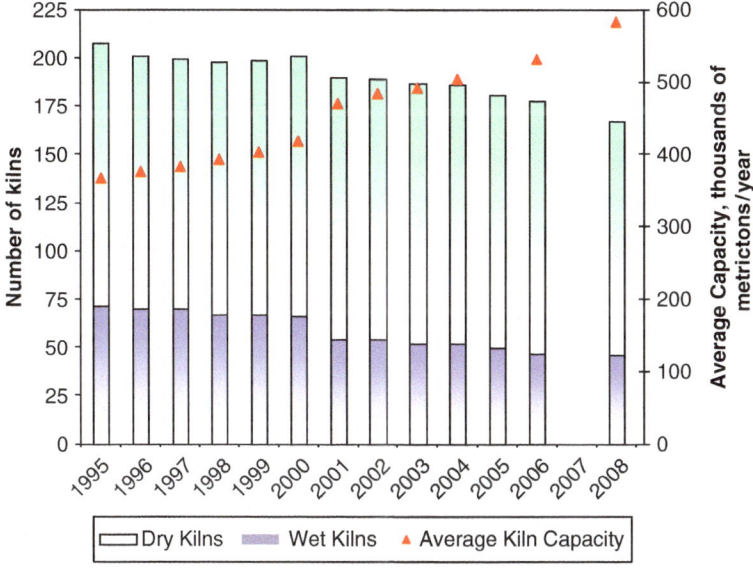

Fig. 8.3 Trends in cement kiln type and capacity in the United States (1995–2008)

The PCA projects a capacity expansion of 27 million metric tons between 2008 and 2013, an 18% increase in existing capacity from 2006. PCA's projected capacity expansions are to come from 23 kilns [11]. The investment in the projected capacity expansions is projected at $6.9 billion.

8.2.4 Cement Production in the U.S.

The cement manufacturing sector remains a vital industry in the U.S. and throughout the world. After China and India, the U.S. is the world's third largest cement producer. Portland cement is a $13 billion industry in the U.S. [12]. In 2005, the U.S. consumed a record 127 million metric tons (140 million short tons) of Portland cement [13] and the U.S. Portland cement industry produced approximately 104 million short tons (94 million metric tons) of cement [14].

In 2005, Portland cement was produced at 115 plants located in 37 States and Puerto Rico. The U.S. cement manufacturing sector is concentrated among a relatively small number of companies. Many U.S. cement plants are owned by, or are subsidiaries of, foreign companies [14]. Together, 10 companies accounted for about 80% of the total U.S. cement production in 2005 [15]. California, Texas, Pennsylvania, Florida, and Alabama are the five leading cement-producing states and accounted for about 48% of the total U.S. production in 2005 [15].

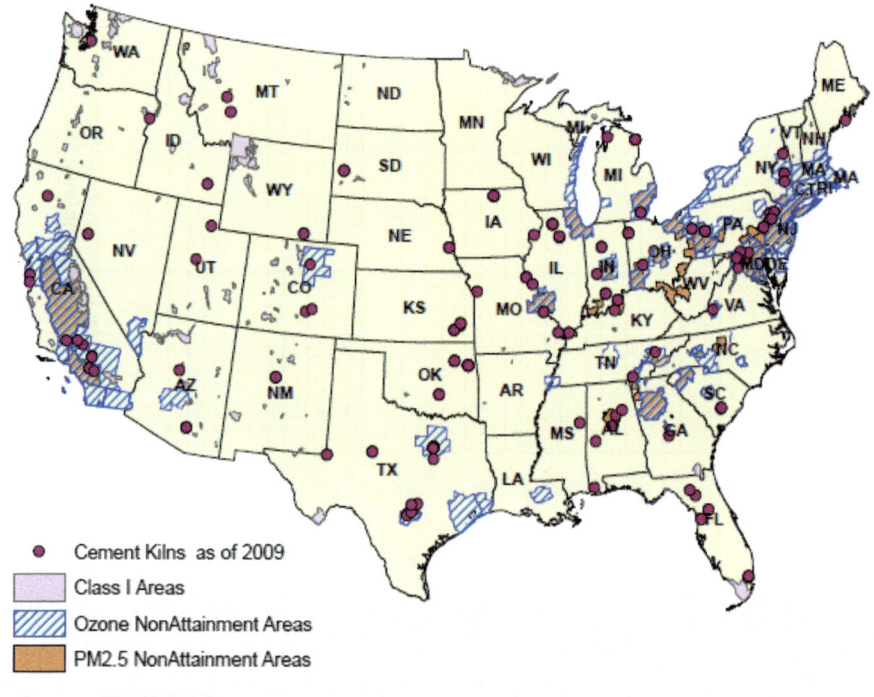

Source: EPA 2009 Data

Fig. 8.4 Portland cement plant locations

The locations of the Portland cement plants in 2005 by state are shown in Fig. 8.4. Many of the plants are located in, or near, ozone and PM nonattainment (NAA) areas and Class 1 areas.[1]

8.2.5   Imports of Cement in the U.S.

Portland cement is not only produced and consumed domestically, but is also traded internationally. In 2005, the U.S. imported 37 million short tons (34 million metric tons) of cement [14]. The level of imports to the U.S. is cyclical with domestic producers importing more when domestic plants are operating at full capacity and cannot meet the remaining demand. Historically, imported cement and clinker have accounted for 20–27% of the domestic consumption of cement. In 2005, total imports of cement and clinker rose owing to continued high demand and accounted for about 24% of the total cement sales in the U.S. [16].

---

[1] Class I areas are those of special national or regional natural, scenic, recreational, or historic value for which the Prevention of Significant Deterioration (PSD) regulations provide special protection.

The ten leading countries supplying cement and clinker to the U.S. in 2005 were, in descending order, Canada, China, Thailand, Greece, the Republic of Korea, Venezuela, Mexico, Colombia, Taiwan, and Sweden. The ten busiest customs districts of entry in 2005 were, in descending order, New Orleans, LA; Tampa, FL; Los Angeles, CA; Houston-Galveston, TX; San Francisco, CA; Miami, FL; Seattle, WA; Detroit, MI; New York, NY; and Charleston, SC [14].

8.2.6   Cement Demand Centers

Because of the relatively high transportation costs involved in delivering cement, the U.S. cement industry is structured around state-specific cement demand centers. PCA reports that the vast majority of cement produced in the U.S. is being transported less than 300 miles by truck due to cement's low value by weight and high cost of transport [17]. However, cement may be transported over longer distances, especially when the less expensive rail and water transportation modes are available [18].

8.3   Emissions from the U.S. Cement Industry and Applicable Regulations

Criteria pollutants, hazardous air pollutants, and CO_2 are released during cement manufacturing. Nitrogen oxides (NO_x) emissions from cement kilns result primarily from the combustion process involving oxidation of fuel nitrogen (fuel NO_x) and oxidation of nitrogen in the combustion air (thermal NO_x). EPA's 2005 National Air Toxics Assessment (NATA) Inventory reports that cement kilns released 181,000 metric tons (200,000 short tons) of NO_x emissions from the combustion of fuels [19].

Sulfur dioxide (SO_2) emissions from cement kilns result from the sulfur in the fuel and the sulfur in the feed materials. Sulfur in the fuel will oxidize to SO_2 during pyroprocessing and a significant amount is likely to be captured in the form of sulfates as the flue gas passes through the calcination zone. Compared to long dry and wet kilns, preheater and precalciner kilns tend to be more effective at capturing fuel-generated SO_2. Accordingly, oxidation of sulfur in the feed materials is likely to be the major component of total SO_2 emissions. The 2005 NATA Inventory reflects that cement kilns released 133,000 metric tons (147,000 short tons) of SO_2 emissions in 2005.

Quarrying operations, crushing and grinding of raw materials and clinker, and the kiln process result in particulate matter (PM) emissions. The NATA Inventory reflects that cement kilns released 10,000 metric tons (11,000 short tons) of PM10 emissions in 2005 [19]. The cement industry also emits hazardous air pollutants (e.g., hydrochloric acid vapor, chlorine, and metals such as mercury, antimony, cadmium, and lead) [20]. Also, the U.S. cement sector emits significant amounts of carbon monoxide. In 2005, these emissions amounted to 148,228 short tons [19].

In a cement kiln, the calcium carbonate in the limestone gets calcined to calcium oxide and in the process releases CO_2. Additional CO_2 is generated from the combustion of fuels in the kiln. Estimates of CO_2 emissions from the U.S. cement industry amounted to 81.4 million metric tons in 2005. Of these, combustion-related emissions were estimated at approximately 35.5 million metric tons of CO_2. Between 1990 and 2005, the process-related emissions resulting from the calcination of limestone increased 38%, from 33.3 million metric tons CO_2 to 45.9 million metric tons CO_2. A relatively small amount of off-site emissions (0.1 million metric tons CO_2) also occur as a result of electricity use [21].

Multiple regulatory requirements currently apply to the U.S. cement industry. The New Source Performance Standards (NSPS) and the National Emissions Standards for Hazardous Air Pollutants (NESHAP) are two of the federal requirements that apply to cement facilities. On the other hand, state and local regulatory requirements might apply to individual cement facilities depending on their locations. In 2008, 44 cement facilities were located within ozone nonattainment areas and 20 facilities were within PM2.5 nonattainment areas. Seventeen facilities were found to be located in, or within 50 km, of Class 1 areas.

8.4 CO_2 Abatement Options for the U.S. Cement Industry

Numerous abatement measures for reducing on-site CO_2 emissions at cement kilns are either already available or under development, and have been discussed and reviewed by other sources [1, 3, 22, 23]. Broadly, the abatement measures fall into four categories: (1) process modifications and upgrades, (2) raw material and/or fuel substitution, (3) product substitution, and (4) mitigation technologies and approaches. These measures are described below.

8.4.1 Process Modifications and Upgrades

Reduction of energy consumption by modification of cement manufacturing process or upgrading the equipment to newer more efficient ones, will reduce the CO_2 emissions either directly through reduction of fossil fuel requirements at the kiln, or indirectly through reduction of electrical demand. Energy efficiency methods are especially attractive because they can provide economic and environmental benefits.

8.4.1.1 Replacement of Kiln Capacity

In 2006, plants with wet and dry process kilns consumed, on average, 6.2 GJ/metric ton of cement (5.8×10^6 Btu/short ton of clinker) and 5.5 GJ/metric ton of cement (5.2×10^6 Btu/short ton of clinker) of fuel-based energy, respectively. In contrast, specific fuel consumption for the overall industry (including dry, wet, preheater and precalciner kilns) averaged 4.3 GJ/metric ton of cement (4.2×10^6 Btu/short ton of clinker), and plants with modern precalciner kilns consumed approximately 3.8 GJ per metric ton [24]. From these statistics, it is apparent that replacement of wet, dry, and older preheater kiln capacity with state-of-the-art precalciner kiln processes can yield substantial reductions in fuel use and commensurate reductions in CO_2 emissions [25]. This observation is also supported by the data presented in Table 8.2 earlier.

Capacity replacements, such as those described above, have taken place continually over the history of the cement industry in response to increased market demand for cement. Generally, replacement of low-fuel-efficiency processes with higher-efficiency processes is done with an expansion in production capacity to take advantage of the economy-of-scale associated with capital costs.

8.4.1.2 Other Energy Efficiency Improvement Options

Opportunities exist within U.S. cement plants to improve energy efficiency while maintaining or enhancing productivity [23]. Energy efficiency improvements may be undertaken in various areas of a cement plant. Energy consuming equipment such as motors, pumps, and compressors require regular maintenance and replacement, when necessary. Consequently, a key component of plant energy management is ensuring efficient operation of crosscutting equipment that powers the production process of the plant. Another important component is ensuring efficient operation of the production process. In this regard, use of process optimization and the most efficient technologies is important.

Broadly, energy efficiency improvement options may be categorized under measures for raw materials preparation, measures for clinker making, measures for finish grinding, and plant-wide measures. Worrell and Galitsky [23] have examined these measures in some detail and have developed cost and performance estimates. Tables 8.3–8.6 list these measures and associated energy savings.

Table 8.3 Energy efficiency measures for raw materials preparation

Energy efficiency improvement method	Electricity consumption change (kWh/short ton clinker)			
	Dry	Wet	Preheater	Precalciner
Efficient transport system	−3.20	n/a	−3.20	−3.20
Raw materials blending	−2.70	n/a	−2.70	−2.70
Process control vertical mill	−0.90	n/a	−0.90	−0.90
High efficiency roller mill	−11.05	n/a	−11.05	−11.05
Slurry blending and homogenization	n/a	−0.35	n/a	n/a
Wash mills with closed circuit classifier	n/a	−12.00	n/a	n/a
High-efficiency classifiers	−5.05	−5.05	−5.05	−5.05

n/a = not applicable or data not available

Table 8.4 Energy efficiency measures for clinker making

Energy efficiency improvement method	Electricity consumption change (kWh/short ton clinker)				Heat input change (10^6 Btu/short ton of clinker)			
	Dry	Wet	Preheater	Precalciner	Dry	Wet	Preheater	Precalciner
Energy Management and Control System (EMCS)	−190	−150	−190	−190	−0.15	−0.21	−0.15	−0.15
Seal Replacement (SR)	n/a	n/a	n/a	n/a	−0.02	−0.02	−0.02	−0.02
Combustion System Improvement (CSI)	n/a	n/a	n/a	n/a	−0.25	−0.35	−0.25	−0.25
Indirect Firing (IF)	n/a	n/a	n/a	n/a	−0.16	−0.16	−0.16	−0.16
Shell Heat Loss Reduction (SHLR)	n/a	n/a	n/a	n/a	−0.20	−0.20	−0.20	−0.20
Optimize Grate Cooler (OGR)	0.90	n/a	0.90	0.90	−0.09	−0.10	−0.09	−0.09
Convent to Reciprocating Great Cooler (CGC)	2.40	2.40	2.40	2.40	−0.23	−0.24	−0.23	−0.23
Heat Recovery for Power Generation (HRPG)	−18.0	n/a	n/a	n/a	n/a	n/a	n/a	n/a
Efficient Mill Drives (EMD)	−2.00	−1.70	−2.00	−2.00	n/a	n/a	n/a	n/a

n/a = not applicable or data not available

Table 8.5 Energy efficiency measures for finish grinding

Energy efficiency improvement method	Electricity consumption change (kWh/short ton clinker)			
	Dry	Wet	Preheater	Precalciner
Energy management and process control	−1.60	−1.60	−1.60	−1.60
Improved grinding media [Ball mills]	−1.80	−1.80	−1.80	−1.80
High-pressure roller press	−16.00	−16.00	−16.00	−16.00
High-efficiency classifiers	−3.85	−3.55	−3.85	−3.85

Table 8.6 Plant-wide energy efficiency measures

Energy efficiency improvement method	Electricity consumption change (kWh/short ton clinker)			
	Dry	Wet	Preheater	Precalciner
Preventative Maintenance (PM)	−2.50	−2.50	−2.50	−2.50
High Efficiency Motors (HEM)	−2.50	−2.50	−2.50	−2.50
Adjustable Speed Drives (ASD)	−6.25	−6.00	−6.25	−6.25
Optimization of Compressed Air System (OCAS)	−1.00	−2.50	−1.00	−1.00

8.4.2 Raw Material and/or Fuel Substitution

Substitution of limestone with raw materials containing non-carbonate calcium and using fuels with lower carbon contents can provide CO_2 reductions. These options are discussed in this section.

8.4.2.1 Non-Carbonate Calcium Containing Raw Materials

Raw mix often contains industrial byproducts such as iron and steelmaking wastes and certain coal ashes, when available. Although the use of slags and fly ashes as an iron source in clinker raw material is commonplace world wide, addition of components for the purpose of minimizing CO_2 emissions has become a focus within the last decade. Alternative raw materials capable of supplying significant amounts of calcium, without requiring calcination, have been used to reduce specific fuel energy consumption and increase kiln system capacity in some cement kilns [26, 27]. One of the more commonly used raw material substitutes is blast furnace slag (BFS), which may contain approximately 40% carbonate-free CaO [26, 27]. In addition, Class C fly ashes, which typically contain 25% CaO, are also a potential source of carbonate-free CaO [28].

Reductions in CO_2 emissions associated with use of reduced carbonate raw materials occur due to reduced calcination and thermal energy requirements. A secondary, but important, related benefit associated with the use of reduced carbonate raw materials is potentially increased kiln system capacity.

A commercially available technology called CemStar has been used in some U.S. kilns. In this technology, a portion of the limestone feed is replaced with BFS. The amount of slag used varies between about 5% and 10% of the clinker output. Use of CemStar potentially can provide several benefits, including reductions in NO_x (about 30%, ton/ton of clinker basis) and CO_2 (about 4%, ton/ton of clinker basis), and increase in kiln output (about 7.5%) [29].

8.4.2.2 Biomass Fuel Substitution

Biomass is "material that comes from plants" [30]. Sources of biomass fuels may include primary wood, wood products, and wood-related wastes. Most of these materials are not widely used in U.S. cement kilns. Another source of biomass is scrap tires. Vehicle tires contain between 14 and 27% natural rubber [31]. As of 2006, approximately one third of all U.S. cement kilns used scrap tire derived fuel (TDF) as a kiln system fuel [15]. TDF represents approximately 5% of the thermal energy consumed by the U.S. cement industry [25]. Also as of 2003, approximately 53 million of the 130 million scrap tires generated in the U.S. were consumed in cement kilns [31]. The extent to which cement kilns use biomass has a bearing on the magnitude of CO_2 emissions from these kilns because of the difference in the carbon contents of biomass materials and traditional fuels like coal.

A technology called midkiln firing (MKF) facilitates the firing of tires (MKF-tires), coal, or other fuels in the mid-kiln region of a kiln. This results in less intense firing at the primary burners and consequently about 35% lower NO_x formation in the kiln. Tires or other waste fuels are often used at the mid-kiln location because high temperatures and residence time in the kiln permit good combustion. In particular, the use of MKF-tires results in reduced fossil fuel consumption (by about 15%) [29].

8.4.3   Product Substitution

Opportunities for the Portland cement industry to reduce CO_2 emissions associated with the finished cement product include replacement of a portion of clinker in Portland cement with materials that do not require the same degree of fuel or electrical energy input for processing, or do not themselves emit CO_2 (as does limestone when calcined as a component of cement raw mix). Certain clinker substitutes may improve cement performance while others may be neutral or detrimental.

Portland cement users and producers have long recognized the ability of natural and synthetic cementitious materials to contribute to cement and concrete performance [32]. These materials have been used directly in concrete, incorporated as a component of cement, or both. Performance specifications have been developed for ground granulated BFS, fly ash, and silica fume in concrete [33–35]. BFS substitution requires grinding of the material to approximately cement fineness, a process that requires electrical energy comparable to, or greater than, that required for clinker grinding [36]. Despite potentially higher grinding energy requirements, CO_2 reductions may still accrue.

Processing requirements for adding non-clinker cementitious components in cement vary depending on the nature of the replacement material. Addition of finely ground materials (fly ash or separately ground BFS) may be made at the point of shipping or distribution. Alternatively, fly ash, BFS, or silica fume may be inter-ground with clinker and other cement components in cement mills.

In 2004, the ASTM revised its standard specification for Portland cement to allow the incorporation of up to a 5% mass fraction of limestone in ordinary Portland cements [6]. CO_2 reductions can accrue from using limestone as a filler replacing an equivalent amount of clinker.

8.4.4   Mitigation Measures

Over the years, the cement industry has utilized the above measures to some degree to primarily improve its operations and reduce costs of supplying cement. In doing so, CO_2 emission reductions have also taken place. While further emission reductions with broader use of above measures should be possible, mitigation measures utilizing CCS, when available, will be able to provide substantial reductions in future CO_2 emissions. Three factors make the case for utilizing carbon capture technologies at cement plants more compelling: (1) cement plants generally are relatively large point sources of emissions, (2) the CO_2 concentration in the flue gas of a cement plant is relatively high (about 25% on a molar basis), and (3) more than 50% of the CO_2 emissions result from limestone calcination. The first two of these factors relate to the cost-effectiveness of CO_2 capture, while the third factor speaks to the potential use of alternative raw material formulations (e.g., replacing a fraction of limestone with blast furnace slag) to reduce CO_2 generation, thereby reducing capture requirements and associated costs.

Two CO_2 capture technologies applicable to cement plants are: (1) post-combustion capture in which CO_2 is scrubbed from the flue gas, and then compressed and transported to the sequestration site; and (2) oxy-fuel combustion in which fuel is combusted with oxygen, instead of air, to generate a concentrated stream of CO_2, which can then be compressed and transported to the sequestration site. A comprehensive coverage of these measures is provided in IEA [3]. Development of these technologies for power plants is ongoing. It is expected that the controls developed for power plants will be translatable to other industrial sectors, including cement manufacturing. Current indications are that CCS is not likely to be available for the cement industry before 2020 [4, 37].

8.5 The Industrial Sectors Integrated Solutions (ISIS) Model

8.5.1 Overview

ISIS, a dynamic linear programming model, facilitates analyses of emission reduction strategies for multiple pollutants, while taking in to account plant-level economic and technical factors such as the type of kiln, associated capacity, location, cost of production, applicable controls, and their costs. ISIS' design allows for incorporating representations of multiple industries within a multi-market, multi-product, multi-pollutant, and multi-region modeling framework. For the emission-reduction strategies under consideration, the model has been designed to provide information on: (1) optimal (maximum profit) industry operation, (2) cost-effective controls to meet the demand for commodities produced by the sectors under consideration, and (3) the emission reductions achieved over the time period of interest.

ISIS has a modular architecture as shown in Fig. 8.5 below. Input data is organized in various spreadsheets of an Excel Workbook. As shown in Fig. 8.5, the inputs are transmitted to the optimization part of the ISIS model, where they are used to solve the selected Business-as-Usual (BAU) and policy cases. Potential policy options may include cap-and-trade, emissions taxes, or emissions limits as emission reduction mechanisms. After solving, the results are post-processed to

ISIS Model and Data Structure

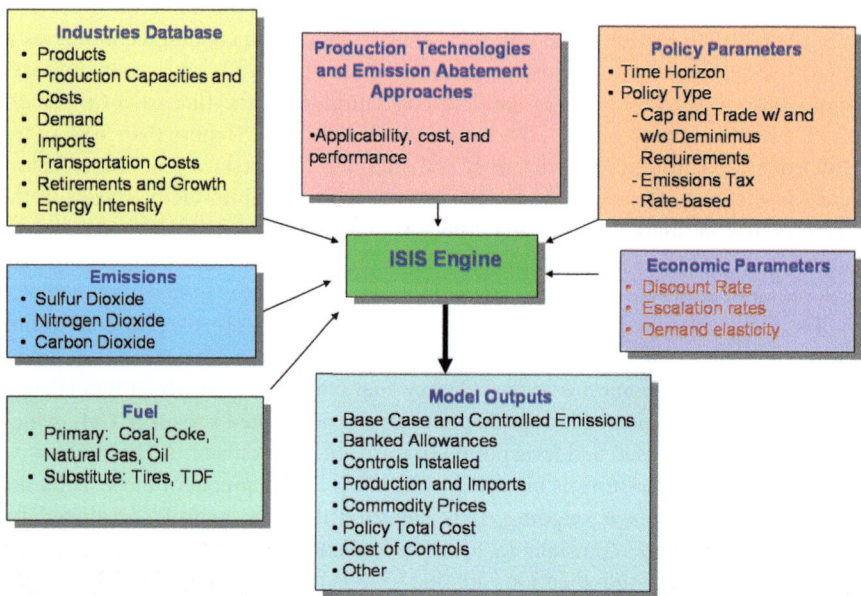

Fig. 8.5 Architecture of the ISIS model

calculate values of various outputs of interest. The output data are exported to Excel spreadsheets for further analyses and graphical representations of selected results.

In the BAU case, the model minimizes the total discounted cost over the horizon of interest while meeting regional market demands, prescribed exogenously, for the applicable commodities. The default discount rate has been chosen to be 7%, as recommended by the U.S. Office of Management and Budget (OMB) for project evaluation [38]. In general, total cost includes the cost associated with operation of production units; cost of endogenous capacity changes resulting from installing new production capacity, replacing or expanding the existing capacity, and retiring or mothballing non-competitive existing units; cost of imports; and the cost associated with transporting the commodities to the pertinent markets.

While evaluating a potential policy option, ISIS maximizes the consumer and producer surplus[2] to account for demand changes associated with price changes relative to the BAU case. An elastic formulation of the demand function is used to estimate the area under the demand curve. Demand for cement is relatively

---

[2] The consumer surplus is the amount that consumers benefit by being able to purchase a product for a price that is less than the most that they would be willing to pay. The producer surplus is the amount that producers benefit by selling at a market price that is higher than the least that they would be willing to sell for.

inelastic and a value of −0.88 is used for the price elasticity of demand [39]. The total cost function is modified to also include the cost of installation and operation of emission reduction measures to meet the applicable emissions reduction requirements.

A detailed description of the ISIS model is available elsewhere [40].

8.5.2 ISIS-Cement and Related Data

ISIS is currently populated with data and information on the U.S. cement industry. This cement industry component of ISIS is hereafter referred to as *ISIS-cement*. The data used in ISIS-cement are described below.

8.5.2.1 Existing and Projected Units and Costs

Currently, the ISIS-cement model contains information on 189 cement kilns that were in existence in 2009 and PCA's projected capacity expansions from 2009 to 2012, as shown in Table 8.7 below [11].

Each kiln in ISIS-cement is characterized by its location, design (i.e., wet, dry, preheater, or precalciner), clinker capacity (short tons per year), vintage, and retirement information when available [15]. In addition, each kiln is characterized by its variable cost (VC) components.

In general, five inputs are required in cement production including raw materials, repair and maintenance, labor, electricity, and fuel. For use in ISIS-cement, kiln-specific VC functions for each of these inputs were developed [41]. The formulation of fuel-specific VC permits each kiln to select a fuel based on the relative costs of available fuels. Under policy, this choice is also influenced by fuel-specific emission factors. In addition to the above VC components, additional cost components included in ISIS-cement are: capital costs associated with use of new and replacement units, indirect labor costs, and applicable overhead costs.

8.5.2.2 Model Markets

As stated previously, the U.S. cement markets are organized in state-specific demand centers. In ISIS-cement, each modeled kiln is located in one of the states as shown in Fig. 8.6. Each state containing at least one kiln is shaded in this figure.

Table 8.7 Summary of kilns modeled in ISIS-cement

Kiln population	Number of kilns
Existing kilns (2009)	189
PCA's projected new kilns (2009–2012)	17

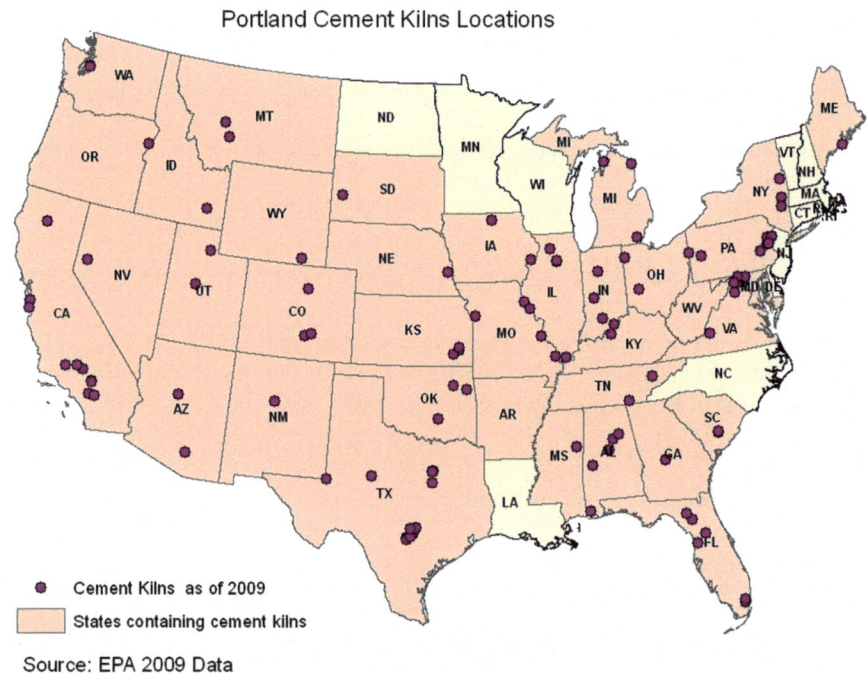

Fig. 8.6 Demand centers for Portland cement in the U.S.

8.5.2.3   Portland Cement Demand

One of the key data inputs for ISIS-cement is the projection of demand for each demand center. In general, the demand is a function of gross domestic product (GDP) growth, interest rates, special construction projects (e.g., highways), and public sector construction spending. Portland cement demand was 128 million metric tons in 2005. PCA expects that the cement demand will reach 192 million metric tons by 2035, which reflects an increase of nearly 64 million metric tons with a compound annual growth rate of 1.4%. Cement demand through the year 2035 is reported in the PCA Long-Term Cement Consumption Outlook [11] and is used in ISIS cement.

8.5.2.4   Transportation-Interregional Trade

In ISIS-cement, a transportation matrix is used to describe the costs for transporting cement from kiln and import district locations to demand centers. To develop these costs, information on distances between supply and demand points and costs of transportation modes (truck, rail, or water transport) was obtained. In particular, the TRAGIS model [42] was used to estimate the origin-destination distances. Also, in the transportation-matrix, the applicable lowest cost transportation mode is used to connect a supply point with a destination. While the cement demand centers are interlinked through a transportation matrix, the competition is generally maintained on a regional level because the cost of transporting cement is relatively high.

8.5.2.5   Imports

U.S. cement markets receive imported quantities of cement and clinker from a number of countries, and these imports arrive at more than 30 import districts [43]. In ISIS-cement, international supplies from exporting countries, excluding Canada and Mexico, to U.S. import districts are modeled using a supply elasticity and then these imports are transported to the demand centers. Supplies from Canada and Mexico are modeled similarly to supplies from domestic kilns.

Excluding Canada and Mexico, the five largest international suppliers of cement and clinker to the U.S. are China, Thailand, Venezuela, South Korea, and Greece. An econometric study was conducted to determine an estimate of international supply elasticity for supplies from these countries and the rest of the world. The results of this study [44] reflected that the best estimate of the international supply elasticity of cement and clinker from China, Thailand, Venezuela, South Korea, Greece, and the rest of the world into the U.S. is 3.94. This indicates that if the price of cement was to increase by 1% within any import district in the U.S., then, *ceteris paribus*, the quantity of cement imported from each of these five supply countries into that district would increase by 3.94%.

8.5.2.6   Capacity Retirement and Growth

Cement plants have relatively long lifetimes of up to 50 years [45]. Various factors including, but not limited to, raw material availability in quarry, technology changes, productivity, efficiency, longevity, reliability, maintenance, and long-term costs can affect the lifetime of a cement kiln. In ISIS-cement, projected kiln retirements of certain existing kilns are based on information from PCA, supplemented with information from individual cement companies on their plans for shutdowns, new construction, and kiln consolidation. Further, as mentioned earlier, ISIS-cement includes algorithms for endogenous capacity growth and retirement of kilns. To determine capital recovery factor for capital costs associated with kiln capacity changes, an economic life of 25 years and an interest rate of 15% are used. Capital costs in 2005 $ per short ton of clinker for new, replacement of wet, and replacement of dry capacity are 208, 296, and 238, respectively [46].

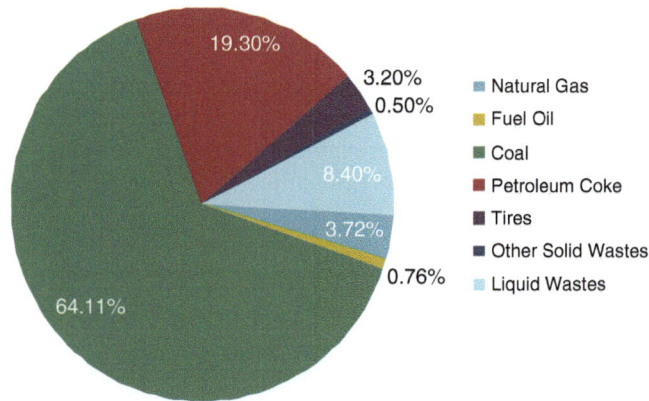

Fig. 8.7 Fuel use profile for the U.S. cement industry in 2005. (Source: EIA 2008 [47])

8.5.2.7   Fuel Use

The Annual Energy Outlook 2008 [47] energy use profile for the U.S. cement sector in 2005 is shown in Fig. 8.7. As shown in this figure, the primary fuel being used in cement kilns is coal. However, there has been an increasing trend towards using other fuels, particularly alternative fuels, such waste tires and oily wastes [48].

In ISIS-cement, information on coal, coke, natural gas, fuel oil, and tire fuels is included. This information includes state-specific fuel prices for the years 2005–2030 (EIA 2008) and fuel-specific CO_2 emission intensity (see next section). Fuel-specific escalation factors from EIA (2008) are used to obtain projected fuel prices for future years. Additionally, based on data from PCA, information on kiln-specific availability of each fuel is also included in ISIS.

8.5.2.8   Emission Intensities

In ISIS-cement, each kiln is characterized by its NO_x, SO_2, PM, hydrochloric acid (HCl), mercury, total hydrocarbon (THC), and CO_2 emission intensities. These emission intensities were developed using available data (Andover Technology Partners [9, 40]).

Tables 8.8 and 8.9 show the NO_x and CO_2 emission intensities for the cement kilns in ISIS.

Kiln specific emission intensities for SO_2 vary by geographic location (states) and kiln type. These intensities (lb/short ton clinker) range from 0.02 to 24.85 [9].

Table 8.8 Estimated NO_x emission intensity (lb/10^6 Btu) for cement kilns

Kiln Type	NO_x Emission Intensity (lb/10^6 Btu)
Wet	1.99
Dry	1.38
Preheater	1.27
Precalciner	0.99

Table 8.9 CO_2 emission intensity (lb/t clinker) for cement kilns

Fuel	CO_2 emission intensity (lb/short ton clinker)
Coal	199.52
Coke	199.52
Natural gas	105.02
Oil	169.32
Tires	187.44

8.5.2.9 Emissions Abatement Approaches

ISIS-cement contains information on abatement approaches for NO_x, SO_2, PM, HCl, mercury, THC, and CO_2 emissions described above. The three categories of abatement approaches included are: process modifications and upgrades, raw material and/or fuel substitution, and mitigation technologies. For each emission abatement approach, where possible, information on the following parameters was developed [29] and included in the model: capital cost, fixed operating cost, variable operating cost, emission reduction performance for all of the pollutants, impacts on fuel and/or raw material use, impact on electricity consumption, byproduct generation and cost, and impact on water use.

To estimate capital recovery factors for capital costs associated with control technologies, economic life of 15 years and an interest rate of 7% are used. Payback periods and technical life for the energy efficiency measures shown in Tables 8.3–8.6 are given in Worrell and Galitsky [23]. Economic life for each of these measures was taken to be the average of the technical life and the payback period. Again, an interest rate of 7% was used for capital recovery.

8.5.2.10 Policy Parameters

The ISIS model framework allows the user to select a variety of potential policy options for evaluation. The user can select from cap-and-trade policy (with or without *deminimus* requirements), emissions charge policy, or rate-based policies. In a cap-and-trade policy scenario, separate caps on pollutants of interest can be specified. The user has the option to run a cap-and-trade policy scenario with or without banking of emissions. Further, a cap-and-trade policy scenario can include *deminimus* requirements, where the user defines a minimum level of emission reduction required for each emission unit. As mentioned before, it is also possible for the user to input an emission charge for pollutants of interest. Furthermore, traditional policy scenarios (rate-based policies) with unit specific emission reduction requirements specified by the user can be modeled in ISIS.

The user can specify the policy horizon (time period) to be used for the model runs. Since climate-related simulation horizons can be long (e.g., 40 years), the user may choose to run ISIS with blocks of years (e.g., 5-year blocks). The simulation horizon and blocks of years can be chosen by the user subject to availability of data.

8.6   An Example Analysis of Potential Reductions in CO_2 Emissions from the U.S. Cement Industry

This section presents an example analysis of the U.S. cement industry, investigating the potential for near-term reductions in CO_2 and other pollutants, associated costs, and industry operation. The ISIS-cement model described above was used to conduct this analysis and the regulatory mechanism selected was cap-and-trade of CO_2. This mechanism has been used in many of the recent policy proposals addressing reduction of CO_2 emissions. The analysis presented is an example of the type of sector-specific analyses that may be needed for developing GHG reduction policies for industrial sectors.

Two broad questions were investigated in this example analysis: (1) what range of CO_2 reduction options may be practicable in the near-term (i.e., the decade ending 2020 selected for this analysis), and (2) for that range, what may be the market characteristics for the U.S. cement industry. The first question is relevant because in the absence of carbon capture technology, the path forward for reducing emissions in the near-term will need to depend on the currently available energy efficiency measures and raw material and product substitution approaches, some of which are described earlier. The second question speaks to industry operation under the potentially practicable options determined while answering the first question.

To investigate the above questions, CO_2 emissions targets corresponding to a range of reductions from projected CO_2 emissions in 2013 were analyzed. Relatively modest reduction levels ranging from 5% to 25% were chosen to be generally consistent with the capabilities of energy efficiency and raw material substitution measures. These targets are summarized in Table 8.10. Only the measures depicted in Tables 8.4 and 8.6 were considered in this example analysis because at this time ISIS-cement does not include calculation procedures for the measures shown in Tables 8.3 and 8.5. In addition to the energy efficiency measures of Tables 8.4 and 8.6, a fuel substitution technology, MKF-tires, and a raw material substitution technology, CEMstar, are also considered in this analysis. As mentioned before, the use of MKF-tires results in reduced fossil fuel consumption (by about 15%) and this is included in ISIS-cement. Relative to Cemstar, a 7.5% reduction in limestone and a 3% reduction in fuel for every short ton of clinker produced is included in ISIS-cement. The costs for energy efficiency measures are described

Table 8.10 CO_2 caps modeled in this analysis

CO_2 cap (short tons)	Reduction from projected 2013 emission level (%)[a]
64,099,115	5
60,725,478	10
57,351,840	15
53,978,202	20
50,604,565	25

[a]2013 emissions were those under the BAU case

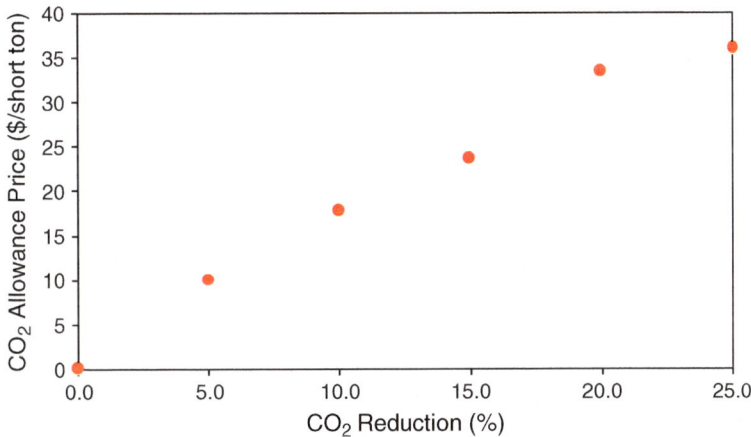

Fig. 8.8 Projected CO_2 allowance prices under a range of CO_2 reduction levels

in Andover Technology Partners [49] and for MKF-tires and Cemstar in Andover Technology Partners [29]. These costs are included in ISIS-cement. Information on the extent to which each of the energy efficiency measures in Tables 8.4 and 8.6 could be applied on existing kiln capacity was obtained [50] and included is ISIS-cement. These penetration numbers ranged from 5% for CGC to 88% for PM. Also for this example analysis it was assumed that each of CEMstar and MKF-tires cannot be applied on more than 25% of the existing and unretired kiln capacity over the horizon 2013–2020.

Runs were made with ISIS-cement for the BAU case (i.e., when no emission reduction requirements are in effect) and with each of the above caps. The corresponding results are presented in Figs. 8.8–8.14.

In a cap and trade framework, allowance price is a primary metric for deciding on emission reduction levels. Figure 8.8 reflects that CO_2 allowance price (in 2013 for example) for various reduction levels range from about 10 to 35 $ per short ton of CO_2. A recent analysis of the American Clean Energy and Security Act of 2009, reflects a price range of 13–24 $ per metric ton in 2015 [51]. Considering this

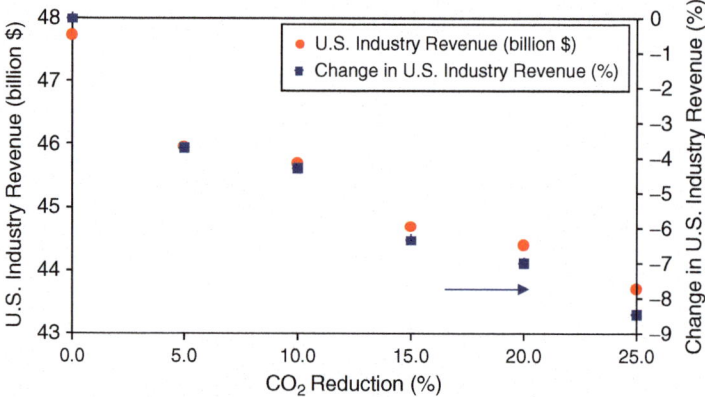

Fig. 8.9 Projected revenue for the U.S. cement industry under a range of CO_2 reduction levels over the horizon 2013–2020

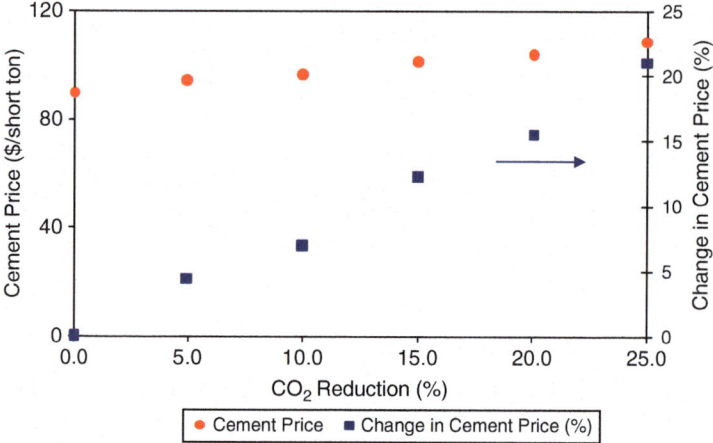

Fig. 8.10 Projected average cement prices in 2013 for the U.S. cement industry under a range of CO_2 reduction levels

indication, a reduction range of 5–15% appears practicable for purpose of this example analysis.

Two metrics for additional evaluation of potential emission reduction levels could be: (1) U.S. industry revenue under policy, and (2) cement price under policy. Arguably, these metrics speak to the interests of both the industry and consumers.

An increase in cement price under a CO_2 reduction policy will cause a drop in demand. Also, if the policy does not impose any requirements on imports, these can increase under policy because production can shift to other countries. Both factors,

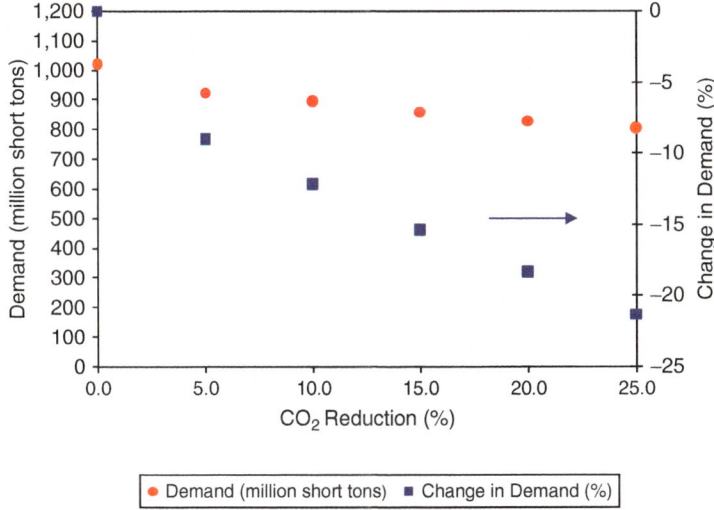

Fig. 8.11 Projected demand for cement under a range of CO_2 reduction levels

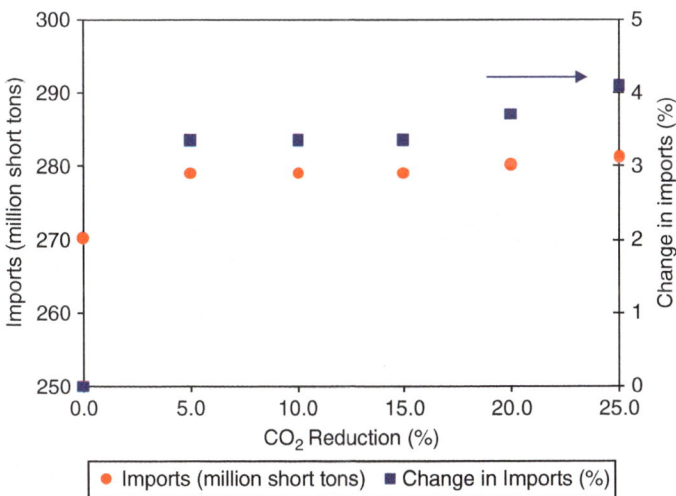

Fig. 8.12 Projected imports for the U.S. cement industry under a range of CO_2 reduction levels over the horizon 2013–2020

drop in demand and increased levels of imports, can cause a reduction in revenue for the cement industry in the U.S.

Increase in price will, in general, result in a reduction in demand because consumers will reduce their needs at higher prices and the drop in demand will contribute to

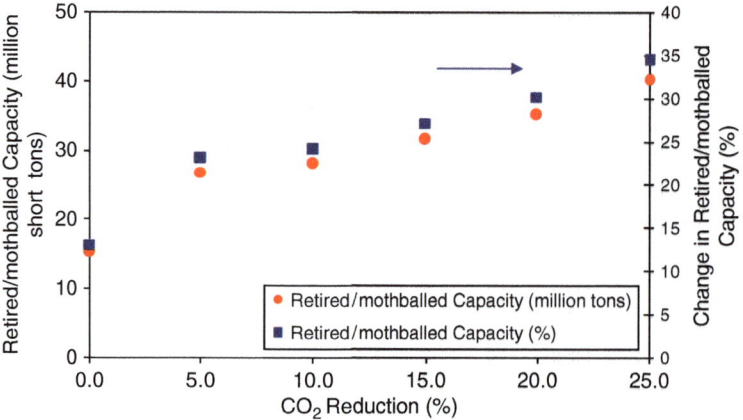

Fig. 8.13 Projected capacity retirement/mothballing

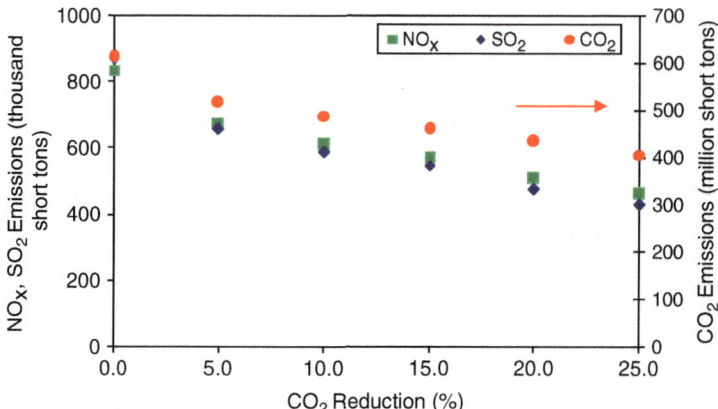

Fig. 8.14 Projected emissions of NO_x, SO_2, and CO_2 from the U.S. cement industry under a range of CO_2 reduction levels over the horizon 2013–2020

reduction in revenue for the industry. As mentioned before, ISIS-cement includes a calculation procedure for calculating drop in demand with increase in price.

Figures 8.9 and 8.10 show the U.S. industry revenue and cement price (in 2013 for example) for the BAU case (0% reduction), and for emission reductions ranging from 5% to 25%. Figure 8.9 shows that the drop in U.S. industry revenue under policy relative to BAU ranges from about 4% to 6.5% for the reduction range of 5–15%. For the same reduction range, Fig. 8.10 reflects that the increase in cement price ranges from about 5% to 12%. Figure 8.11 shows the drop in demand due to rising prices under increasing CO_2 reduction levels. This drop contributes to the loss of revenue seen in Fig. 8.9. Notably, the drop in demand relative to BAU is less than about 16% if the reduction levels are kept at or below 15%.

Figure 8.12 reflects that under example policy options imports increase relative to the BAU case. In addition, as the cap is tightened from the 15% reduction level to the 25% reduction level, the industry resorts to increasing imports in a monotonic fashion. Since imports come with zero emissions in the U.S. and result in reduced domestic fuel and raw materials processing, they can help comply with a policy requiring reductions in domestic emissions. While the impact of increases in imports in the U.S. on emissions in exporting countries is beyond the scope of this analysis, it is recognized that such increases in imports will generally result in increases in CO_2 and other emissions in exporting countries.

A concern related to drop in revenue can be retirement/mothballing of kiln capacity in the U.S. cement industry. Figure 8.13 reflects the potential for retirements/mothballing under the example reduction options. For this work, a unit is considered retired or mothballed if it does not produce in any year of the horizon 2013–2020. The figure shows that even under the BAU case, 12–13% of the existing capacity may be retired or mothballed. This is because in each year of the horizon 2013–2020, the U.S. cement industry has excess kiln capacity relative to BAU demand. For the reduction range 5–15%, the retired or mothballed capacity ranges from 23% to 27% of the existing capacity, or about 10% to 15% points over the BAU projection.

Significant collateral reductions in other pollutant emissions may be possible under the range of practicable CO_2 reduction levels arrived at in this example analysis. Figure 8.14 shows such collateral reductions in NO_x and SO_2 emissions. At the 15% CO_2 reduction level, each of NO_x and SO_2 emissions may be reduced by more than 200,000 short tons over the selected time horizon. These reductions result from use of energy efficiency measures (see below) and reduced production.

For the cap at the 15% reduction level, which is in the range of practicable reductions for purpose of this example analysis, additional detailed results are presented in Figs. 8.15 and 8.16.

Figure 8.15 shows the projected CO_2 emissions under the BAU and policy cases, emission caps corresponding to 15% reduction from CO_2 emissions in 2013, and banked allowances. As seen in this figure, the industry engages in relatively modest levels of CO_2 allowance banking and related trading activity. This is consistent with the relatively modest level of reduction required.

Figure 8.16 reflects that, in response to the CO_2 reduction requirement, the industry installs and operates measures with multi-pollutant, NO_x, SO_2, and CO_2, reduction benefits. The majority of these measures are installed in 2013 to help comply with the reduction requirement. The energy efficiency measures reflected in the legend in this figure are defined in Tables 8.4 and 8.6. In addition, CEMstar and MKF-tires are briefly described in the section on raw material and/ or fuel substitution. Note that Fig. 8.16 reflects that CEMstar will need to be applied on about 40% of operating capacity in 2013. Consequently, a reliable and adequate supply of BFS will need to be ensured for this level of CEMstar application.

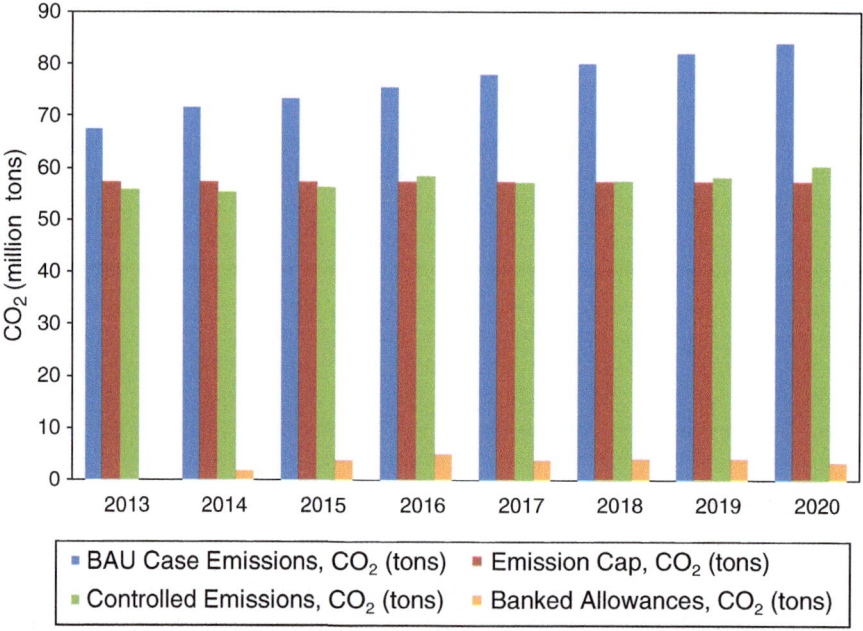

Fig. 8.15 Projected CO$_2$ emissions and reductions from the U.S. cement industry under the BAU and the 10% CO$_2$ reduction cases

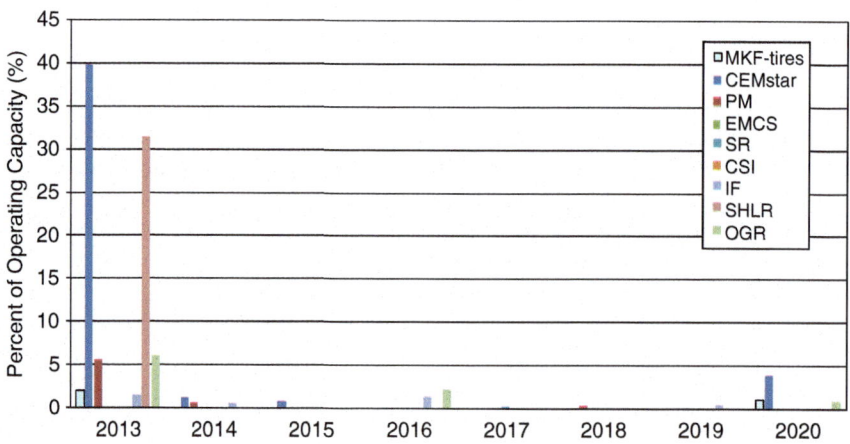

Fig. 8.16 Projected control technology applications in the U.S. cement industry under the 10% CO$_2$ reduction case

8.7 Summary and Thoughts for Future Work

Carbon dioxide accounts for more than 90% of worldwide CO_2-eq greenhouse gas (GHG) emissions from industrial sectors other than power generation. Amongst these sectors, the cement industry is one of the larger industrial sources of CO_2 emissions. Notably, cement manufacturing was responsible for about 6% of the global anthropogenic CO_2 emissions in 2005. Further, global production of cement has been growing steadily, with the main growth being in Asia. China, in particular, now accounts for almost half of the global cement production [2]. Considering these trends, the worldwide cement industry is a key industrial sector relative to CO_2 emissions.

The development of policy options for managing emissions and air quality can be made more effective and efficient through sophisticated analyses of relevant technical and economic factors. Such analyses are greatly enhanced by the use of an appropriate modeling framework. Accordingly, the ISIS model for industrial sectors is under development at the U.S. Environmental Protection Agency (U.S. EPA). Currently, this model is populated with data on the U.S. cement-manufacturing sector and efforts are underway to build representations of the U.S. pulp and paper and iron and steel sectors. The cement industry representation in the ISIS framework is referred to as ISIS-cement.

In this chapter, ISIS-cement was used to conduct an example analysis of the U.S. cement sector to gain some insights relative to two broad questions: (1) what range of CO_2 reductions options may be practicable in the near-term, and (2) for that range, what may be the market characteristics for the U.S. cement industry. These questions are relevant because in the absence of carbon capture and sequestration technology, the path forward for reducing CO_2 emissions in the near-term (decade ending 2020 selected for this work) will need to depend on the available energy efficiency measures and raw material and product substitution options. Only the energy efficiency measures for clinker making and plan-wide application were considered in this example analysis because calculation procedures for other energy efficiency measures associated with raw material preparation and finish grinding still need to be included in ISIS-cement. In addition to the energy efficiency measures, a fuel substitution technology, MKF-tires, and a raw material substitution technology, CEMstar, were also included in the analysis.

The example analysis reflects that, based on CO_2 allowance prices, a range of CO_2 reductions, 5–15%, from the projected 2013 emissions appears to be practicable. For this range, the drop in U.S. industry revenue under policy ranges from about 4% to 6.5% of the revenue in the BAU case and the increase in cement price (in 2013 for example) ranges from about 5% to 12% relative to the BAU. Significant collateral reductions in other pollutant emissions may be possible under the range of 5–15% CO_2 reductions. For example, at the 15% CO_2 reduction level, each of NO_x and SO_2 emissions may be reduced by more than 200,000 short tons over the selected time horizon. To meet a CO_2 reduction level of 15% from the projected emissions in 2013, the U.S. cement industry would potentially use multi-pollutant energy efficiency improvement options and reduce domestic production.

The results of the example analysis presented above are indicative only for several reasons. First, as mentioned before, the analysis did not include the full suite of abatement options potentially available in the near-term. In particular, it did not take in to account the energy efficiency measures of Tables 8.3 and 8.5 and feasible raw material and product substitution approaches. Second, some assumptions associated with the abatement measures considered need additional evaluation. In particular, availability of materials (e.g., blast furnace slag, tires) and the degree to which a specific abatement measure could be applied across the industry need further assessment. Finally, extra-U.S. CO_2 emissions associated with imports need to be taken in to account while developing policy options. These areas need to be addressed to permit more comprehensive evaluation of the CO_2 reduction potential relative to the U.S. cement industry. Work is being initiated to address these areas and modify ISIS-cement as needed.

The focus of the example analysis presented is on modest CO_2 reductions potentially possible with use of energy efficiency approaches in the near-term. It is recognized, however, that in the longer-term more significant CO_2 reductions (e.g., greater than 50%) perhaps can only be achieved with use of CO_2-specific mitigation technologies, which are currently under development.

References

1. Bernstein L, Roy J, Delhotal KC, Harnisch J, Matsuhashi R, Price L, Tanaka K, Worrell E, Yamba F, Fengqi Z (2007) Industry. In: Metz B, Davidson OR, Bosch PR, Dave R, Meyer LA (eds) Climate change 2007: Mitigation, Contribution of Working Group III to the fourth assessment report of the intergovernmental panel on climate change. Cambridge University Press, Cambridge/New York
2. IEA (2008) Energy technology perspectives 2008: scenarios and strategies to 2050. International Energy Agency, Paris
3. IEA (2008) CO_2 capture in the cement industry, IEA Greenhouse Gas R&D Programme (IEA GHG), 2008/3, Jul 2008
4. Mckinsey & Company (2009) Pathways to a low-carbon economy – version 2 of the global greenhouse gas abatement cost curve
5. WBCSDA (2009) A sectoral approach – greenhouse gas mitigation in the cement industry. World Business Council for Sustainable Development, Geneva, June 2009
6. ASTM C150-07 (2007) Standard specification for Portland Cement. ASTM International, West Conshohocken
7. PCA (2008) History and manufacture of Portland cement. Portland cement association. http://www.cement.org/basics/concretebasics_history.asp
8. van Oss HG (2008) Personal communication: Hendrik G. van Oss, USGS, with Elineth Torres, U.S. EPA, 7 July 2008
9. Andover Technology Partners (2008) NO_x, SO_2 and CO_2 emissions from cement kilns, memorandum to U.S. EPA, 23 Sept 2008
10. PCA (2004) U.S. and Canadian Portland Cement Industry: plant information summary. Portland Cement Association, Skokie, 31 Dec 2004
11. PCA (2009) Forecast report: long-term cement consumption outlook, Portland Cement Association. Portland Cement Association, Skokie, 28 Oct 2009

12. PCA (2008) Practical application of PCA economic forecast and market assessments. Portland Cement Association, Education and Training, Skokie, 12–13 Aug 2008
13. PCA (2008) Forecast report: long-term cement consumption outlook, Portland Cement Association, 31 Jan 2008. http://www.cement.org/econ/pdf/Long-TermFlashwinter2007nonmember.pdf
14. USGS (2007) 2005 Minerals yearbook: cement. U.S. Geological Survey, Feb 2007, p 16.2. http://minerals.usgs.gov/minerals/pubs/commodity/cement/cemenmyb05.pdf
15. PCA (2006) U.S. and Canadian Portland Cement Industry: plant information summary. Portland Cement Association, Skokie, 31 Dec 2006
16. USGS (2007) Mineral commodity summaries: cement, U.S. Geological Survey, Jan 2007, pp 40–41. http://minerals.er.usgs.gov/minerals/pubs/commodity/cement/cemenmcs07.pdf
17. PCA (2005) Letter from David S. Hubbard, Director, Legislative Affairs, Portland Cement Association. RE: hours of service of drivers; proposed rule (Docket Number FMCSA-2004-19608). http://www.cement.org/exec/DHOS%20Comments%203.10.05.pdf. Accessed 10 Mar 2005
18. APCA (1997) Comments on EPA's draft economic analysis of air pollution regulations for the Portland Cement Industry (May 1996), prepared for American Portland Cement Alliance by Environomics, 29 Jan 1997
19. EPA (2009) EPA 2005 national air toxics inventory. U.S Environmental Protection Agency, Research Triangle Park, NC
20. EPA (2008) AP 42, Compilation of air pollutant emission factors, Volume I, 5th edn. Chapter 11: Mineral products industry. http://www.epa.gov/ttn/chief/ap42/ch11/final/c11s06.pdf
21. EPA (2008) Inventory of U.S. greenhouse gas emissions and sinks: 1990–2006, U.S. Environmental Protection Agency, Washington, DC. http://epa.gov/climatechange/emissions/downloads/08_CR.pdf. Accessed Apr 2008
22. Martin N, Worrell E, Price L (1999) Energy efficiency and carbon dioxide emissions reduction opportunities in the U.S. cement industry, Sep 1999 LBNL-44182 Nathan Martin, Ernst Worrell, Lynn Price; http://www.osti.gov/energycitations/servlets/purl/751775-bfyH4x/webviewable/751775.PDF
23. Worrell E, Galitsky C (2008) Energy efficiency improvement opportunities for cement making; Lawrence Berkeley National Laboratory, LBNL-54036, Jan 2004. An ENERGY STAR® Guide for Energy and Plant Managers. http://www.osti.gov/energycitations/servlets/purl/821915-Re2kcK/native/
24. EPA (2008) Air pollution controls and efficiency improvement measures for cement kilns (Final Report). Prepared for: U.S. Environmental Protection Agency. Prepared by: ARCADIS U.S., Inc. Contract No.: EP-C-04-023, Project No.: RN990234.0030, Mar 2008
25. PCA (2006) U.S. and Canadian labor-energy input survey. Portland Cement Association, Skokie
26. Young RD (1995) Method and apparatus for using steel slag in cement clinker production. U.S. Patent No. 5,421,880. Issued 6 June 1995
27. Young RD (1996) Method and apparatus for using blast-furnace slag in cement clinker production. U.S. Patent No. 5,494,515. Issued 27 Feb 1996
28. Scheetz BE (2004) Chemistry and mineralogy of coal fly ash: basis for beneficial use. In: Proceedings of state regulation of coal combustion by-product placement at mine sites: a technical interactive forum, Harrisburg, 4–6 May 2004
29. Andover Technology Partners (2008) Cost and performance of controls, memorandum to U.S. EPA, 25 Sept 2008
30. Wright L, Boundy B, Perlack B, Davis S, Saulsbury B (2006) Biomass energy data book: 1st edn. Sept 2006, ORNL/TM-2006/571
31. Rubber Manufacturers Association (2004) U.S. scrap tire markets 2003 edn. https://rma.org/publications/scrap_tires/index.cfm?PublicationID=11302&CFID=8029519&CFTOKEN=25261291. Accessed Jul 2004
32. Livingston RA, Bumrongjaroen W (2004) Optimization of silica fume, fly ash and cement mixes for high performance concrete. In: World of coal ash. Lexington, KY. http://www.flyash.info/2005/79liv.pdf

33. ASTM C 1157-03 (2003) Standard performance specification blended hydraulic cement. ASTM International, West Conshohocken
34. ASTM C 618-05 (2005) Standard specification for coal fly ash and raw or calcined natural pozzolan for use as a mineral admixture in concrete. ASTM International, West Conshohocken
35. ASTM C 595-08 (2008) Standard specification for blended hydraulic cements. ASTM International, West Conshohocken
36. Fortsch DA (2005) Modern slag grinding. World cement. http://www.flsmidth.com/NR/rdonlyres/2E77A28A-F359-4C92-93A3-0427457339A9/26270/ReviewNo158.pdf. Accessed Sep 2005
37. WBCSDA (2009) Cement technology roadmap 2009. World Business Council for Sustainable Development, Dec 2009
38. OMB (1992) Guidelines and discount rates for benefit-cost analysis of federal programs. OMB circular no.A-94 (Revised). U.S. Office of Management and Budget. http://www.whitehouse.gov/omb/circulars/a094/a094.html#8. Accessed 29 Oct 1992
39. EPA (1998) Regulatory impact analysis of cement kiln dust rulemaking. U.S. Environmental Protection Agency, Washington, DC. http://www.epa.gov/osw/nonhaz/industrial/special/ckd/ckd/ckdcostt.pdf. Accessed Jun 1998
40. EPA (2010) Industrial Sectors Integrated Solutions (ISIS) model for the Portland Cement Manufacturing Industry. U.S Environmental Protection Agency, Research Triangle Park, NC
41. Depro BM (2007) RTI International. Documentation for Portland Cement kiln cost functions (2005), memorandum to Keith Barnett, U.S. EPA, 31 Aug 2007
42. TRAGIS (2003) Transportation Routing Analysis Geographic Information System (TRAGIS) user's manual, revision 0, Oak Ridge National Laboratory. https://tragis.ornl.gov/TRAGISmanual.pdf. Accessed Jun 2003
43. USGS (2009) 2005 Minerals yearbook: cement. U.S. Geological Survey (Table 17), pp 16.21. http://minerals.usgs.gov/minerals/pubs/commodity/cement/myb1-2007-cemen.pdf
44. Burtraw D (2010) Supply elasticity estimation, memorandum to Ravi Srivastava, U.S. EPA, 10 Mar 2010
45. FLSmidth & Co. A/S (2007) Q2 Report 2007. http://hugin.info/2106/R/1148414/219358.pdf. Accessed Aug 2007
46. PCA (2009) PCA Comments on the NESHAP from the Portland Cement Manufacturing Industry; Proposed rule (Docket Number: EPA-HQ-OAR-2002-0051). Appendix 13 review and comment on EPA's ISIS model, 4 Sept 2009
47. EIA (2008) Annual Energy Outlook 2008, DOE/EIA-0383(2008). http://www.eia.doe.gov/oiaf/aeo/pdf/0383(2008).pdf. Accessed Jun 2008
48. USGS (2005) Background facts and isues concerning cement and cement data. U.S. Geological Survey, open-file report 2005–1152. http://pubs.usgs.gov/of/2005/1152/2005-1152.pdf
49. Andover Technology Partners (2009) GHG mitigation methods for cement, memorandum to U.S. EPA, 10 Jul 2009
50. Xu T (2010) Personal communication: Tengfang Xu, Environmental Energy Technologies Division, Lawrence Berkeley National Laboratory with Ravi Srivastava, U.S. EPA, May 2010
51. EPA (2009) EPA analysis of the American Clean Energy and Security Act of 2009 HR 2454 in the 111th Congress. U.S Environmental Protection Agency, Washington, DC. http://www.epa.gov/climatechange/economics/economicanalyses.html. Accessed Jun 2009

Chapter 9
Geoengineering: Direct Mitigation of Climate Warming*

Brooke L. Hemming[†] and Gayle S.W. Hagler

Abstract With the concentrations of atmospheric greenhouse gases (GHGs) rising to levels unprecedented in the current glacial epoch, the earth's climate system appears to be rapidly shifting into a warmer regime. Many in the international science and policy communities fear that the fundamental changes in human behavior, and in the global economy, that will be required to meaningfully reduce GHG emissions in the very near term are unattainable. In the 1970s, discussion of "geoengineering," a radical strategy for arresting climate change by intentional, direct manipulation of the Earth's energy balance began to appear in the climate science literature. With growing international concern about the pace of climate change, the scientific and public discourse on the feasibility of geoengineering has recently grown more sophisticated and more energetic. A wide array of potential geoengineering projects have been proposed, ranging from orbiting space mirrors to reduce solar flux to the construction of large networks of processors that directly remove carbon dioxide from the atmosphere. Simple estimates of costs exist, and some discussion of both the potentially negative and "co-beneficial" consequences of these projects can be found in the scientific literature.

---

*The findings included in this chapter do not necessarily reflect the view or policies of the Environmental Protection Agency. Mention of trade names or commercial products does not constitute Agency endorsement or recommendation for use.

[†] © US Government 2011

B.L. Hemming (✉)
Air Pollution Prevention and Control Division, National Risk Management Research Laboratory, Office of Research and Development, United States Environmental Protection Agency, Research Triangle Park, NC
Currently with: Global Change Research Program, National Center for Environmental Assessment, Office of Research and Development, United States Environmental Protection Agency, Washington, DC
e-mail: hemming.brooke@epa.gov

G.S.W. Hagler
Air Pollution Prevention and Control Division, National Risk Management Research Laboratory, Office of Research and Development, United States Environmental Protection Agency, Research Triangle Park, NC

F.T. Princiotta (ed.), *Global Climate Change - The Technology Challenge*, Advances in Global Change Research 38, DOI 10.1007/978-90-481-3153-2_9, © Springer Science+Business Media B.V. 2011

The critical, missing piece in the discussion of geoengineering as a strategy for managing climate is an integrated evaluation of the downstream costs-versus-benefits inter-comparing all available climate management options, including geoengineering. Our examination of the literature revealed a number of substantial gaps in the knowledge base required for such an evaluation. Therefore, to ensure that the decision framework arising from this analysis is well founded, a focused program of scientific research to fill those gaps is also essential. As with any sound engineering plan, international decisions on how to address human-induced climate warming must be founded on a thoughtful and well-informed analysis of all of the available options.

9.1   Introduction

With the concentrations of atmospheric greenhouse gases (GHGs) rising to levels unprecedented in the current glacial epoch, due in large part to human activities, the earth's climate system appears to be rapidly shifting into a warmer regime [1] . Studies by Raupach et al. [2] and Canadell et al. [3] have concluded that global GHG emissions are growing at a rate in excess of 3% per year (Princiotta, this book). The human and ecosystems costs associated with the consequent climate change are expected to be great.

Many in the international science and policy communities fear that the fundamental changes in human behavior, and in the global economy, that will be required to significantly reduce GHG emissions in the very near term are unattainable. This fear has stimulated discussion in the scientific and policy communities, and in the popular press, of a radical alternative strategy for arresting climate change – intentional, direct manipulation of the earth's energy balance. The term coined to describe this strategy is "geoengineering" [4].

The notion that weather and climate could be modified through technological means was proposed early in the twentieth century, and evolved to include schemes ranging from cloud seeding to stimulate rainfall to strategies for warming the climate in the northern latitudes [5]. The concept of geoengineering greatly enlarges upon these more local- and regionally-focused ideas to consider the earth's climate system, as a whole. Directly and intentionally modifying the global climate system to achieve specific ends was initially dismissed as science fiction. More recently, the sensational aspects of geoengineering have made the topic an attractive subject for the popular press and television [6–11].

With the consensus conclusion by the Intergovernmental Panel on Climate Change (IPCC) that climate change is underway, publications concerning geoengineering in prominent scientific journals have proliferated, and the level of discourse on the subject has rapidly grown more sophisticated and more energetic. The range of opinion concerning geoengineering as a legitimate policy option is broad. Geoengineering appears to be a plausible means of "buying time" while humanity undergoes the profound changes in its use of energy and choices in transportation that are needed to eliminate GHG emissions. Many fear, however, that the adoption of geoengineering as a climate management strategy will lessen the sense of urgency needed to drive the global-scale changes critical for reducing future climate change.

While the concepts underlying geoengineering strategies are simple, given our current level of knowledge, the downstream consequences of their implementation are not at all clear. A critical, missing piece in the discussion of geoengineering as a strategy for managing climate is a systematic evaluation of the downstream costs versus benefits, from a global perspective, based on rigorous scientific study. Should the geoengineering approach become necessary for minimizing the risk of a major climate catastrophe, such an evaluation will be essential for the design of a successful climate management program.

This chapter introduces the basic concepts underlying the geoengineering strategy for mitigating climate warming, provides an overview of the more widely discussed project proposals, and highlights the critical uncertainties regarding the implementation of each proposal. Also included is a discussion of potential co-benefits along with the undesirable consequences of the implementation of these proposals that should be factored into the risk assessment of a global climate management program. The chapter concludes with a discussion of the need for an integrated risk assessment/risk management framework that incorporates the available quantitative information on the risks associated with the intentional manipulation of the climate system – should direct manipulation of the climate system become a policy choice. Recommendations for research targeted at minimizing some of the larger uncertainties are included in this discussion.

9.2 The Planetary Energy Balance: Levers Available for Climate Manipulation

The earth's climate system is a dynamic and very complex system, but it obeys the basic principles of physical and chemical thermodynamics and kinetics. The climate system at equilibrium is a balance between the incoming solar radiative energy and outgoing long-wave terrestrial radiative energy. The "levers" that have the capacity to substantially shift the earth's energy balance are: (1) altering the flux (quantity) of solar radiation entering the earth's atmosphere; (2) altering the fraction of solar radiation that is reflected, unchanged, back into space, and; (3) altering the radiative emissivity of the earth's atmosphere, e.g., its capacity for absorbing infrared radiation. Geoengineering proposes to apply these large levers to intentionally shift (manage) the earth's energy balance.

The Stefan-Boltzmann Law succinctly describes the relationship among these levers,

$$\left(1-\mathbf{a}\right)\mathbf{S_0} = \varepsilon\sigma T^4 \qquad (9.1)$$

Here, S_0 is the average solar irradiance outside Earth's atmosphere; **a** is Earth's average albedo, ε is the emissivity of Earth's atmosphere; σ is the Stefan-Boltzmann constant (equal to 5.67×10^{-8} W m^{-2} K^{-4}), and T is Earth's "black body" temperature. The left side of Eq. 9.1 provides a simple estimate of the solar energy that is absorbed, while the right side of the equation is an estimate of the quantity of energy that radiates back to space by the earth system. When the solar energy absorbed is equal to the energy re-emitted by the earth system, Earth's climate system is at thermodynamic equilibrium. The scientific understanding concerning Earth's energy budget on decadal to century timescales is incomplete. However, over the long term, this fundamental thermodynamic balance will hold.

Geoengineering, as a strategy for mitigating climate warming, involves purposefully altering some of the values in this energy balance equation – modifying the solar irradiance (S_0), Earth's average albedo (**a**), and the emissivity of Earth's atmosphere (ε).

Solar irradiance, S_0, is the average quantity of solar energy that falls upon one square meter of Earth's atmosphere per second. If the earth were flat, S_0 would average 1,366 W/m^2 [12]. However, the earth's curvature combined with the fact that only half of the planet is illuminated by the sun at any time substantially reduces the actual energy absorbed – down to a mean of 343 W/m^2. A number of astrophysical processes, from the size and ellipticity of the earth's orbit around the sun, to cyclic changes in the sun's magnetic fields, lead to variability in S_0 [13]. Over the past 50 years, the 11-year solar sunspot cycle has been the chief cause of this variability. Analyses of the available solar luminosity data over this period show a variance in the range of 2 W/m^2 or less than 0.1% [14, 15]. While these cycles in luminosity can be seen in the temperature record, there has been no upward trend that might explain the observed climate warming (IPCC 2007).

Earth's *albedo* (**a**) is a measure of the reflectivity of its surface and atmosphere. Multiplying the solar irradiance (S_0) by (1-**a**) gives the fraction of solar energy that is absorbed into the system. Albedo is a function of the fraction of white or light-colored surfaces, such as sea ice, glaciers, or deserts, or in the atmosphere, clouds and scattering (non-absorbing) aerosol. For simple calculations like Eq. 9.1, the earth's albedo is usually estimated to be around 0.3, indicating that approximately a third of the sun's radiation is reflected into space by these light scattering surfaces. However, in practice, these reflective features over or at the earth's surface vary over time scales ranging from seconds to millennia. Shifting continents, and ice ages have altered global albedo on geological time scales. Over centuries, and more recently, on decadal and annual time scales, human agriculture and development has altered the albedo of the earth's land surfaces. Recent rapid deforestation has dramatically altered surface albedo (as well has the biosphere's capacity for carbon sequestration). The cloud and aerosol composition of the atmosphere changes on scales from years down to seconds. Thus, in practice, the earth's albedo varies widely and changes constantly.

If Earth were a true "black body," the energy it re-emits to space would simply be a function of its absolute temperature. The ideal version of the Stefan-Boltzmann equation, $j = \sigma T^4$, where j is the energy flux per surface area, σ is the Stefan-

Boltzmann constant, and T is Earth's black body temperature in Kelvin units, quantifies this energy flux. However, Earth's atmosphere interferes to a degree with the emission of terrestrial (IR) radiation. The emissivity (ε), in this context, is a measure of our atmosphere's opacity to infrared radiation. If the atmosphere were perfectly transparent to IR radiation, ε would equal unity (1). Values of ε that are less than unity indicate that the atmosphere absorbs IR radiation to some degree.

Earth's black body temperature, as measured above the atmosphere by satellite, is −19°C – well below the freezing point of water – while the average surface temperature is 15°C. The strong thermal gradient that exists between the surface and the top of the earth's troposphere is a consequence of the non-ideal emissivity of the atmosphere. We know this as the familiar "Greenhouse Effect." The greater the atmospheric concentration of greenhouse gases, the lower the emissivity of the atmosphere. The lower the emissivity of the atmosphere, the higher the earth's surface temperature (T, in Eq. 9.1) must rise to achieve thermal equilibrium.

A number of factors complicate the absorption and dispersion of energy at the earth's surface. These factors include the planet's near-spherical shape; the heterogeneous mix of surface types with a wide range of albedo values; the earth's photochemically active atmosphere; the vast world ocean, and; a biosphere that absorbs solar radiation and CO_2, and subsequently re-emits CO_2 and other GHGs (CH_4, methane; and N_2O, nitrous oxide). The ocean not only absorbs solar radiation, but also contains 50 times the concentration of CO_2 as the atmosphere. On geologic time scales, changes in any of these climate system components have induced ice ages and warm interglacial periods, along with climate variability at shorter time scales.

The perturbation represented by rapidly increasing atmospheric GHG concentrations due to human activities has disrupted the earth's energy balance. While the global atmospheric burden of GHGs continues to grow, the equilibrium state towards which the earth system is moving will include higher average surface temperatures. The climate science community has projected increasingly dramatic changes in meteorological patterns for the future, as the additional heat energy trapped by the enhanced Greenhouse Effect disperses within the Earth system [1] .

The geoengineering strategy is to apply the climate system levers implied by the variables in Eq. 9.1: reducing solar irradiance; increasing the earth's albedo; increasing the emissivity of the atmosphere. In the following sections, we describe several examples of *how* geoengineering proponents have proposed to apply these levers. While the list of proposals is not exhaustive, from a physical standpoint, all geoengineering proposals seek to employ one of the three available levers. Any difference amongst them is in the details. A summary of the options discussed here is provided in Table 9.1, which qualitatively compares the current level of understanding on the potential environmental outcomes (side effects) associated with successful implementation of each of the suggested geoengineering strategies. Other metrics included for comparison include implementation cost and the expected timeframe when action would be required to maintain climate forcing. Each proposal's timeframe-to-maintain can also be interpreted as a reversibility-timeframe, assuming the geoengineering action would only be halted by natural mechanisms.

Table 9.1 Proposed methods to geoengineer climate

Proposal	Forcing mechanism	Cost to implement[a]	Timescale to maintain	Potential co-benefits	Potential undesirable consequences	Level of understanding
Stratospheric aerosol injection	Reflecting shortwave radiation	$–$$$[b]	Years	• Brilliant sunrises and sunsets	• Loss of stratospheric ozone • Whitening of the sky • Unknown effects of manufactured novel aerosols • Altered hydrological cycles	Feasibility: High Co-benefits: High Consequences: Low
Tropospheric marine cloud-seeding	Reflecting shortwave radiation	$	Weeks	None	• Increased aerosol pollution of ocean and ocean-bordering regions • Altered hydrological cycles	Feasibility: Low Co-benefits: High Consequences: moderate
Space-based reflectors	Reflecting shortwave radiation	$$$	Decades	None	• Altered primary productivity due to altered PAR flux • Altered hydrological cycles	Feasibility: Low Co-benefits: Low Consequences: Low
Surface albedo modification	Reflecting shortwave radiation	$$–$$$	Months	• Reduced urban heat island effects	• Significant interference with ocean ecosystem • Ocean surface modification likely to have poor aesthetics and affect shipping industries	Feasibility: Low Co-benefits: Low Consequences: Low

Ocean pH modification	Greenhouse gas drawdown	Unknown	Years	• Mitigation of CO_2-driven ocean acidification	• Large-scale manipulation of ocean surface chemistry likely to impact ocean ecosystems	Feasibility: Low Co-benefits: Low Consequences: Low
Phytoplankton fertilization	Greenhouse gas drawdown	Unknown	Months	• Restoration of depleted fish populations	• Algae overgrowth and creation of dead zones in ocean • CH_4 and N_2O production below fertilized euphotic zones	Feasibility: Low Co-benefits: Low Consequences: Low
Reforestation and afforestation	Greenhouse gas drawdown	Unknown	Years	• Replenishment of depleted forests • Urban heat island mitigation for urban forestry	• Impacts of increased water demand, soil run-off, and fertilizer. • Changes in biodiversity with wide-scale forest monoculture	Feasibility: Moderate Co-benefits: High Consequences: Low
Chemical weathering	Greenhouse gas drawdown	Unknown	Years	Unknown	• Environmental impacts of large-scale reagent and waste production.	Feasibility: Low Co-benefits: Low Consequences: Low

[a] Cost approximated using existing published estimates and assuming scale of implementation would be enough to compensate for forcing equivalent of doubling CO_2 (from preindustrial leaves) and held for 50 years. $ = 10–100 billion USD, $$ = 100 billion – 1 trillion USD, $$$ = 1–10+ trillion USD

[b] Sulfate aerosols have projected costs at 2–7 trillion USD [16], while novel manufactured aerosols optimized to reflect solar radiation were projected to be considerably less costly at 15–40 billion USD [17]

9.3   Geoengineering Strategy: Reduce Solar Irradiance

Modeling studies have found that a reduction of a small percentage (estimated at 1.6–1.8%) in absorbed solar radiation would compensate for the temperature-warming induced by CO_2 doubling from pre-industrial levels (280 ppm) to future elevated concentrations (560 ppm) [18, 19]. Proposed approaches to geoengineer climate by reflecting solar radiation all depend upon some form of scattering mechanism situated at locations ranging from Earth's surface to outer space, as illustrated in Fig. 9.1. Several geoengineering proposals seek to reduce the quantity of solar radiation entering the earth's atmosphere by placing light scattering objects either outside of Earth's atmosphere in low-earth orbit (LEO; 222 km) or at the Lagrangian 1 point (L1, 1.5×10^6 km from Earth) where the gravitational fields of Earth and the Sun are in balance and allow a small mass to remain stationary relative to Earth. This concept was first introduced by Early [20], who

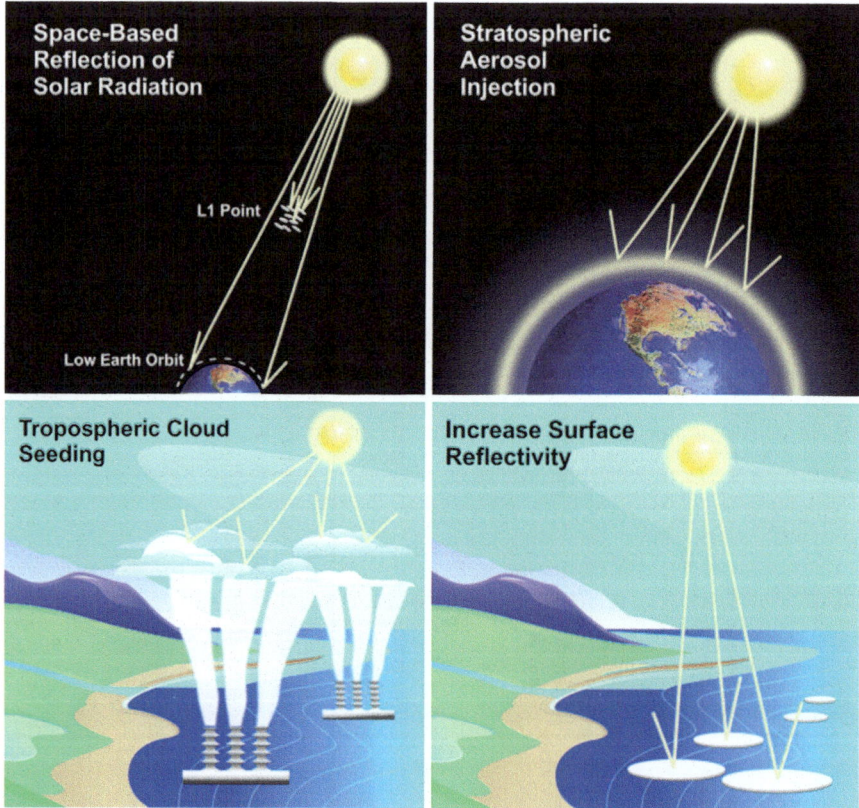

Fig. 9.1 Geoengineering proposals to deflect incident solar radiation: placing scattering materials in outer space; injecting aerosols into the stratosphere; brightening tropospheric clouds, and; increasing the reflectivity of Earth's surface

proposed to loft into outer space a 10 μm thick and 2,000 km in diameter shield constructed of lunar materials and locate it at the L1 point. A more recent study proposed to launch 800,000 m-sized reflective objects manufactured on Earth to the L1 point, creating a 100,000 km diameter reflective "cloud" to deflect solar energy [21]. Both studies discussed the challenge of maintaining the delicate balance between the opposing gravitational fields, centripetal acceleration from orbiting the sun, and the forcing associated with the deflection of solar photons. In the [22] report, the list of proposed objects to be placed into LEO included a large solar-reflective screen, thousands of mirrors, and clouds of dust. Due to instability in orbit, the NAS committee ruled out the dust cloud proposal as impractical.

A project to reduce solar flux situated outside of the Earth's atmosphere has advantages. The strategy avoids the need to disturb Earth's surface or the atmosphere, alleviating the concern for potential damage to the environment or human health. In addition, launching reflective objects into outer space requires less maintenance, as objects placed into outer space have lifetimes as long as decades. This strategy is therefore less vulnerable to the wide number of factors that may interrupt other geoengineering strategies, such as international financial crises or conflict.

Mitigation Option: Launch reflective material either into LEO or to the L1 point, where the reflective material would remain to block incident solar radiation for decades.

Feasibility: The technical challenges involved with sending reflective material to block solar radiation in outer space are enormous. Current proposals involve yet undeveloped technologies, such as using lunar material to construct reflective objects in outer space [20]. Currently, cost estimates for the various proposals are very preliminary and range from $1 to 10 trillion current USD.

A crucial design consideration for a system of space-based reflectors will be the need for "fine-tuning" the quantity of solar radiation blocked by the system, as information about the climate system response to reduced S_0 emerges. Each device will require a remote control mechanism for reducing or increasing the active reflective area. This will add significantly to the cost of implementing the scheme. The remoteness of the reflectors makes removal or other significant modification difficult at best.

Co-benefits and undesirable consequences: Geoengineering a decrease in solar irradiance from outer space is the most "clean" option, involving no direct interaction with Earth's surface or atmosphere. However, as for other proposals to limit solar radiation, the outer space options would also likely reduce the amount of photosynthetically active radiation (PAR, or wavelengths of 400–700 nm) available to support photosynthetic organisms. This effect is of note both in terms of its impact on ecosystem viability but also as a climate feedback mechanism – to what degree might a suppression of photosynthesis induce a positive (i.e., warming) feedback on climate?

9.4   Geoengineering Strategy: Increase the Planetary Albedo

On June 15, 1991, Mt. Pinatubo, a volcano in the Philippines, explosively erupted. The volcanic plume lofted approximately 20 Tg of gaseous sulfur dioxide into the upper troposphere and stratosphere. During the subsequent weeks, the volcanic sulfur dioxide dispersed across the globe, while photochemically oxidizing to form sulfate aerosols. These aerosols had both significant "direct" and "indirect" effects on the Earth's radiative balance – directly scattering solar radiation to outer space and indirectly affecting climate by enhancing cloud reflectivity in the upper troposphere. Two months after the eruption, the Earth's albedo had increased, on average, by 6%. Cloud-free regions, which otherwise absorb a large fraction of the earth's incoming solar energy, saw a 20% increase in albedo due entirely to direct aerosol scattering. In the year to follow, global temperatures dropped by an estimated 0.7°C (relative to 1991). As the volcanic aerosols settled out of the stratosphere and processed out of the troposphere by wet and dry deposition, the cooling effect faded. By 1993, the temperature anomaly had decreased by half ([23] and references therein).

Many in the scientific community view the 1991 eruption of Mt. Pinatubo as a natural demonstration of the geoengineering strategy for managing climate warming through the enhancement of planetary albedo. The cooling effect of volcanic eruptions has inspired proposals to inject aerosols into the stratosphere and, as with the volcanic emissions, increase the amount of back-scattered solar radiation at the top of the atmosphere. Other proposals to increase the Earth's albedo include brightening tropospheric clouds overlying the ocean and widespread modification of the ocean and/or land surface.

9.5   Tropospheric Cloud Albedo Enhancement

Atmospheric particulate pollution, which consists of airborne solid or liquid particles generated from a variety of sources, is most often associated with a visible haze and adverse impacts on human health. Particles, also referred to as aerosols, in the lower atmosphere have natural sources such as volcanic emissions, wild fires, and sea spray, as well as anthropogenic sources such as fossil fuel combustion, prescribed forest fires, and road dust. Once generated, tropospheric particles usually reside in the atmosphere for days to weeks before being removed by precipitation or dry deposition to surfaces. It is well-known that particles impact climate [1] , inducing both warming and cooling forcings depending on the aerosol composition, meteorological conditions, and location over the Earth surface. Particulate matter can directly affect climate by scattering (cooling effect) or absorbing (warming effect) solar radiation. Particles may also indirectly impact climate by serving as cloud condensation nuclei (CCN) and altering cloud formation. Increased CCN leads to smaller cloud droplets, creating brighter and longer-lasting clouds that would increase planetary albedo [24].

The study of atmospheric particles is an active field of research, with a diversity of scientific disciplines studying the chemical and physical mechanisms of particle formation, impacts on human health and the ecosystem, and climate linkages. Particulate pollution in the lower atmosphere is currently regulated in the United States under the EPA National Ambient Air Quality Standards (NAAQS) for specific size fractions known as "fine" particulate matter ($PM_{2.5}$ or particles smaller than 2.5 μm) and "coarse" particulate matter (PM_{10} or particles smaller than 10 μm). Fine and coarse particle standards are also in place in other countries, although the ambient concentration limits vary.

Ambient air pollution and its associated climate effects are unintentional by-products of fossil fuel powered industrialization. In addition to this inadvertent climate forcing, the long-standing practice of weather modification *purposefully* injects CCN into already existing clouds in an attempt to control precipitation. First introduced in the 1940s by researchers in the USSR and in the United States ([5] and references therein), weather modification is still in use today worldwide. Weather modification, also commonly referred to as "cloud seeding", is usually practiced by dispersing a chemical compound to serve as CCN using aircraft or ground-based rockets. Two different particle types have been used – glaciogenic particles (e.g., silver iodide) which are used as ice nuclei in cold clouds and hygroscopic particles (e.g., salt) which serve as water droplet nuclei in warm clouds. Over its history, cloud seeding was used and then banned as a military strategy [25] and has also been peacefully applied worldwide to meet specific regional goals. Cloud seeding has been applied as a strategy to enhance snow fall for ski resorts, boost precipitation for agriculture or drought-stricken regions, prevent hail formation, and even reportedly to reduce pollution and control rainfall during the 2008 Olympics in Beijing, China [26].

While cloud seeding has been widely practiced to induce precipitation, results have been mixed. Particle-cloud interactions are complex and the study of these interactions has been limited until recently. Geoengineering climate using a cloud seeding process has a nearly opposite goal of traditional cloud seeding operations – to reduce cloud droplet size and extend the lifetime of a cloud that overlies a region with low solar reflectivity, leading to higher planetary albedo. Rather than enhance precipitation which is the general goal of weather modification, this strategy would lower the chance of precipitation by preventing the growth of large cloud droplets.

Mitigation Strategy: Tropospheric cloud-seeding would be implemented over the ocean, increasing the reflectivity of marine stratocumulus clouds [27–306, 22]. As ocean-covered portions of Earth have low albedo, increased reflective cloud coverage would have a significant cooling effect.

Feasibility: Several methods have been proposed to support long-term seeding of marine stratocumulus clouds. One suggested method would be to continuously emit sulfur dioxide over the ocean, which would convert to sulfate particles and seed clouds overlying the ocean [22]. A more recent proposal is to aerosolize seawater using unmanned wind-powered vessels, resulting in salt particles to operate as CCN [30, 31]. Both studies projected implementation costs, which would equate to approximately \$10–100 billion if operated at a level sufficient to offset a doubling of CO_2 and maintained for a time horizon of 50 years (providing time to lower greenhouse gas emissions). However, it should be noted that these cost estimates depend on underlying assumptions about the fraction of Earth's surface covered with marine stratocumulus clouds and the required amount of CCN to change cloud albedo, which are of debate [19]. Cloud seeding for climate application is only a concept at this point, with little data existing to show whether extensive cloud-seeding over the ocean would substantially alter the planetary albedo and sustain a climate forcing on timescales of years to decades.

Co-benefits and undesirable consequences: No co-benefits have been identified for this strategy. Potential consequences include the reduction of PAR and subsequent suppression of ocean phytoplankton growth, deposition of acidic particles to the ocean, modification of precipitation patterns, altered storm frequency, and intensity, and human exposure to respiratory pollutants. Human exposure risk is of substantial concern for particulate matter injected into the troposphere, as particles can be transported up to thousands of kilometers from the point of injection before removal from the atmosphere by wet or dry deposition. While these consequences have not been well studied for the application of large-scale cloud-seeding operations, research on anthropogenic particulate pollution may provide some understanding of the potential impacts. One important aspect of a geoengineering cloud seeding program is the capability to rapidly "turn off" the climate forcing, as tropospheric particles have a short lifetime of days to weeks. This can be seen as a benefit in allowing greater control over the process and capability to respond to unintended consequences. However, the need for continuous operation also implies that a major interruption in large-scale geoengineering via cloud seeding may lead to a sudden change in global climate.

9.6   Stratospheric Aerosol Injection

The most frequently discussed geoengineering option for mitigating climate change is the injection of aerosols (solid or liquid particles) into the stratosphere, to mimic the cooling effects of volcanic eruptions. The majority of effort by those interested in geoengineering, to date, has been applied to understanding whether purposeful injection of sulfate aerosol would be a viable strategy to control Earth's surface temperature. This idea was originally introduced by Budyko [32] and was recently revitalized in an editorial by Nobel Laureate Paul Crutzen. Crutzen [16] noted that tropospheric particles will decrease in concentration in response to public health measures, a fact which has been supported by recent trends in sulfur emissions [33]. As anthropogenic particulate pollution has an overall cooling effect [1] , the lowering of particle concentrations while greenhouse gas concentrations continue to increase would shift the energy balance towards accelerated warming. In order to counteract both this effect and that of poor progress regarding greenhouse gas emission reductions, Crutzen [16] advocates that stratospheric sulfate aerosol injection be given serious consideration and further research.

Sulfate aerosol has been the most well-studied and popular choice for strato-spheric aerosol injection proposals, due to its ease of production (in fact, the coal and oil power industry invests in flue gas desulfurization to be rid of it) and the ability to observe natural volcanic events which demonstrate the results of large-scale sulfate aerosol injection. Other aerosol types have also been proposed, including dust [22], soot [34], and the development of novel aerosols optimized to scatter solar radiation [17]. As the sulfate option has been more thoroughly analyzed, in comparison to other aerosol types, the following discussion will be confined to sulfate injection.

Mitigation Strategy: Inject sulfate aerosol into the stratosphere, with continued injections on an annual or biennial basis to sustain climate forcing.

Feasibility: Using past volcanic eruption events as a guide, it has been estimated that annual injections of 5.3 Tg sulfur would meet the goal of 1.8% reduction in incident radiation and compensate for the temperature effects of doubling CO_2 levels [16]. Rasch et al. [35] calculated that less sulfur may be needed annually (1.5–3 Tg) if the particle size were optimized. Using the cost metrics developed by [22] and extending the implementation out to 50 years, the deployment of this option was estimated to range $2–7 trillion (1992 USD). It was estimated that the amount of sulfur needed is likely attainable, with 5.3 Tg sulfur equal to less than 5% of inad-vertent global sulfur emissions in the year 2000 [33]. Several mechanisms for loft-ing sulfate aerosols into the stratosphere have been proposed, including dispersal by aircraft, ground-based artillery, or balloons. The optimal generation of sulfate aero-sol and mechanisms of dispersal are areas of needed research and development.

Co-benefits and undesirable consequences: The only known co-benefit of aerosol addition to the stratosphere is aesthetic – sunrises and sunsets would likely be intensely colorful, as often observed after volcanic eruptions. A major area of concern is the fate of the stratospheric ozone layer, which blocks biologically harmful ultra-violet radiation. After Mt. Pinatubo injected tens of teragrams of sulfate aerosol in the stratosphere, the ozone hole over Antarctica grew 25% greater in geographical extent [36]. A recent modeling study by Tilmes et al. [37] also demonstrated that sulfate aerosol addition to the stratosphere may lead to a loss of ozone, predicting that this geoengineering scheme could delay the ozone hole recovery by 30–70 years. An inverse relationship with particle size was also found, with the higher surface area associated with smaller particles enhancing ozone loss through heterogeneous chemical reactions. Although smaller aerosols may provide a cost savings [35], it appears they also increase the risk of stratospheric ozone destruction.

The human health, ecosystems, and climate effects of air pollution-related and stratospheric sulfate aerosols have been extensively studied by the atmospheric sciences community, as discussed for tropospheric cloud enhancement strategy. Known consequences include reduced PAR, altered hydrological cycles leading to shifts in regional precipitation, and potential human health effects. The knowledge that now exists on these effects can assist us in assessing the environmental impacts from boosting sulfate aerosol loadings in the atmosphere. For manufactured aerosols, however, we are in uncharted territory. Proposals to spread novel aerosols [17] in the atmosphere at a global scale need careful evaluation prior to any level of implementation. As we learned the hard way with manufactured chlorofluorocarbon compounds (CFCs), what is an inert compound within the troposphere can be reactive in the stratosphere and cause long-term global impacts.

9.7 Increase Surface Reflectivity

The conversion of a natural landscape to an urban environment often induces a local warming trend due to the introduction of impervious and light-absorbing surfaces, such as dark-colored rooftops and pavements. This is known as the "urban heat island" effect. Warmer temperatures in urban environments have a number of negative impacts, such as higher energy demand for air conditioning and related emissions (pollutants and greenhouse gases), thermal stress to local ecosystems, and an increased risk of high temperature episodes [38]. Strategies for reducing urban surface temperatures in cities include the use of "cool" building materials, e.g., light-colored pavement, to increase the surface albedo.

Strategies already in place to mitigate urban heat islands will likely also serve to slow global climate change to some extent by lowering fossil fuel use and increasing surface reflectivity. However, geoengineering projects to substantially alter global climate by modifying Earth's surface reflectivity pose a significant engineering challenge. A major physical limitation is location – only half of the incident solar radiation actually penetrates to the surface of the planet. A second major limitation is the availability of suitable regions to modify. Many areas of the Earth have naturally high albedo, such as snow- or ice-covered regions and deserts. Covering these areas of earth with solar-reflective material would have little effect on the planetary albedo.

Mitigation Option: Cover low albedo portions of Earth with reflective coverage. Proposals include covering large areas of the ocean with highly reflective material, such as manufactured material, films, foams, biological organisms, or sea ice. Land-based proposals include use of reflective building materials, genetically-modified high albedo vegetation, or spraying vegetation with bright material [34, 39, 40].

Feasibility: Flannery et al. [30] calculated that covering 10% (or approximately 36 million km^2) of Earth's ocean with highly reflective material would compensate for a doubling of CO_2. This basic theoretical calculation alone illustrates the extensive surface coverage that would need to be constructed and maintained in order to mitigate warming. Other proposals to mitigate warming by boosting land albedo do not reach the target of offsetting a doubling of CO_2, but in combination could cause substantial cooling effect [19]. An important factor to consider in all cases is the potential diminishing reflectivity of the surface coverage under aging, which may require continued maintenance and/or larger area coverage to sustain the desired albedo. While the use of reflective surface materials in urban areas is a well-established method for combating urban heat islands, large-scale modification of Earth's albedo is currently restricted to the point of a theoretical proposal. The feasibility of implementing such large-scale modification and expected costs are unknown at this time.

Co-benefits and undesirable consequences: Large-scale modification of either land or ocean surface properties is bound to cause substantial disturbances to ecosystems and in regional and global hydrological cycles. In particular, blocking solar radiation to the ocean or land surface will likely reduce the amount of PAR available to photosynthetic organisms beneath the modified surface. Covering the ocean surface to the extent suggested by Flannery et al. [34] can also be expected to reduce the evaporation of water into the atmosphere, altering precipitation patterns to an unknown degree. While not an intentional change, reduced solar flux, due to the presence of a dense plume of air pollution particles over the Indian Ocean, appears to have weakened the Asian monsoon, leading to drought in Northern China and southern flooding [41]. To the authors' knowledge, co-benefits of large-scale surface modification beyond improving urban building materials are nonexistent.

9.8 Geoengineering Strategy: Increasing the Emissivity of the Atmosphere by Direct GHG Capture

A geoengineering strategy that involves changes to the thermal emissivity of the atmosphere would be constrained by the practical aspects of extracting GHGs at trace-level concentrations. Carbon dioxide is not the only GHG and, on a molecule-by-molecule basis, it is not the gas with the highest global warming potential (GWP). However, its atmospheric concentration makes it the most important of the anthropogenic GHGs. The anthropogenic greenhouse gases with higher GWPs, such as methane, nitrous oxide, and the hydrofluorocarbons, are present at concentrations that are lower by orders of magnitude [1] .

The capture of CO_2 at its point of emission – typically at very large combustion sources such as coal-fired power plants – is a strategy under serious consideration by energy systems analysts as a means of reducing further GHG emissions, while allowing the continued use of fossil fuel (Princiotta, this book). However, many anthropogenic sources of CO_2 exist that are not amenable to carbon capture and storage (CCS), such as fossil-fueled mobile sources, and those emissions related to forest clearing and agriculture. In addition to its limitations concerning the types of sources it can effectively control, carbon capture at the point of emission does not affect the existing excess concentrations of anthropogenic CO_2 in the atmosphere. Several geoengineering proposals offer approaches for reducing overall atmospheric CO_2 concentrations.

Carbon dioxide is readily removed from the atmosphere by natural chemical and biological processes. For example, naturally occurring alkaline minerals react with CO_2 when exposed to the atmosphere to form carbonate compounds. Photosynthesis, the process by which green plants synthesize carbohydrates, is the natural route of biological CO_2 sequestration. These processes seem suitably inexpensive and to have fewer negative environmental side effects than those involving the reduction of solar insolation. Nevertheless, CO_2 exists at parts-per-million concentrations in the atmosphere, implying the need for very large scale processing of the atmosphere to ensure a meaningful reduction in its atmospheric concentrations.

Basic thermodynamics predicts, and geochemical models have shown, that removal of atmospheric CO_2 by a terrestrial sink, such as tree farming or chemical weathering, will result in the out-gassing of CO_2 from the oceans [42]. Assuming humanity ultimately succeeds in eliminating its carbon emissions, the global oceans will require centuries to off-gas the excess dissolved CO_2 before the ocean-atmospheric system begins to approach thermodynamic equilibrium [1] . Therefore, the success of any direct capture project is not only tied to the very long-term stability of the method used to store the captured carbon once it is removed from the atmosphere, but the capacity for storing quantities equivalent to current CO_2 emissions, *plus* past emissions that have accumulated as dissolved carbon in the oceans.

The proposed techniques for direct capture, illustrated in Fig. 9.2, include altering the pH of the ocean surface, stimulating biological-photosynthetic sequestration at the ocean's surface or through reforestation or afforestation, or artificial chemical weathering projects that employ materials that react with and sequester CO_2. Unlike

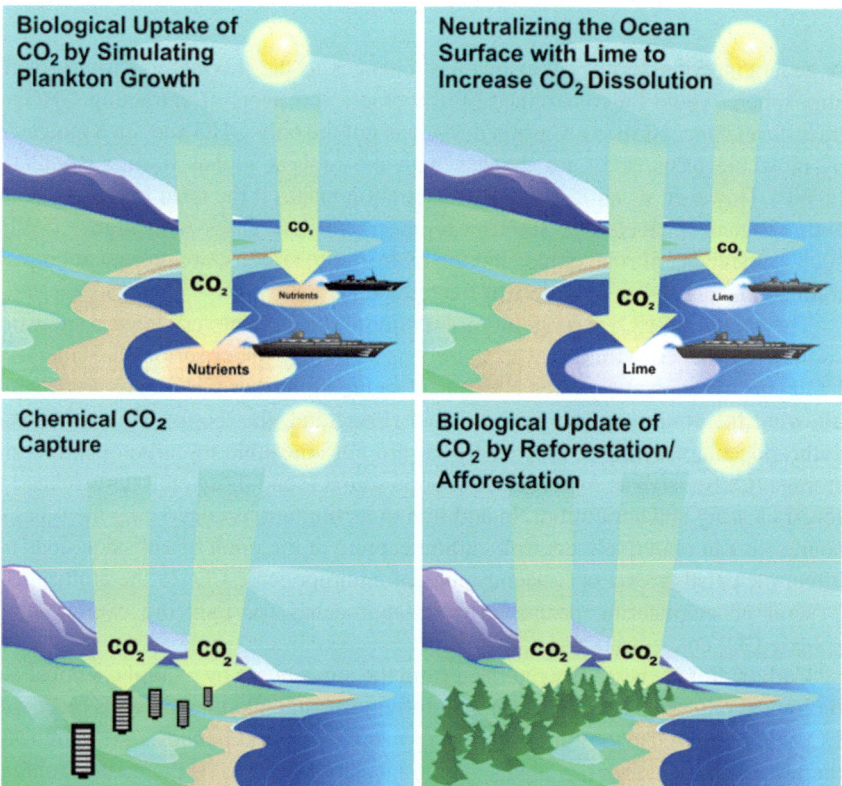

Fig. 9.2 Proposed approaches to the direct capture of CO_2 from the atmosphere: enhancing phytoplankton growth; reducing ocean surface pH to increase CO_2 dissolution; the construction of artificial "trees" that use alkaline chemical compounds to absorb atmospheric CO_2, and; large-scale expansion of managed forests

the geoengineering strategies involving changes in solar flux, the costs to implement direct GHG capture methods relying on components of the natural cycle are not easily quantified. Further exploration is required to determine the capacity of the natural carbon sinks that are suitable for engineered enhancement.

9.9 Ocean-Based Geoengineering Proposals: pH Modification and Feeding Phytoplankton

Among the natural carbon sinks in earth's biogeochemical system, the world's oceans play the dominant role in buffering changes in atmospheric CO_2 concentrations. Covering 70% of the earth's surface, the oceans contain approximately 50 times the carbon present in the atmosphere. The annual carbon flux between the atmosphere and the oceans is approximately 100 Pg (100 billion metric tons) [42]. The net uptake of CO_2 by the global oceans, under non-El Nino conditions, has been estimated to be approximately 2 Pg C year^{-1} (2 billion metric tons) [43].

Atmospheric CO_2 partitions into the ocean by dissolution and is, then, removed by either chemical ("solubility pump") or biological processes ("biological carbon pump") [44] . Amplifying these natural CO_2 sequestration processes by changing the chemistry of the ocean surface has been proposed as a geoengineering strategy.

9.10 Ocean pH Adjustment

The first step in the ocean solubility pump is the reaction between CO_2 and water (Eq. 9.2). Water and CO_2 combine to form a weak acid known as carbonic acid (H_2CO_3). Depending on the local pH at the ocean surface, some fraction of the carbonic acid molecules dissociates to form bicarbonate ion and hydrogen ion. Bicarbonate can then further dissociate to form carbonate and another hydrogen ion. This series of reactions is responsible for the increasing acidification of the world's oceans.

$$CO_{2(aq)} + H_2O \longleftrightarrow H_2CO_3 \longleftrightarrow HCO_3^- + H^+ \longleftrightarrow CO_3^{2-} + 2H^+ \quad (9.2)$$

Mitigation Option: Broadcast spreading of alkali compounds, such as limestone or soda ash, to neutralize the ocean surface, and amplify carbonate production [45, 46]. A substantial fraction of the carbonate that is formed in this manner is expected to deposit in deep ocean sediments that, according to geological observations, tend to remain stable for thousands of years.

Feasibility: Estimates of the quantity of CO_2 that can be sequestered using this approach must account for the variability in ocean surface temperature, wind speeds and phytoplankton, along with the uncertainties associated with quantifying the role of biology and ocean dynamics in carbon sequestration. The equilibrium distribution of CO_2 between the ocean and the atmosphere is highly sensitive to the temperature, alkalinity and salinity of surface waters [42]. CO_2 concentrations vary widely from season to season, by as much as 60% above and below the mean atmospheric concentration. Wind speed also substantially alters surface CO_2 concentrations [43]. Increasing the temperature of the ocean surface drives CO_2 out of solution and back into the atmosphere.

The biological pump competes with chemistry in the sequestration of CO_2. Whether CO_2 reacts and precipitates as carbonate, is utilized for coralline algae and corals for constructing shells, or is converted to organic carbon through photosynthesis, sequestration in ocean sediments depends upon the physical processes driving vertical, downward transport to the ocean floor, including both sedimentation and ocean currents.

Each of the factors described will influence the estimated mass of alkali mineral required for effective CO_2 sequestration. The costs associated with locating sufficient supplies of the mineral, mining, and transporting and then distributing it to the designated ocean regions cannot be estimated without first knowing the resource quantity required.

Co-benefits and undesirable consequences: Large-scale manipulation of ocean surface chemistry can be expected to affect the ocean food web, but the potential impacts have not been investigated. Depending upon local conditions, the response by the resident ocean organisms may be positive or negative.

9.11   Phytoplankton Fertilization

Marine phytoplankton play a very large role in the global carbon cycle, accounting for about 50% of the global biological (photosynthetic) uptake of CO_2 [47]. However, the efficiency of this process is sensitive to the availability of several essential macronutrients required by the various species comprising phytoplankton. Phosphorus, nitrogen, silicon, iron, and zinc are critical for the growth and reproduction of these organisms. In areas limited in one or more of these nutrients, fertilization of the sunlit euphotic zone with the limiting element can stimulate transient phytoplankton blooms to enhance CO_2 uptake.

Iron (Fe) is the most practical of these nutrients, from a geoengineering perspective, for stimulating phytoplankton growth and photosynthesis. Three regions of the world ocean – the eastern equatorial Pacific, the subarctic Pacific, and the Southern (Antarctic) Oceans – are extremely limited in nutrient Fe, an element critical for phytoplankton photosynthesis ([48] and references therein). Areas suitable for fertilization have very specific characteristics: they are up-welling zones that are limited in an easily obtained and dispersed macronutrient.

Mitigation Option: Continuous broadcast spreading of phytoplankton macronutrients in nutrient-limited regions of the world's oceans, stimulating the biological carbon pump.

Feasibility: With careful screening of Candidate Ocean regions for their suitability vis-à-vis local ocean currents, biological composition and prevailing meteorology, fertilization can result in sequestration of atmospheric CO_2.

The fraction of carbon bound by biological processes is dependent upon an array of factors. Phytoplankton are a heterogeneous lot, comprising diatoms, cyanobacteria, dinoflagellates and other types of algae – all photoautotrophic organisms at the base of the ocean food web. Each species has preferred living conditions, responds differently to fertilization, and utilizes CO_2 to differing degrees. The presence of predators (zooplankton) which can eat the phytoplankton increases the possibility that the carbon will be quickly recycled back into the atmosphere. Vertical advection of phytoplankton detritus is quite variable.

Calcification, the formation of insoluble forms of carbonate, is a necessary process for reproduction by coralline algae, an important species of phytoplankton present in coastal waters and the open ocean – and is highly pH dependent [49]. Acidity drives the chemical equilibrium away from carbonate to favor the formation of bicarbonate (see Eq. 9.2). The current excess of atmospheric CO_2 is driving the acidification of ocean surface layers. This suggests that the co-implementation of pH adjustment with Fe fertilization might be necessary for optimal CO_2 drawdown.

Assuming the infrastructure existed to permit ready dispersion of the appropriate nutrient, identifying the best candidate areas appears to require a great deal of exploration and *in situ* experimentation.

Co-benefits and undesirable consequences: Effects on the ocean food web can be expected. Phytoplankton feed zooplankton (krill) which, in turn, feed a number of larger ocean species. The effects could favor the growth of currently depleted fish populations, or could alter the food web to favor other, less desirable species [50]. Stimulating phytoplankton growth, sufficient to draw down a meaningful quantity of CO_2, increases the risk of overgrowth of harmful algae and the creation of dead zones that could deplete fish stocks. Another unintended side effect could be the production of N_2O – a gas with a high global warming potential – in the O_2-depleted depths below the fertilized euphotic zone [51]–[52].

9.12 Land-Based Geoengineering: Reforestation/Afforestation

Reforestation/afforestation has already been incorporated into current international efforts to create carbon "offsets," along with renewable energy and energy efficiency programs, and projects that qualify under the Kyoto Protocol Clean Development Mechanism. However, current implementation of forest-based carbon sequestration is limited to private enterprise engaged in selling carbon-offsets, and to some nations with commitments under the Kyoto protocol. As a geoengineering strategy, reforestation/afforestation would be far more extensively and systematically employed and managed to sustain a larger-scale drawdown – and sustained sequestration of CO_2.

Mitigation Option: Very large-scale reforestation of previously cleared land, reclamation of mine land, afforestation of marginal lands, agroforestry, forest preservation, and urban forestry has been proposed [53]. With investment in inventory management and monitoring, development of appropriate technologies and decision support systems, an estimated additional 100–200 Tg C/year of forest carbon sequestration is, for example, achievable within the US. New forestry technologies including improvements in thinning and better utilization of products from thinning, low-impact harvesting, optimizing rotation length, species or genotype selection, and forest biotechnology will be required to achieve significant drawdown [53].

Feasibility: The success of forestry for CO_2 direct capture and sequestration depends on a number of variables: harvesting and soil disturbance rates, future forest productivity, site characteristics, and the local environment. Quantification and verification of sequestered CO_2 requires measurement of the exchange of CO_2 between forests and the atmosphere – this is by no means trivial, given the number of individual carbon pools present in a forest ecosystem. An added complication is that the stability of forest carbon sequestration is vulnerable to natural disturbances. Stored carbon in forests can be unexpectedly released back to the atmosphere due to, for example, wildfires, and insect infestations.

Estimates of the feasibility of forest management as a mechanism for reducing atmospheric CO_2 levels must factor in the climate-induced changes that will occur in forest ecosystems, due to climate warming currently underway [1] . Warming can be expected to negatively affect forest productivity.

An important cost consideration is the requirement that forest management technologies, along with the necessary resources, must be available in a sustained manner to international carbon managers in order to ensure the success of this strategy at the global scale.

Co-benefits and undesirable consequences: Restored forestlands, and the cultivation of new-forested areas may be beneficial for wildlife and regional ecosystems. However, silviculture on marginal and reclaimed lands implies the need for extensive fertilization and irrigation, with coincident risk of water pollution, nutrient runoff impacts (such as the extensive dead zone in the Gulf of Mexico) and depletion of fresh water supplies.

9.13 Geoengineering with Chemical Weathering

One of the natural mechanisms responsible for stabilizing Earth's atmospheric CO_2 concentrations, on geologic time scales, involves the weathering of soil-based silicate minerals. As CO_2 concentrations increase in the atmosphere, climate warming affects weather patterns that, in turn, enhance the weathering of these minerals. Alkaline metal ions (calcium, magnesium) are freed from the soil and are washed into surface water bodies and the ocean, where they can participate in the sequestration of dissolved CO_2. Technologies are under development that would use this chemical approach to capturing CO_2 at the point of emission, particularly at power generation facilities [54]. One form of geoengineering that has been proposed involves the expansion of this approach to draw CO_2 directly from the atmosphere.

Mitigation options: Various alkaline materials (sodium hydroxide, amines), have been proposed for use in wet-scrubbing devices (artificial trees) that would be deployed on a massive scale or spread over agricultural fields and in forests (olivine), for CO_2 direct capture [55–57]. Cost estimates for the application of this technique vary widely, depending upon the chosen technology, estimated capital costs and operating assumptions, from \$15/t- CO_2 for olivine application, \$30/t-CO_2 for one sodium hydroxide-based wet scrubbing approach, to as much as \$500/t-CO_2, using other currently available technologies [56, 57].

Feasibility: Deployment on the scales needed to meaningfully affect atmospheric CO_2 concentrations appears to be decades away. Critics have argued that the energy demands for operating the proposed scrubbers will substantially reduce the net carbon sequestered [58] . For efficient high volume-high applications, it appears that scrubbing technologies are not yet in reach.

Co-benefits and undesirable consequences: Assuming stable depositories for the products of the scrubbing process could be found, this method may have fewer potentially negative environmental consequences than other geoengineering approaches. Any impacts related to the long-term storage of the carbon sequestered would depend upon the chemical form in which the CO_2 is sequestered, as well as where and how that material is stored. As with windmills for energy production, however, the installations used for processing the atmosphere may be viewed as having a negative visual impact on the landscape.

9.14   Geoengineering Prerequisites

No examples exist of previous attempts to manage environmental problems at a global scale using technology-based strategies from which we may learn about potential pitfalls. However, the lessons we've learned from the unintended global-scale changes resulting from human activities and technologies apply to the use of geoengineering. Anthropogenic greenhouse gas-induced climate change is one example. Other unintended changes include long-term damage to Earth's protective stratospheric ozone layer through the use and emission of synthetic chlorofluorcarbons and the global ecosystems damage induced by wide-spread use of persistent organic pesticides and herbicides (persistent organic pollutants or POPs). We have learned from the consequences of these activities that the benefits of a new technology must be balanced against potential environmental costs.

The implementation of any of the proposals discussed here will be costly and their capacity, on an individual basis, for mitigating climate warming without significant, negative global consequences appears limited. However, the implementation of an ensemble of these proposals, at scales that limit harmful downstream impacts, may yet be an option available to the global environmental policy, if circumstances demand it. However, before serious attempts are undertaken to directly manage the earth's climate system, an integrated risk assessment/risk management decision framework inter-comparing the management options available, is needed.

9.15 An Integrated Risk Assessment/Risk Management Decision Framework

Prudent practice in the engineering of large, public projects calls for rigorous, systematic evaluation of the risks associated with the project design and, in the case of large civil engineering projects, environmental impact statements. The potentially far-reaching consequences of any project meant to alter the global climate system imply the need for a multi-stage assessment, enumerating the immediate and down-stream implications associated with changes in the major climate system components.

This risk framework should inter-compare each climate management-policy option, e.g., candidate geoengineering activities, other mitigation actions such as those described elsewhere in this book, versus inaction. Quantification of the probabilities, uncertainties and related costs (or benefits) of immediate, near- and long-term negative and positive impacts, along with the cost of implementation and long-term maintenance of each option will be needed. The global climate and ecosystems research communities must be called upon to populate this framework with the details necessary to enable well-founded decision-making by international climate management decision-makers.

Furthermore, any risk management framework, especially in the context of climate change, must be periodically re-evaluated to account for the change in risk with time. For example, new information, if aggressively sought, may meaningfully alter the relative uncertainties associated with possible courses of action; changing human behavior alters the potential impacts of inaction. Finally, the complete risk framework will not only incorporate the essential questions regarding the feasibility and risks associated with implementation of the various climate-modifying schemes, but also the risks associated with inaction.

Of course, with even the best possible decision framework, surprises may still be possible. The enormous effort underway within the international scientific community to understand the consequences of human activities has yielded the valuable insight that the earth's climate system is complex and laden with positive and negative feedbacks [1] . Perturbations to one component of the climate system propagate to the other components in unpredictable ways. Nevertheless, the responsible course dictates that due diligence be applied to the assembly of an integrated risk assessment/risk management framework and that such a framework be utilized in decision-making with such far-reaching consequences.

9.16 Support for Research

Geoengineering, as an approach for managing climate change, is still very much in the realm of theory. The best-studied option in the scientific literature – the proposal to inject sulfate aerosols into the stratosphere – has only a small handful of modeling studies that predict the climate response along with the potential side effects. Continued vigorous research, experimental as well as theoretical, on the candidate geoengineering approaches is required to quantify and narrow the existing uncertainties, and to establish the probabilities for negative, down-stream side effects.

Targeted research may quickly improve the informational basis upon which to weigh the risk of action to the risk of inaction. The fundamental questions that must be answered for each approach include:

- What is the expected timeframe and magnitude of the global climate response to the proposed geoengineering action?
- What are the immediate, ancillary (positive and negative) consequences associated with the action?
- What are the long-term risks associated with sustained implementation of the strategy for greenhouse warming mitigation, e.g., from years to decades?
- What are the immediate and long-term financial, material, and personnel demands for this strategy?
- How scalable and reversible is this strategy?
- Is the risk of unintended consequences distributed evenly or would some regions of the world pay a higher price? The complexity of the climate system ensures that there will be winners and losers if climate is intentionally managed using the schemes proposed to date. How will the losers be identified and compensated?
- How do each of the suggested geoengineering strategies compare on a cost/tonne-CO_2 basis? The current literature regarding global climate modeling studies to evaluate the effects of reduced solar flux and the proposed projects to extract CO_2 from the atmosphere does not provide a common metric for comparing costs.

9.17 Conclusion

Across the centuries, humans have adapted to the challenges presented by nature through the development of increasingly sophisticated technologies. In all cases, these technologies have altered the environment to some degree. As the global human population has grown into the billions, these impacts are now so substantial that the resources, including the services provided by earth system upon which humans depend for their livelihood, health, and well-being, are in jeopardy. Arguably, the most important of the earth's systems, global climate, is changing in a manner that is difficult to predict – due to human activities.

Seen as a delaying tactic or as a possible "last resort" action to limit catastrophic climate change, geoengineering is receiving increasing attention among academic, government, and commercial groups. The claim that geoengineering may become necessary as a means of mitigating the consequences of our continuing global dependence upon fossil fuels reflects the philosophy that the earth's climate system can be explicitly and successfully managed with human technology. This philosophy has yet to be evaluated in a *systematic* analysis that integrates existing knowledge of the earth system response to significant perturbations of the type and scale that would result from geoengineering projects. Given the scale of the potential effects of attempts to directly manage global climate, the international community must engage in just such a thorough analysis.

Should the global community conclude that some form of geoengineering may be both feasible and effective at rapidly mitigating climate warming; the immediate and long-term costs of implementation of the proposed projects must be evaluated. Cost, in the context of a geoengineering project assessment, must include an accounting for the complex, potentially undesirable side effects of altering large components of Earth's climate system. It will be these costs, in addition to the costs of implementation and maintenance of the scheme, that must be weighed against the assessed costs of the alternatives, such as aggressive GHG mitigation policies, as well as the costs associated with continued reliance on fossil fuels without strict GHG emissions controls, our "business as usual" status.

The existing scientific literature is, however, insufficient for the task of judging the efficacy and global impacts of current geoengineering proposals. Therefore, the essential first step that must be taken in advance of serious consideration of large-scale geoengineering is targeted, thorough scientific evaluation of the proposals under discussion.

While the degree to which geoengineering should or can be employed as a climate management strategy is the subject of debate, the scientists participating in these discussions agree that aggressive action towards cutting GHG emissions should not be supplanted by geoengineering. The earth system may not recover from the sustained implementation of some of the proposed strategies. Geoengineering, at best, should be regarded as a short-term strategy in the long-term management of Earth's climate system.

References

1. Intergovernmental Panel on Climate Change (2007). Climate change 2007: the physical science basis. Contribution of Working Group I to the Fourth Assessment Report of the Intergovernmental Panel on Climate Change. Cambridge University, Press, Cambridge/New York
2. Raupach MR et al (2007) Global and regional drivers of accelerating CO_2 emissions. Proc Natl Acad Sci USA 104:10288–10293

3. Canadell JG et al (2007) Contributions to accelerating atmospheric CO_2 growth from economic activity, carbon intensity, and efficiency of natural sinks. Proc Natl Acad Sci USA 104:18866–18870
4. Marchetti C (1977) On geoengineering and the CO_2 problem. Clim Change 1:59–68
5. Keith DW (2000) Geoengineering the climate: history and prospect. Annu Rev Energy Environ 25:245–284
6. Broad WJ (1988) Scientists dream up bold remedies for ailing atmosphere. New York Times, New York
7. Broad WJ (2006) How to cool a planet (Maybe). New York Times, New York
8. Discovery Chanel (2011) Project Earth, http://dsc.discovery.com/tv/project-earth/project-earth.html. Retrieved March 4, 2011
9. Economist T (2008) A changing climate of opinion? The Economist
10. Svoboda E (2008) The sun blotted out from the sky. Salon.com
11. Walsh B (2008) Future revolutions: geoengineering. Time Magazine, Time Inc
12. NASA (2010) Earth fact sheet. http://nssdc.gsfc.nasa.gov/planetary/factsheet/earthfact.html. Retrieved 13 June 2010
13. Nissen KM, Matthes K, Langematz U, Mayer B (2007) Towards a better representation of the solar cycle in general circulation models. Atmos Chem Phys 7:5391–5400
14. Foukal P, Frohlich C, Spruit H, Wigley TML (2006) Variations in solar luminosity and their effect on the Earth's climate. Nature 443:161–166
15. Solanki SK, Fligge M (1998) Solar irradiance since 1874 revisited. Geophys Res Lett 25:341–344
16. Crutzen PJ (2006) Albedo enhancement by stratospheric sulfur injections: a contribution to resolve a policy dilemma? Clim Change 77:211–219
17. Teller E, Hyde T, Wood L (2002) Active Climate Stabilization: Practical Physics-Based Approaches to Prevention of Climate Change." National Academy of Engineering Symposium, Washington, D.C. April 23–24, 2002. Lawrence Livermore National Laboratory, Livermore, CA. US DOE Report Number: UCRL-JC-148012
18. Govindasamy B, Caldeira K (2000) Geoengineering Earth's radiation balance to mitigate CO_2-induced climate change. Geophys Res Lett 27:2141–2144
19. Lenton TM, Vaughan NE (2009) The radiative forcing potential of different climate geoengineering options. Atmos Chem Phys Discuss 9:2559–2608
20. Early JT (1989) Space-based solar shield to offset greenhouse effect. J Br Interplanet Soc 42:567–569
21. Angel R (2006) Feasibility of cooling the Earth with a cloud of small spacecraft near the inner Lagrange point (L1). Proc Natl Acad Sci USA 103:17184–17189
22. National Academy of Sciences, Committee on Science, Engineering and Public Policy, Panel on Policy Implications of Greenhouse Warming (1992). Policy implications of greenhouse warming: mitigation, adaptation, and the science base. National Academy Press, Washington, DC
23. McCormick MP, Thomason LW, Trepte CR (1995) Atmospheric effects of the Mt Pinatubo eruption. Nature 373:399–404
24. Charlson RJ et al (1992) Climate forcing by anthropogenic aerosols. Science 255:423–430
25. United Nations (1976) Convention on the prohibition of military or any other hostile use of environmental modification techniques
26. Springer S (2008) No time to be under a cloud. The Boston Globe
27. Khan E et al (2001) White paper: response options to limit rapid or severe climate change, assessment of research needs. U.S. Department of Energy, Washington, DC
28. Latham J (1990) Control of global warming? Nature 347:339–340
29. Latham J (2002) Amelioration of global warming by controlled enhancement of the albedo and longevity of low-level maritime clouds. Atmos Sci Lett 3:52–58
30. Latham J et al (2008) Global temperature stabilization via controlled albedo enhancement of low-level maritime clouds. Philos Trans R Soc A Math Phys Eng Sci 366:3969–3987

31. Salter S, Sortino G, Latham J (2008) Sea-going hardware for the cloud albedo method of reversing global warming. Philos Trans R Soc A: Math Phys Eng Sci 366:3989–4006
32. Budyko MI (1977) Climatic changes (Translation of 1974 Russian edition). American Geophysical Union, Washington, DC
33. Stern DI (2005) Global sulfur emissions from 1850 to 2000. Chemosphere 58:163–175
34. Flannery BP, Kheshgi H, Marland G, MacCracken MC (1997) Geoengineering climate. In: Watts RG (ed) Engineering response to global climate change. CRC Lewis Publishers, Boca Raton
35. Rasch PJ, Crutzen PJ, Coleman DB (2008) Exploring the geoengineering of climate using stratospheric sulfate aerosols: the role of particle size. Geophys Res Lett 35:6
36. Hofmann DJ, Oltmans SJ (1993) Anomalous Antarctic ozone during 1992: evidence for Pinatubo volcanic aerosol effects. J Geophys Res 98:18,555–18,561
37. Tilmes S, Muller R, Salawitch R (2008) The sensitivity of polar ozone depletion to proposed geoengineering schemes. Science 320:1201–1204
38. U.S. Environmental Protection Agency (2008) Reducing urban heat islands: compendium of strategies. U.S. Environmental Protection Agency, Washington, DC
39. Hamwey R (2007) Active amplification of the terrestrial Albedo to mitigate climate change: an exploratory study. Mitig Adapt Strateg Glob Change 12:419–439
40. Akbari H, Menon S, Rosenfeld A (2009) Global cooling: increasing world-wide urban albedos to offset CO_2. Clim Change 94(3–4):275–286
41. Ramanathan V, Agrawal M, Akimoto H, Aufhammer M, Devotta S, Emberson L, Hasnain SI, Iyngararasan M, Jayaraman A, Lawrance M, Nakajima T, Oki T, Rodhe H, Ruchirawat M, Tan SK, Vincent J, Wang JY, Yang D, Zhang YH, Autrup H, Barregard L, Bonasoni P, Brauer M, Brunekreef B, Carmichael G, Chung CE, Dahe J, Feng Y, Fuzzi S, Gordon T, Gosain AK, Htun N, Kim J, Mourato S, Naeher L, Navasumrit P, Ostro B, Panwar T, Rahman MR, Ramana MV, Rupakheti M, Settachan D, Singh AK, St Helen G, Tan PV, Viet PH, Yinlong J, Yoon SC, Chang W-C, Wang X, Zelikoff J, Zhu A (2008) Atmospheric brown clouds: regional assessment report with focus on Asia. United Nations Environment Programme. Nairobi, Kenya
42. Raven JA, Falkowski PG (1999) Oceanic sinks for atmospheric CO_2. Plant Cell Environ 22:741–755
43. Takahashi T, Sutherland S, Sweeney C, Poisson A, Metzl N, Tilbrook B, Bates N, Wanninkhof R, Feely R, Sabine C (2002) Global sea air CO2 flux based on climatological surface ocean pCO2, and seasonal biological and temperature effects. Deep Sea Research Part II: Topical Studies in Oceanography 49(9–10):1601–1622
44. Gruber N, Gloor M, Sara E. Mikaloff Fletcher, Scott C. Doney, Stephanie Dutkiewicz, Michael J. Follows, Markus Gerber, Andrew R. Jacobson, Fortunat Joos, Keith Lindsay, Dimitris Menemenlis, Anne Mouchet, Simon A. Muller, Jorge L. Sarmiento, and Taro Takahashi (2009). Oceanic sources, sinks, and transport of atmospheric CO2 . GLOBAL BIOGEOCHEMICAL CYCLES, VOL. 23, GB1005, doi:10.1029/2008GB003349
45. Harvey LDD (2008) Mitigating the atmospheric CO_2 increase and ocean acidification by adding limestone powder to upwelling regions. J Geophys Res Oceans 113
46. Kheshgi HS (1995) Sequestering atmospheric carbon-dioxide by increasing ocean alkalinity. Energy 20:915–922
47. Arrigo KR (2007) Carbon cycle – marine manipulations. Nature 450:491–492
48. Pollard RT, Salter I, Sanders RJ, Lucas MI, Mark Moore C, Mills RA, Statham PJ, Allen JT, Baker AR, Dorothee CE, Bakker MA, Charette SF, Fones GR, French M, Hickman AE, Holland RJ, Alan Hughes J, Jickells TD, Lampitt RS, Morris PJ, Nédélec FH, NIelsdótir M, Planquette H, Popova EE, Poulton AJ, Read JF, Seeyave S, Smith T, Stinchcombe M, Taylor S, Thomalla S, Venables HJ, Williamson R, Zubkov MV (2009). Southern Ocean deep-water carbon export enhanced by natural iron fertilization. Nature 457:577–581.
49. Gattuso JP, Buddemeier RW (2000) Ocean biogeochemistry – calcification and CO_2. Nature 407:311–313

50. Trick C, Bill B, Cochlan W, Wells M, Trainer V, Pickell L (2010) Iron enrichment stimulates toxic diatom production in high-nitrate, low-chlorophyll areas, PNAS 107(13):5887–5892
51. Suntharalingatn P, Sarmiento IL, Toggweilcr JR (2000) Global significance of nitrous-oxide production and transport froin oceanic low-oxygen zones – a modeling study. Global Biogeochem Cy 14(4):1353–1370
52. Law CS (2008) Predicting and monitoring the effects of large-scale ocean iron fertilization n marine trace gas emissions. Mar Ecol Prog Ser 364:283–288
53. Birdsey RA, Pregitzer K, Lucier A (2006) Forest carbon management in the United States: 1600–2100. Journal of Environmental Quality 35:1461–1469
54. Metz BD, Davidson O, De Coninck H, Loos M, Meyer L (eds) (2005) IPCC special report on carbon dioxide capture and storage. Cambridge University Press, Cambridge/New York, p 442
55. Keith DW, Ha-Duong M, Stolaroff JK (2006) Climate strategy with CO_2 capture from the air. Clim Change 74:17–45
56. Lackner KS (2002) Carbonate chemistry for sequestering fossil carbon. Annu Rev Energy Environ 27:193–232
57. Schuiling RD, Krijgsman P (2006) Enhanced weathering: an effective and cheap tool to sequester CO_2. Clim Change 74:349–354
58. Velasquez-Manoff M (2007) Giant carbon vacuums could cool Earth, Christian Science Monitor, 4/19/2007. http://www.csmonitor.com/2007/0419/p13s01-sten.html. Retrieved 4 Mar 2011

Chapter 10
Research, Development, Demonstration and Deployment Issues in the Power Sector*

Bruce Rising

Abstract In this chapter we explore the challenges in developing and deploying technology for mitigation of CO_2 emissions associated with power generation. Past successes with controlling other pollutants (notably SO_2) provide insight as to the difficulty of extrapolating those successes to applications for carbon capture and control. Technology innovations that have yet to reach commercial fruition are noted, but for the near term we can make effective use of commercial processes readily available and achieve significant reductions in carbon emissions. These reductions can be obtained by fuel switching, efficiency upgrades introduced fleet-wide, and expanded use of lower CO_2 emitting technologies, all of which should be done in parallel with a robust R&D program to develop new technologies for extraction of CO_2 from exhaust gases or strategies for fuel decarbonization.

10.1 Introduction

Here we consider the research and development needs that can effectively mitigate human induced impacts on climate change, while balancing the competing requirements of energy supply and security. For mitigation of CO_2 emissions from coal combustion, technical innovations are discussed for Integrated Gasification Combined Cycle (IGCC), oxygen-fuel combustion, and CO_2 reduction through de-carbonization or post-combustion extraction. But there are other effective alternatives that can be implemented in the near term (<5) years using state-of-the technologies available today. Such readily available technologies have broad appeal because they can yield positive effects with minimum technical and economic risk.

---

*The findings included in this chapter do not necessarily reflect the view or policies of the Environmental Protection Agency. Mention of trade names or commercial products does not constitute Agency endorsement or recommendation for use.

B. Rising (✉)
Siemens Energy, Orlando, FL, USA
e-mail: bruce.rising@siemens.com

F.T. Princiotta (ed.), *Global Climate Change - The Technology Challenge*,
Advances in Global Change Research 38, DOI 10.1007/978-90-481-3153-2_10,
© Springer Science+Business Media B.V. 2011

Technologies such as renewables, nuclear power, and high efficiency facility upgrades represent the tools that are can yield effective rapid, near-term reductions in CO_2 emissions.

To avoid the potential for adverse effects from global warming, a major increase in R&D resources is proposed, since cost effective carbon capture technologies are currently not available nor are they being developed at a pace consistent with proposals to offset potential anthropogenic effects on global climate. The methods (or solutions) to be implemented from an intensive R&D campaign could differ geographically since the relationship between uptake and release of GHG varies on a global scale. For developed economies, market-based energy pricing on carbon bearing fuels is likely to continue to be a tool for affecting man-made CO_2 emission, at least in the short run. As the price of fuel increases, consumers will respond by taking actions that will minimize the economic impact on them: they will use less fuel, or maximize the use of available fuels (e.g. drive cars with greater fuel efficiency). Not everybody can alter their energy consumption rates equally.

As noted in Fig. 10.1, energy efficiency usage varies widely across the globe, and so will the opportunities for saving energy and reducing fossil fuel consumption. Energy consumption per capita is typically greater in those countries situated at extreme latitudes (e.g. Canada), where winter seasons can be extreme.

In subsistence economies, where biomass burning is a significant source of CO_2, the price of oil and gas is likely to have minimal impact on emissions of combustion related GHG. Rather, for agrarian based economies, anthropogenic emissions are more strongly correlated to land conversions (forestland to cropland, cropland to grazing, etc.). These two extremes suggest that there is an enormous gulf between the desire to mitigate anthropogenic CO_2 released to the atmosphere through combustion and the tools, both technical based and policy based, that can be employed to reach this goal.

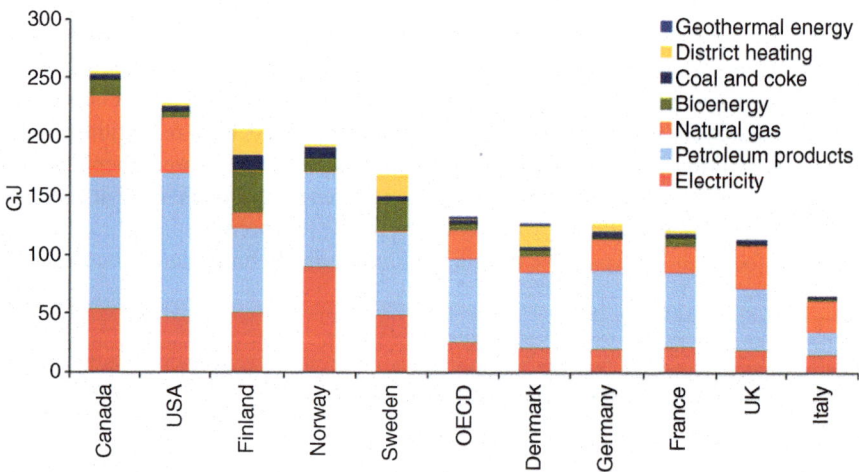

Fig. 10.1 Per capita energy consumption in OECD countries

Focusing on large, stationary combustion sources of CO_2 alone, there are more than 8,000 major point sources around the world, emitting 60% of global CO_2. Coal-fired power plants make up more than half of those point sources. Coal fired generation, including use of coal in residential applications, releases approximately ten billion metric tonnes of CO_2 annually, 38% of the world's yearly CO_2 emissions, leading many to the conclusion that coal represents all or most of the carbon challenge. However, the global economies depend heavily on such coal-fueled facilities for the bulk of electricity generation. While the U.S. is slowly adding to its complement of coal-fired power plants (15,000 MWe under construction as of late 2009), with some exceptions, there are virtually no new coal additions in the planning stages. Meanwhile, China has accelerated their pace of re-industrialization. Just since 2001, China placed on order in excess of 345,000 MWe of coal-fired generation (these are vapor power cycle units larger than 300 MWe in size), roughly equal to the entire fossil coal capacity of the United States.

Lacking significant technical innovations, carbon reduction proposals made in today's context will require adopting power intensive environmental control methods to a fleet of generation equipment not designed to accommodate them. As of now, carbon capture technologies (those capable of processing most of the exhaust gases in a power plant) have yet to reach full commercialization, except for the very few applications where a small fraction of source carbon dioxide is recovered for very specific purposes. Those examples demonstrate that CO_2 recovery from exhaust gases is certainly possible [1]. They also reveal the impracticality of recovering a substantial fraction of CO_2 emissions using that same technology; and that is partly the point, finding methods of recovering CO_2 will be research intensive—finding ways to make it affordable will require more than a few full-scale demonstration projects.

It will take years for carbon capture and sequestration to reach a level of maturity (and cost effectiveness) that permits widespread industry acceptance. However, there are CO_2 mitigation steps that can be implemented much quicker. Near term emission reductions can be readily achieved by consuming less fuel (or improving fuel utilization), switching to less carbon intense fuel sources, or increased use of renewables. Perhaps just as compelling, there is emission control technologies that are commercially available today that could begin to correct the impacts of other pollutants (NO_x, CH_4, particulates, etc.) believed to play a major role in global climate change.

10.2   Power Generation Technology Status

The United States inventory of power generation includes a stable of over 1,000 GW of generation, with half of the total generation (2,000 billion kW h in 2008) produced from coal. This comprises about one-third of the total power generation capacity globally, with expansion continuing over the next few decades. It took almost 50 years to build this infrastructure, a generation portfolio based essentially on coal, natural gas, and nuclear fuel (CO_2 emission factor rankings for the major power

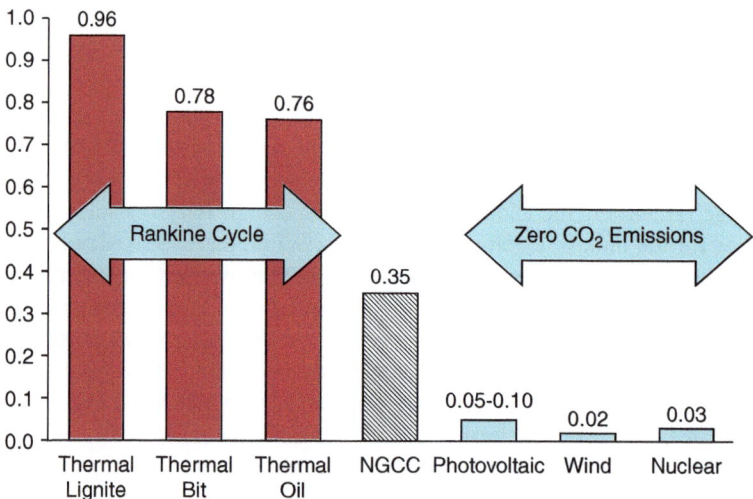

Fig. 10.2 CO_2 emission index (kg/kWh, including construction/manufacturing) for various generation types and fuel sources

generation sources are shown in Fig. 10.2). By 2008, 50-year-old power plants, operating at relatively poor heat rates, were still generating electricity, many of them using coal or heavy oil feedstock. In the United States, regulatory pressure for hazardous air pollutants could result in the replacement of approximately 30,000 MWe of residual oil fired capacity, and in all likelihood replacing it with gas-fired generation that is less CO_2 intensive. Because of its sheer size, replacing the power generation system in its entirety would require decades, if indeed replacement (or upgrading) of the entire system is the correct (or most immediate) strategy to pursue.

10.3   Upgrading of the Power Sector

How long would it take to change the complete generation landscape? Let's consider how long it takes to implement high-efficiency off-the-shelf technology currently available. Thermal steam plants, using coal as fuel, produce the bulk of electricity in the US and worldwide. These plants may require a minimum of 4 years to construct, although schedules in the United States can be as long as 6 years. Add time for permitting (and the likelihood of legal challenges on various fronts) along with shakedown trials to the construction and negotiation phases, and the whole process could take up to 8 years before a facility begins producing power for dispatch. In those parts of the country where water cannot be extracted from rivers or underground aquifers, water restrictions can either delay the project significantly, or kill it entirely. Even restarting a nearly complete nuclear plant, such as *Watts Bar*, could require 2–4 years. These are just a few of the challenges that new project development face in the United States. Such hurdles can be sufficiently daunting that some developers opted to construct facilities just beyond the US border to reduce the administrative barriers and reduce total construction costs.

Switching to an entirely new power generation approach is expected to add significantly to the lead-time. Front End Engineering Design (FEED) work may require 1 or 2 years to complete for such a new product design, a product, (or process) that is expected to be a substantial departure from power generation designs commercially available now. By 2009 a number of FEED studies to evaluate various carbon capture technologies were underway in the US. The punch line here is that deploying a new and untested technology to replace one that works reasonably well will take many years to bring to service, and many more before the installed base of power generation and energy production begins to take an appearance markedly different from what it does today. In 2008, the retirement rate of existing power generation in the United States was less than 1%. Meanwhile the total capacity growth rate is projected to be over 1.5%, requiring on the order of 15–20 GW of new generation annually. Under this scenario it could be decades before the older power generation fleet, as well as the newest fleet additions, could be replaced, or upgraded with completely new environmental systems capable of removing CO_2 gases.

Globally, the prospects are more challenging. China is adding about two new 500 MWe coal power plants each week [10]. In 2008, China reached a landmark point by releasing more CO_2 from fossil fuel burning than the United States (even with a smaller inventory of privately owned motor vehicles, a fleet that is rapidly increasing and could grow to release as much CO_2 to the environment as any other segment of the economy). Both China and India are on an accelerated rate of CO_2 production, making it difficult to envision a rapid turnaround in CO_2 emissions and even in Italy, new fossil coal plants are under construction in a race to meet growing demand.

10.4 Architecture of the Power Generation System

The backbone of the power generation system (as well as nearly all transport systems) is rooted in one of two basic thermodynamic cycles: The Rankine, or vapor power cycle, and the Brayton, or gas power cycle (combined cycle, combustion turbine, steam turbine boiler, and nuclear components depicted in Fig. 10.3). Improving the efficiency of these cycles is one of the key steps to achieving CO_2 reductions in the near term. Together, these two cycles are responsible for about 80% of the electricity generated in the United States. In France, where nuclear power comprises 70%+ of the electricity, the Rankine steam cycle is still the primary basis of power generation. Since these two processes are the heart of most power generation, improvements here are probably the best place to start with a strategy for carbon mitigation at the source. Improving efficiency, especially of those facilities that have the greatest carbon intensity, requires that the energy conversion processes

2008 U.S. Operating Plant Capacity by Technology
Type

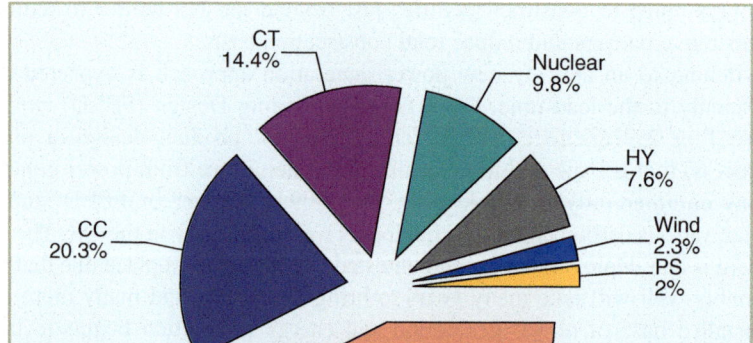

Fig. 10.3 Breakdown of the US power generation fleet (*PS* photovoltaic solar, *HY* hydroelectric, *CT* combustion turbine, *CC* combined cycle, *STB* steam turbine boiler) (Source: www.snl.com)

operate at extreme thermal conditions, and that typically means operating at higher temperatures and pressures (or both) to achieve maximum thermal efficiency. Today, the United States coal turbine fleet (thermal steam) currently averages about 33% efficiency when weighed across all coal types and all plant designs.

10.4.1 Fossil Steam Plants: Rankine Cycle (Sub-critical Steam)

The Rankine cycle (or vapor steam cycle) represents the basic energy conversion cycle providing 70% of the electricity in the US economy. Fundamentally, it is the conversion of water into steam, followed by the extraction of work, which today is usually accomplished by conversion into electricity (or in the case of cogeneration, steam for industrial processes). The first design was developed in the late 1712, yielding a cycle efficiency that was less than 2%. It has taken almost 300 years to reach an efficiency of over 40%, the current state-of-the art for fossil steam plants. Today's facilities operate with steam temperatures on sub-critical plants range from 537°C/1,000°F to 565°C/1,050°F.

In the United States, 50% of the electricity is generated using coal as the primary source of energy, releasing about 1 tonne CO_2/MWh. The energy extracted from coal is converted into heat, converting water to steam, and fed into a rotating steam turbine. Overall efficiency of a sub-critical unit (one that operates well below 3,200 psig) is on the order of 10,000 Btu/kW h, about 33% (even less if the fuel contains a large quantity of water as is found in some coals such as lignite or Powder River Basin coal).

10.4.1.1 Rankine Cycle (Supercritical)

Increasing the steam pressure and temperature increases the amount of energy carried with the steam, and more energy can then be extracted by the steam turbine. Operating at supercritical conditions (above 3,200 psig), the cycle efficiency can increase by an additional 5% (depending on plant size, complexity, and fuel).

Supercritical steam units are more efficient to operate, and if fuel costs are high, they are likely to be the preferred thermal system. The first commercial supercritical plant in the United States went into service in 1957 (operated by AEP), the second in Philadelphia by 1961. Both units were relatively small by today's standards—125 and 325 MWe respectively. By any standards, these units were groundbreaking, and pioneering, and not without early problems.

By the mid-1960s, about half of all United States units under order were super-critical, suggesting that the initial teething problems with the new technology had been largely solved. Some of the key problems were materials related, leading to decisions to derate the units to improve reliability. As of 1986, 15% of the United States fleet operated at supercritical steam conditions.

In Japan and Europe, supercritical units were continuously built throughout the 1980s and onward. It took years after the first wave before problems with materials, startup, controllability, and reliability could be solved. In the United States much of the supercritical fossil steam construction came to a halt in the 1980s partly, because of the introduction of nuclear power generation and partly because the low cost of domestic coal didn't demand the more extreme operating conditions.

By 2001, a new wave of supercritical fossil steam plant construction began. Nearly all new power plants under consideration will be supercritical (that is, operate above the critical conditions for steam). Worldwide, more than 400 supercritical plants are in operation.

10.4.1.2 Ultra-supercritical

Operating steam and gas turbines at higher pressures and temperatures improves efficiency. Pushing thermal plant operating conditions beyond 5,000 psig and 700°C may allow operation at efficiencies approaching 50%+, about 10% points higher than current commercial supercritical power plants. There are extreme material challenges to be overcome as the metals weaken at the high temperatures and pressures proposed (perhaps as high as 4,000–5,000 psig). It is another facet of the research effort to find materials capable of operating long term under these conditions.

Once metal temperatures exceed 620°C (1,150°F) threshold, steels comprised of substantially iron (ferrite) must be replaced with a more exotic, expensive, difficult to weld and machine nickel-based alloys. Some of the materials proposed for these extraordinary conditions do not conduct heat as well as those used in less severe conditions (i.e., conditions found in the most advanced boilers and steam turbines produced today). At some of the more extreme conditions (approaching 700°C/1,290°F), very expensive alloys such as Inconel or Hastelloy may be required. This will require the high-pressure steam piping, steam turbine blades, vanes, and casing to be fabricated from materials to handle these conditions, materials that are also quite expensive.

Specialty materials will be required for ultra-high efficiency plants still on the drawing boards, and high efficiencies will be needed if fossil plants must adopt energy intensive post combustion emission controls for carbon capture. All known carbon capture processes degrade unit performance substantially. Therefore, it is critically important to push the core efficiency of the power station to its maximum so that the efficiency penalty is offset once additional controls are put in place. If the loss in efficiency is not offset, it will require a larger facility, consuming even more carbon fuels to produce the same energy production as a unit without controls. Consider it from a different perspective: If we use existing established boiler technology, it would require mining and consuming even more coal, since a significant part of the energy in the fuel would be required to operate carbon capture system.

We clearly have the tools to model, design, and even construct facilities that can produce power while extracting CO_2. The impacts on power and efficiency, and ultimately on the cost to the consumer, have yet to be resolved. This is one reason why improving plant efficiency (increased MWh per tonne of CO_2) or using a lower carbon fuel (natural gas) provides a convenient short-term solution to a very challenging problem. A major benefit of selecting the NGCC option is that one realizes both a lower emission rate of CO_2 and improved efficiency, significantly higher than any competing technology. This benefit can be quickly lost by adding emission controls to a source that already exhibits a substantially lower CO_2 emission profile compared to any other fossil-fueled power generation. The natural gas option only has long-term viability if supplies of gas are robust. Fortunately, that appears to be the case. Supplies of unconventional gas (gas from shales and coal bed methane) and offshore supplies are making the gas option more viable going forward. While the United States appears to lead the world in the "unconventional gas expansion," these geological formations are not unique to North America and can be found in virtually every continent, making this a long-term global option. Thus what we have learned in solving the issue here in the US could readily be adapted to other major economies wrestling with the same problem.

10.4.1.3   Combined Cycle

Combining both a Rankine and Brayton cycle yields the largest efficiency step increase in the last few decades. While steam cycles have topped out in the range of 40–45%, and air cycles are close to that, combining the two cycles is capable of producing an overall cycle efficiency in the range of 55–60%. Since the vast majority of these units operate on natural gas as the primary fuel, the mix of cycle efficiency and fuel carbon content is the primary reason for CO_2 emission levels achieved in the 0.3–0.35 tonne/MWh range. Reaching these performance levels also required decades of research and development. In the 1970s operating temperatures for the gas turbine component were limited to 760°C, with efficiency in the range of 28–30%. Using state-of-the-art materials, design methods, and cooling strategies, operating temperatures now are in the range of 1,300–1,400°C. This evolved over a period of almost 40 years, a process that is still on going.

10.5 Post Combustion Emission Controls

Flue Gas Desulfurization (FGD) technology provides an example of both the scale and challenge of solving an environmental control comparable to carbon capture. Virtually all of the technologies for carbon capture will require conditioning the gas stream to eliminate the presence of sulfur (to prevent degradation of the solvents). FGDs were first installed in England to combat the environmental impacts associated with SO_2 from coal burning. Using a process developed by ICI (Imperial Chemicals Industries), three large fossil steam plants were retrofitted early in the twentieth century, although work on FGDs came to a halt during the Second World War [2]. The first full-scale FGD in the US was installed in 1967 on a plant operated by Union Electric. Despite a several decade head start on the technology, many of the "first" adopters experienced significant challenges in plant reliability with what was perceived as a relatively simple emission control adaptation. A large number of fossil units constructed in the 1960s and 1970s, would later be retrofitted with FGDs.

Some of the early problems, which would take a decade or more to resolve include:

- Efficiency: FGDs inherently reduce the efficiency of a plant. Plant performance deteriorated by as much as 3–5% with the addition of the post combustion sulfur control. One impact is the need to reheat the flue gases to obtain buoyancy; and this usually meant extracting thermal energy from the base unit. At the time the retrofit program began, there was sufficient excess generator capacity that the loss was not a problem in terms of sufficiency of electricity supply.
- Corrosion: Early systems were plagued by corrosion on a large scale. Materials ranging from rubber to gunite pastes to stainless steel and titanium were tried over the years. Today, many "wet" designs have settled on using a tile and mortar coating that can stand up to the corrosive environment. In the early days, materials selection was more or less haphazard.
- Dewatering: Water is used to cycle sorbent and product through the system, and saturates the stack with water vapor in the process. This requires handling and disposition of the enormous quantities of water consumed in the process.

- Waste disposal: At the largest plants hundreds of tons of calcium sulfite and sulfate are generated each day. It was a major challenge to process these enormous quantities of waste. Entire valleys were filled with the sludge from FGDs before a substantial recycle route was used. Eventually the sulfate would become the primary product. Today, nearly 30% of all gypsum wallboard is recycled from sulfur captured at power plants.[1]

At some plants, those original FGDs are no longer in place, having been replaced by better designs that evolved out of this process. It would take until the late 1980s before most of the technical challenges (plugging, erosion, corrosion, water handling, additive usage, etc.) would be resolved, creating a jumping off point for the next phase that would become Title IV (the Acid Rain Program). This would allow for the technology expansion indicated in Fig. 10.4, although the expectations from the CAIR rule may come through a different regulatory pathway. The US EPA played a major role in deploying pilot scale demonstrations, catalyzing technology transfer via reports, papers, and symposia. This helped resolve many of the operability and environmental impacts that were revealed during these early years.

After the Clean Air Act of 1990, Title IV of that act initiated a large retrofit campaign to capture SO_2 at the point of origin—the coal-fired power plant. Title IV was able to capitalize on the technical innovations and experience from nearly 20 years of trial and error. Even with that experience and knowledge, the graphic hints at the sheer length of time it can take to complete such a process—some 25 years. Earlier attempts to mitigate the impact of acid rain by treating rivers and lakes

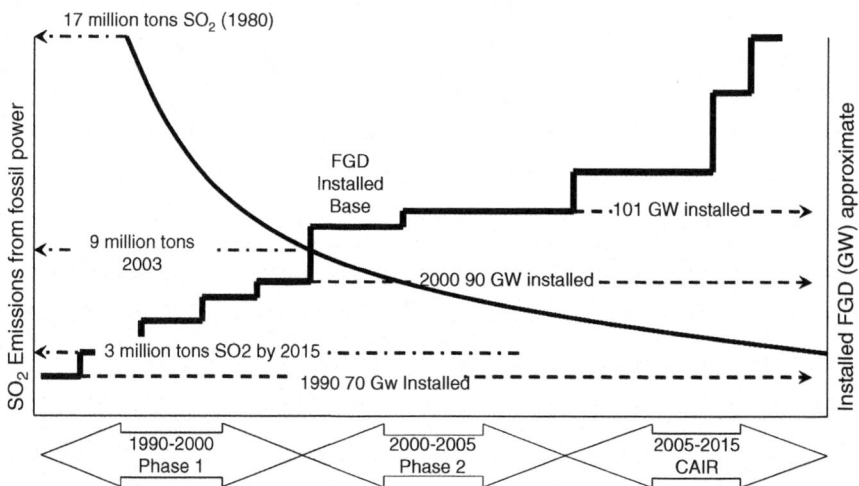

Fig. 10.4 Growth of installed FGD in the United States

---

[1] The 2008 coal ash pond failure at the TVA plant in Kingston, Tennessee injected a new level of complexity in the concern over coal waste disposal.

had proved futile. Addressing the problem at the source achieved significant SO_2 emissions reductions, as well as solving problems of acid deposition on streams and lakes, particulate haze, corrosion of infrastructure (including highways and bridges), and eventually resulting in significant health benefits. While these benefits were achieved with reduced plant output and degraded performance efficiency (and higher electricity rates), the benefits are widely acknowledged to have been worth the investment.

10.6   Carbon Capture and Control Systems

Unlike the SO_2/acid rain program, there is no established process in place to mitigate CO_2 at the scale being proposed. The technology that is well understood now and could be deployed is expensive (nearly doubling the price of a gas fired facility) [3] and is accompanied by substantial energy penalties [4] (the primary application is treatment of pressurized gases to pipeline specifications in the Oil & Gas industry). The United States Department of Energy (DOE) has outlined objectives to advance carbon capture technologies that can achieve 90% reduction while minimizing the electricity cost increase to 35% [5]. That's a major improvement from the estimated 85% price increase expected to result with the current stable of technologies.

Lacking a significant technical breakthrough, CO_2 reductions of 50–80 % are more likely only to be achieved with a shift towards non-fossil fixed energy resources, use of more renewable sources, or just reduced usage of fossil fuel. The benefit here is that these technical choices are commercially available, but it would not be possible to switch the entire generation mix quickly. Alternatively, on the time scales under consideration (next 20 or 30 years), there may be serious concerns about the adequacy of some fossil reserves, which could invoke an entirely different dynamic that would mitigate the rapid increase of CO_2 emissions—the price of energy.

There is working experience base for recovery of CO_2 from energy sources. In 1999, approximately 20 carbon capture systems were operating globally, most using the *Fluor-Economine®* process which is based on liquid amine solvent. By 2008, this base had expanded to 24 plants. To yield a high quality product (usually high purity CO_2), with minimal contamination of the solvent, CO_2 is extracted from a relatively clean exhaust (often natural gas is the fuel, although there are coal plants operating with this process). The product is a commodity for food and beverage processes [6]. In most of the commercial CO_2 exhaust recovery processes, the total volume fraction of flue gas processed is small, usually less than 1%; small enough to limit the parasitic loads on the overall plant.

In these operating plants the carbon capture process here is not tightly integrated into the power production phase of operations, so that failure of the CO_2 recovery process has virtually not impact on the overall power generation capability of the unit. In addition, the carbon capture component of these facilities is not installed as environmental control, but rather devoted to product recovery (CO_2).

As a starting point, this can serve as a template for future process designs, but it is a very long way from a full-scale commercial offering to broadly treat emissions from power generation sources.

10.6.1 Chemistry of Capture

Full-Scale carbon capture systems are expected to be the most complicated new environmental control technology introduction we have undertaken. The reasons are subtle, but ultimately based on the simple fact that carbon dioxide is not a very chemically reactive gas. In power generation, carbon capture (or recovery) is likely to demand a large chemical processing plant installed at the outlet/exhaust of a power plant, substantially raising the cost of what has been a relatively inexpensive source of energy. Power demands (or the loss of equivalent generation output) of 100–400 kW h/tonne of CO_2 capture are expected.

The current stable of post-combustion carbon capture systems differ primarily in choice of solvents and the energy required for regeneration. Amine-based chemicals like monoethanol amine (MEA) rapidly absorb CO_2 over a range of pressures. It is at higher pressures where volumes of gas are reduced due to compression where the process becomes most economic. Because the solvent is also relatively inexpensive, it has become an industry standard for treating natural gas (which is usually processed at pressures well above 1,000 psig). This experience of treating pressurized gas is one of the reasons for considering these solvents in an IGCC application. Capturing the CO_2 at pressure could significantly reduce the capital equipment expense compared to a facility attempting the same feat near atmospheric conditions, but that's a big part of the problem—post combustion gases are not at high pressure, but rather at atmospheric conditions.

Amine-type solvents are widely used in the gas processing industry where natural gases are sweetened (i.e., sulfur is removed) by extracting both CO_2 and H_2S with solvents. This experience has led to MEA being the solvent of choice for most of the power plants considering carbon capture today. MEA binds tightly with CO_2 tightly; incurring a generation loss of 200–400 kW h for each tonne of CO_2 recovered [7]. Ammonia-based solvents previously mentioned are being explored for post combustion exhaust gas treatment because much less energy is required to separate the bound CO_2 from the solvent, with the added benefit that the carrying capacity of ammonia is greater than that of amine solvents. There is a give and take here, because ammonia's volatility demands very low operating temperatures. This too will require large energy expenditure to reduce the exhaust gas temperature, an energy penalty that is not easily missed in the accounting of plant output.

Compared to existing technologies, based on what we know today, carbon capture systems will be much more complex than the emission control systems designed to reduce emissions of priority pollutants (SO_2, NO_x, etc.). Conceptually, the idea is to use design features that include readily available solvents (either liquids or solids) and to duplicate the desirable features of a post combustion control system such as

Flue Gas Desulfurization (FGD). In contrast, CO_2 specific controls will be required to process over 100 times the mass of material as the next largest emission control system (FGDs), while attempting to chemically treat a gas that is almost 1,000 times less reactive.[2] It will require significant technical innovation to overcome five orders of magnitude in reactivity with an economically viable process. This is a critical point missed in most comparisons of post combustion environmental control technologies. The opportunities for innovation and the risks of failure are both enormous, suggesting that it could be incredibly difficult to determine in advance what the most effective method for CO_2 mitigation would, or could, be. Twenty-five years from now, a power generation system equipped with CO_2 separation capability may be entirely different from anything proposed today or constructed within the next decade, placing the initial wave of such facilities in the category of large-scale research facilities, with the potential of full-scale obsolescence in less than 10–15 years. We are still exploring the technology demonstrations: One large utility is experimenting with chilled ammonia for carbon capture, although an initial public announcement for a commercial CO_2 capture plant will be based on a more conventional amine application. Oxy-fuel demonstration plants are continuing to be studied, although none has really made it to a stage of commercial availability. Numerous university laboratories are experimenting with even more developmental processes like chemical looping.

In addition to the process primary components, there are hidden environmental features that have not been widely discussed. Usually these aspects fall into the category of "unintended consequences." For example, water availability is expected to be a critical factor. This, during a period when supplies of water are expected to be declining, not increasing. A recent study by North American Electric Reliability Council (NERC) suggested that reducing the open loop cooling associated with much of the existing thermal fleet (about 400 GW) could result in a reduction of capacity on the order of 50 GW [8]. Inclusion of a post combustion carbon capture plant is expected to increase water consumption to operate the regeneration steps in the process. Where water access is limited, or in short supply, the carbon capture process may be forced rely on less efficient mechanisms for heat transfer (e.g. air cooled condensers or hybrid systems), and potentially increase both the size and cost of the facility.

As we understand it today, simply adding a carbon capture system to an existing power generation unit will substantially lower its efficiency. Studies suggest in the range of 20–30% reduction in plant output [9]. Meanwhile, demand for energy (MWhrs) and capacity (MWe) is increasing virtually everywhere [10]. Summer peak demand increases annually, as does winter peak demand. While it may be counterintuitive, adding this level of emission controls to the fossil fleet would require mining and burning more fossil fuels to produce the same energy output. Identifying these unanticipated responses will be a daunting task for policy makers more accustomed to promoting local, regional, or national based incentives.

---

[2] Assuming a gas concentration of SO_2 of 1,000 ppm at 10% CO_2.

However challenging it may be to capture CO_2 from a fossil plant, it is even more difficult to place that burden on gas-fired generation based on the Brayton (gas turbine) cycle. First, the natural gas combined cycle typically has a much lower CO_2 content in the exhaust than is found in a fossil steam plant (4% vs. 10–12%). The lower concentration implies a reduced uptake rate of CO_2, suggesting that the overall capture system (spray towers or beds) would be physically larger (even though there would be less cleanup of sulfates required prior to carbon extraction). The additional back-pressure on the turbine will exact a large energy penalty.

Perhaps a more effective strategy using a gas turbine would be to increase the gas fired capacity of the entire system, possibly even by converting coal into substitute natural gas, another technology that is already well proven. As noted earlier, the CO_2 emission profile of a gas-combined cycle is about one-third of a fossil steam plant (refer to the lower emission profile noted in Fig. 10.2), a 60% reduction with no penalty associated with add-on emission controls.

10.7 Current Options for CO_2 Reduction

From most perspectives, an obvious place to consider capturing CO_2 is at the source. This is the approach successfully used for SO_2, NO_x, and particulate emission controls, and one could extrapolate that such an approach is applicable to CO_2. At the exhaust stack, the concentration of the CO_2 is greatest, and it should be most amenable to recovery. There are several thousands of concentrated sources in contrast to hundreds of billions of CO_2 receptors in the environment. CO_2 presents its own unique chemical challenges. As already noted, CO_2 is not as reactive as other pollutants (SO_2 or SO_3, for example). It's not easily transitioned to another chemical state. Some have suggested taking CO_2 and converting it to carbon and oxygen, without realizing that energy accounting demands nearly as much energy released from the fuel is required to produce such a conversion. If one-tenth of 1% of CO_2 was required to be converted in such a way, it might be viable but not on the scale of CO_2 reduction being proposed. Some proposals call for 80–90% reduction in CO_2 economy wide. Capturing this much CO_2 from stationary and mobile sources would be as challenging as attempting to alter the global temperature directly. An alternative to capture, or an intermediate step, might be to reduce the CO_2 emissions, or eliminate them entirely from the source. Various technical innovations to achieve this were extensively reviewed only a few years ago [11].

10.7.1 Fuel Switching

Fuel switching is perhaps the easiest choice for reducing CO_2 emissions. Using a fuel with less carbon and more hydrogen in the fuel results in a reduce CO_2 emission profile. In this respect, carbon is the mechanism of delivering hydrogen to the end-user. Fuel switching is not always as simple as it may appear. Fossil coal plants can easily substitute burning natural gas instead of coal or oil, but they do so at efficiencies far less than the most efficient combined cycles or even the simplest high compression diesel engines. Most combined cycles and peaking units are gas fired, providing limited options for fuel substitution. If fuel switching to natural gas is the option, that option should be used only in the most efficient systems available. That usually implies natural gas combined cycles, combined heat and power systems, or some comparable cycle. Because of its cost (and its high inherent fuel quality), the use a high quality fuel like natural gas in a low efficiency system is to be discouraged. Coal, on the other hand, is one of the least expensive fuels because it is so abundant. The reason for the low cost is that fuel is typically of poor quality (due to the presence of water and ash), as well as a high carbon content. The critical technology focus for coal utilization has been to find ways of extracting the thermal energy within the coal, while minimizing the release of pollutants to the environment, or damage to the hardware.

Another fuel switch option is the use of biomass (directly or blended with a fossil fuel). Biomass is derived from CO_2 (and water vapor) absorbed from the atmosphere. Burning biomass may be one approach to reducing the CO_2 emission factor from a conventional fueled power plant. In fact, there are quite few power plants around the world currently operating on biomass as a feedstock, although most of these are under 100 MWe in size (in the US, the average is closer to 50 MWe).

10.7.2   Efficiency Enhancements

In parallel, another option to be considered for CO_2 mitigation involves something as mundane as improving the efficiency of processes we already use. Whether the fuel is gas or coal, efficiency improvements always work to reduce CO_2 emissions. Market forces are already inherently at work here, albeit for different reasons (demand for commodities such as oil and natural gas sometimes quickly outpaces supply). Fuel-efficient hybrid vehicles and diesel engines are preferred over less efficient, larger, and heavier motor vehicles, particularly where fuel price is an issue. Airlines will optimize flight schedules to reduce fuel consumption, which commensurately reduces CO_2 emissions. In many cases, an aircraft may be replaced entirely with a more fuel-efficient version—saving money but also mitigating CO_2 emissions.

There are significant efficiency enhancements which can yield CO_2 reduction in power generation. The results noted in Fig. 10.5 are based on using the 2004 average US fossil fleet (coal) performance figures. By improving overall fleet efficiency by just 1% (from the average 33% to 34%) could reduce CO_2 emissions by about 60 million tonnes per year. Making minor changes to a fleet of 1,400 fossil units is no small feat, but clearly the opportunities for quick returns on reduction of CO_2 are possible, and almost irresistible not to mention the savings related to reduced fuel consumption.

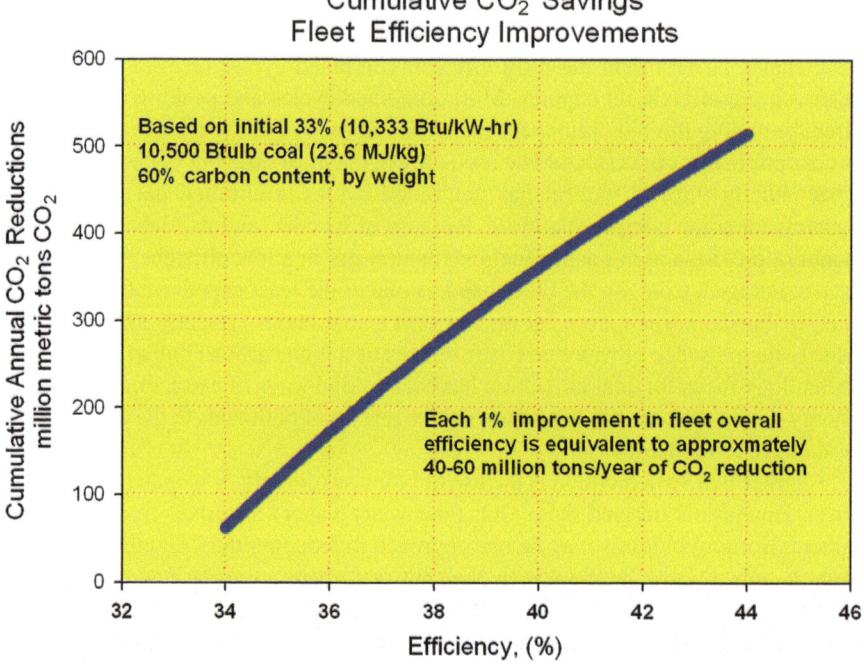

Fig. 10.5 CO reduction potential from fossil coal utility fleet with efficiency enhancements

Going further, we could consider operating the entire United States coal fleet at or near supercritical conditions where thermal efficiencies of 40% HHV (Higher Heating Value) are considered the benchmark for new fossil steam plants. This could yield 400+ million tonnes of CO_2 reduction, which is about half the CO_2 increase in the United States economy since 1990 and what would work in the US markets could be readily deployed globally. A similar exercise with natural gas shows that efficiency improvements reduce CO_2, but because the CO_2 emission factor for natural gas is so much lower initially, the reductions obtained in the gas fleet are not nearly as large as with the fossil coal fleet.

Even though new fossil plants can be rated at 40% efficiency and greater, the fully integrated system (generation, transmission, and distribution, storage, etc.) does not perform at this level all the time. One reason for this is that load and demand must always be in balance, and since demand changes throughout the day (and the season), the system must respond accordingly. A plant operating at maximum efficiency at 4 PM may only be required to support a much-reduced load at night, and invariably this comes with reduced efficiency as the output drops. Virtually all peaking plants will be offline in the late hours. The net result is that the yearly average of the fleet drops from the design maximum continuous rating of the individual components that make up the fleet.

Conflicting plant operational requirements can make improvements in efficiency difficult, especially in older facilities where open loop cooling (for the condenser heat exchanger) has been the norm. A recent study by Tetratech examined the impact of switching from open loop cooling for a conventional thermal power station. Cooling water extracted from offshore can be reduced by 94+%, but for a gas-fired thermal station in California it also worsens the plant heat rate by 1.5–1.7%, and reduces total plant output by about 2.7%. The importance of what might appear to be a minor issue shouldn't be minimized: changing the facility operating characteristics to solve one problem (reducing the amount of water consumed for cooling) comes at a price of reduced efficiency and output, a tack that takes the facility away from improving the CO_2 emission footprint [12].

10.7.3 CO_2 Extraction

Current CO_2 extraction methods are variants of gas separation—isolation of one gaseous component from a blend of other gases or vapors. If it were easy to separate CO_2 from exhaust gas products, and if the market price of the gas justified the separation, a large industrial market for the product would probably already exist. But as it turns out, it is not easy and industrial processes attempting to extract CO_2 don't survive without some unique set of financial conditions or incentives. The largest CO_2 capture plant in the world is only 2,000 tonnes per day, and most of the plants are extracting CO_2 from a relatively clean natural gas exhaust flow. In contrast, a 500 MWe fossil plant can generate 500 tonnes of CO_2 per hour and any new plant will come equipped with SO_2, NO_x, and PM controls already placed in the exhaust flow path.

Overcoming the energy penalties associated with gas separation could be considered a monumental technology prize. Extracting CO_2 from a fossil steam plant may exact a penalty as high as 400 kW h per tonne of CO_2. For coal, each MW h produces a tonne of CO_2 [13], thus, this level of energy penalty could result in efficiency decrease as high as 40%. Optimizing plant performance and selecting solvents more appropriate for the process conditions (e.g. physical solvents for pressurized systems, chemical solvents for atmospheric systems) could reduce that energy penalty, but it would probably still require 20–25% of plant output. Such a large reduction in performance is completely counter to the first principle of carbon reduction—improve plant efficiency to reduce the fuel consumption (which on its own merits results in a significant CO_2 emissions reduction).

Gas separation, as a technology improvement, is not limited to extracting CO_2. Oxygen production is critical to some of the most novel technologies being explored—gasification and oxy-fuel. Gasification, one of the most important steps in an IGCC facility, consumes so much additional power that the overall net efficiency may even be less than that of a modern supercritical fossil steam plant—in the range of 38–41% net efficiency. Reducing that load, or changing the cycle in such a way that oxygen separation is achieved in an entirely different way, could recover a significant amount of that loss.

10.7.4   Cycle Design

It has taken centuries to improve on the 2–4% efficiencies of the very first steam cycles. Both the nuclear and fossil steam cycles are not very different from the very first designs some 300 years ago. Water is boiled at one part of the cycle (where the fuel is consumed) and condensed in another cluster of hardware (with the optimal efficiencies occurring with the lowest condenser temperatures). Cycle efficiencies are found in the range of 30–40% net efficiency for vapor power cycles (plant auxiliary equipment, including emission controls hardware can quickly degrade the plants output and overall performance). Combine the vapor and air cycles together, and efficiencies reach almost 60%. Pushing cycle performance to higher levels is a serious materials and process design challenge. One thought might be to consider changing the cycle completely, and this approach is central to the idea of Chemical Looping.

Table 10.1 summarizes how these technologies compare with and without the ability to chemically isolate the CO_2 component, starting with the most widely used technologies (fossil steam), and ending with consideration of technologies not yet on the market (e.g. Chemical looping and Oxy-Fuel). Chemical looping comprises a series of controlled oxidation-reduction reactions, with pathways that incur lower entropy losses than encountered with the combustion process. If this technically challenging process could be deployed, power systems might reach operating with efficiencies in the range of 50–70%, and without the requirement for specialized carbon separation technologies (or air separation). However, the most advanced processes are only on the drawing boards, or using laboratory scale reactors. Demonstrations using calcium as the working medium are underway at universities on the kilowatt scale [14].

10.7.5   Design Challenges

In a fossil steam or combined cycle power plant, the steam's primary function is to drive a turbine to generate electrical power. However in a carbon capture plant, steam

Table 10.1 Technology requirements for possible future power generation options[a]

	2008: No CO_2 separation			Future: with CO_2 separation[b] (coal feedstock)			
	SCPC	NGCC	IGCC	IGCC	SCPC	Oxy-Fuel	Chemical looping
Efficiency, %, LHV	42–44	58%+	40–42	30–35	30–35	40–45	50–75%
Gasifier	No	No	Yes	Yes	No	Yes	No
ASU for oxygen	No	No	Yes	Yes	No	Yes	No
Solvent recovery of CO_2	No	No	No	Yes	Yes	No	No

[a] Efficiencies derived from various sources [1, 4, 7, 11]
[b] The term "separation" is used as opposed to capture to indicate that the CO_2 can be isolated from the other flow streams. Compression not included

will also be siphoned away to regenerate the solvent, separating the CO_2 as a gas (although at much higher concentration than it was extracted from the exhaust gas). Most of the energy penalty is "paid-for" with low-pressure steam. Steam that should be used to generate electricity is instead sent to regenerate solvent. This situation creates a major design quandary: If the steam turbine is designed to handle the smaller fraction of the total flow (because some will be shifted to the regeneration loop), what happens to the steam when the carbon capture system shuts down? Is it vented to atmosphere or by-passed and sent to a condenser? If so, should the condenser be sized to handle additional steam in case of a failure, a feature that makes the plant that much more expensive. Another option is to design the turbine to handle the occasional additional steam flow and a larger capacity generator. This scenario implies that both a larger steam turbine and generator would be installed, but operated at reduced load while the regenerator is in operation. This too adds to the cost of the plant and the ultimate cost to the consumer.

10.8   Innovations for Carbon Capture

Finding a technical solution to a problem that is as broad in scale as CO_2, and climate change in general, while simultaneously maintaining a secure energy supply, is challenging indeed. Some of the technologies in the pipeline are improved versions of older approaches, albeit vastly improved over their predecessors. Others are completely new, and in that regard, will require significant time and effort to successfully validate them.

10.8.1   Gas Separation

Gas separation is one of the great success stories in industry. However, separating carbon dioxide from other gases at the scale proposed, especially other combustion gases, presents a monumental challenge. One technical solution that is needed is a method of separating gases with minimal energy penalty. If one or more of the gases exhibits properties that are substantially different from the others, it greatly simplifies the problem. Nitrogen and carbon dioxide are relatively similar in their properties and are both relatively inert. As a result, a significant amount of energy must be expended to separate them. (Hydrogen is somewhat different because its relatively small molecular diameter allows it to diffuse rapidly through membranes). Areas of research need include:

- Enhanced Solvents: Improve the efficiency of CO_2 uptake, reduce the parasitic energy losses found in the current stable of chemical agents
- Membranes: Develop permeable or semi-permeable membranes to selectively diffuse CO_2 across the membrane.

- Oxy-combustion: Reduce the pollutants emitted when burning low quality fuels. Produce an exhaust gas rich in CO_2 to permit direct sequestration.
- Solid Sorbents: Develop solids that can absorb CO_2, comparable to that found in liquid solvents. Carbonates are a likely choice. Like the enhanced solvents, reduce the energy penalty for solvent regeneration.

These studies have already begun, but they are quite a long way from commercialization.

10.8.1.1 Solvents for CO_2 Extraction

The current stable of solvents is liquid amines, such as monoethanol amine (MEA) or Methyl-Diethanol Amine (MDEA). They are effective in recovering CO_2, but their use results in a substantial decrease in operating efficiency. These solvents may be considered first generation solvents and can degrade plant performance by up to 12% points. Continued investigation of the traditional stable of solvents shows that there is still some room for improvement. One approach is to use the solvent as a multi-pollutant control; it doesn't reduce the energy burden to zero, but it's another incremental step in the research process [15].

Improving the performance (i.e., maximizing the facilities final overall efficiency) could be achieved by using solvents that require less energy to release the CO_2 from the solvent. These are considered second-generation solvents; chemicals that bind the CO_2 less tightly than current solvents or these results may be obtained with chemical additives that can accelerate the rate of absorption of CO_2 by the solvent. One of the more intriguing materials to come along is a unique matrix that has the capacity to absorb CO_2 readily, even at ambient conditions. These are called Metal Organic Frameworks, or MOF's.

10.8.1.2 Metal Organic Frameworks

Like catalysts, it is the surface and its morphology that affect the performance of Metal Organic Frameworks (MOF). These compounds are porous crystalline materials composed of metal clusters that are organically linked together. First noted in 1999, they exhibit unusually high surface areas. While activated carbon can reach surface area of 300–1,000 m^2/gm, MOF materials have been found with three to five times larger. They have unique properties that can be exploited in gas separation applications such as extraction of CO_2 from exhaust gases. One framework, MOF-177 reportedly could contain 320 volumes of CO_2 per unit volume of solid sorbent [16]. Beyond CO_2, these organic structures can also be used to increase the energy density of hydrogen carriers. A team at UCLA led by chemistry Professor Omar M. Yaghi identified an organic framework that can store up to 7.5 wt % of hydrogen with a volumetric capacity of 32 kg/m^3 at 77 K [17], with a unique framework that exhibits a surface area of over 5,000 m^2/gm. The chemical manufacturer BASF demonstrated that MOF's can also be used to increase the energy density of methane stored in canisters at high pressure. Methane has greater market options if the fuel energy density is increased beyond what is available from compressed natural gas, and it creates new opportunities for utilization of this fuel with the lowest carbon content.

10.8.1.3   Membranes

Some of the properties noted in MOF's are also found in membranes. Membranes preferentially diffuse one species over others. For molecules with small molecular diameters and high diffusivity, like hydrogen or helium, simple materials such as latex and rubber can be used to effectively separate the lighter gas from air. However, the molecular size of CO_2 is large relative to the gases normally found with it. Membranes with high selectivity for CO_2 could reduce both the energy penalty and the capital costs for carbon capture.

10.9   Materials and Process Engineering

Materials engineering is a broad subject that includes virtually ever aspect of R&D required to address the issues at hand. More specifically in this section, we are concerned with specialty alloys and metals used to handle gases or fluids at very high temperatures and pressures.

10.9.1   High Temperature Materials

Virtually all of the technical innovations in power conversion and energy storage are expected to face serious materials/metallurgy limitations. Materials breakthroughs on several levels are required to bring to market some of the most potent technologies. Vapor cycle systems that can operate above 700°C and 35 MPa are but one example. Achieving efficiency levels of 45%, as a minimum, will require applications of specialty materials. Most of the supercritical plants built today use ferritic steels (with costs less than $1,000 per tonne of steel, see the top of Fig. 10.6). To reach ultra-supercritical conditions (those beyond 620°C) will require materials not widely used in the vapor power cycles nickel-based materials that can reach 70x the cost. Also revealed in Fig. 10.6, current technology (or perhaps better described as current affordable technology) has stayed below the critical 620°C threshold for a reason—cost being one of the biggest factors.

   Many of these specialty materials are extensively used in gas turbines; hence, there is a good knowledge base to start with. However, gas turbines have several key attributes that make the use of specialized metallurgy possible—for example,

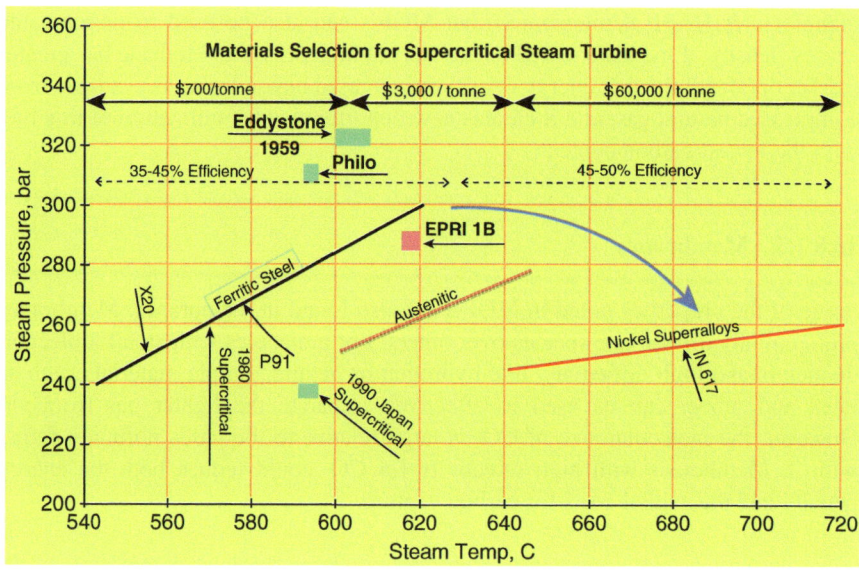

Fig. 10.6 Material requirements for power generation. At temperatures above 620°C, material costs increase rapidly

convective cooling of hot section components that are much smaller in size. This approach to thermal management is not practical in the vapor power cycle. While the highest-pressure ratio gas turbines operate at 20–30 atm, vapor power cycles are expected to operate in a range five to ten times greater to achieve the desired efficiency objectives. Specialty alloys will be required to continue to improve the thermal performance of the most widely used thermal cycles—the Rankine and Brayton cycle.

There are serious barriers to continuous operation above the 620°C limit for ferritic steels. Material properties such as bulk modulus, melting points, eutectics, conductivity need to be mapped out. Because of these limitations in the available materials, much of the world's steam fleet operates well below the 620°C limit, and nearly all of it operates below metal temperatures of 565°C.

Perhaps even more challenging than employing the right materials is to find materials that are affordable. Nickel based alloys could cost as much as 70x the current price of ferritic steels. HP and IP rotors weighing 30–40 tonnes would become very expensive if they were to be manufactured in the same configurations as today's high performance gas turbine units. Economic factors, namely the high price of super-alloys, will play a dominant role in the design approach of these advanced cycles. Piping headers and tubing would have to be made from similar materials, and require a greater level of skill in welding and assembly. All of these are acting in direction of substantial increases in plant costs, even above the seemingly inflationary rates for construction costs being experienced in 2007–2008.

10.9.2   Energy Storage and Transmission of Power

Materials development can also play a significant role in the transmission and storage of power. Nearly 10% of the power generated at the source is lost in transmission to the end-user. Stated another way, the power generation base must be approximately 10% larger in order to overcome the losses. Material improvements that reduce these losses would be the equivalent of building new construction, which translates to about 100 GW of installed capacity.

Superconductivity offers one pathway to minimize energy losses through resistive heating in a metal or conductor. While we've known about the superconductivity effects in metals since its discovery in 1911, we've not been able to expand this understanding much beyond the laboratory. Initially superconductivity was only found at temperatures below 30 K, a temperature that required liquefied hydrogen or helium to reach. It was not until 1986 that superconductivity was observed in the laboratory at 77 K, making it practical to use liquid nitrogen as a coolant. It would be 1993 before that temperature reached 138 K using a ceramic material comprised of thallium, mercury, copper, barium, calcium, and oxygen [18].

The first superconducting power transmission system was brought on line in 2008 in Long Island, almost a century after the discovery superconductivity. While it was a feat of engineering, so were the costs. For 2,000 ft of conducting cable, $58.5 million in capital was expended or about $30,000 per foot—$154 million per mile. By comparison, the most expensive gas pipeline built in the United States in 2007 was less than $10 million per mile; and the average is typically less then $4 million per mile.

Reducing the thermal losses associated with energy transmission can yield significant gains. As noted previously, the energy loss between the point of generation and use is about 10%, and that translates to about 150 million tonnes of CO_2 per year for the US alone. If CO_2 were assigned a market value of $50 per tonne, the savings would be equivalent to $7.5 billion. Of course, the thermal losses can never be eliminated, but there is the possibility of reducing the resistive losses, perhaps by 3–4%.

Several years ago, IBM researchers suggested a dual approach to energy and power. Energy, in the form of electricity, would be delivered through a superconducting grid. Power, in the form of hydrogen, would be delivered in the same grid, while it would also be used as a coolant to maintain the superconductivity effect. Currently, a scan of the literature reveals only a limited number of technical papers on this subject. However, such an innovative concept may only find its way to markets when substantial development and deployment have been undertaken [19].

Materials performance improvements may offer one of the paths forward to reaching high energy density, while at the same time providing high output power. It's expected that in the near term, this could be achieved with a lithium ion battery coupled with a super capacitor (a device that can supply power for very short periods, typically less than 60 s). Successful exploitation of energy storage could result in significant reductions in CO_2 emissions due to the increase in efficiency at the point of end-use. Net reductions in CO_2 of 25–45% would result using this strategy even if more coal plants were constructed to supply the additional electricity. This is due to the lower efficiency of the gasoline IC engine that could be replaced by more efficient electric motors. To maximize the potential of these candidate approaches, materials improvements for energy storage should focus on:

1. Increasing energy density. Reduction in size and mass of the energy storage device.
2. Use of non-toxic and non-hazardous materials. Some of the current battery designs use acids which are corrosive, and can generate explosive gases. High surface area carbons that can readily oxidize on exposure to air.
3. Operation over a wide range of temperatures. Most battery energy storage devices exhibit rapid performance deterioration below −20°C and above 30°C. Extending these ranges will substantially increase opportunities to improve applications of energy storage devices.
4. Development of hybrid systems: energy storage devices that can offer performance of either a battery or a super-capacitor in a single low cost device.

Materials development will be a center of excellence in developing new technologies that can be used to achieve a combination of increased energy efficiency, high energy density storage, and gas separation without the overwhelming parasitic energy costs of today's solvents.

10.9.3 Catalysis

Catalysis is a key component of the chemical process for almost every important industrial reaction. It is used in environmental controls to destroy CO and VOC's (as it does in the catalytic converter in an automobile), and is essential to the production of specific chemical commodities (high-octane gasoline being one of the more valuable ones). Catalysts are used extensively in refining and chemical processing, and in the gasification process they are used in the **Water Gas Shift** (WGS) reaction to produce hydrogen from a mixture of CO and H_2O. While hydrogen is widely used in the processing industry, there is little application for it beyond refining and petrochemicals. There is no distribution network, and its low energy density makes it an enormous challenge to substitute directly as a fuel for the transport sector. For hydrogen, this has been, and remains, a major barrier.

In the 1920s it was discovered that several catalysts (Iron and Cobalt) could combine CO and H_2 to produce long chain hydrocarbons into synthetic oil. Substantially paraffin in structure, they could readily substitute for the straight chain hydrocarbons needed in the diesel reciprocating engine. In catalysis, particle size, uniformity, and dispersion are key factors. The greater the surface area is, the more reactive the catalyst, assuming the material shows catalytic activity. Combining catalyst chemistry with new nanostructures has the potential to yield significant improvements in catalyst reaction rates. In one novel process, the electronic properties of iron can be controlled by confining metallic nanoparticles inside carbon nanotubes. This alters the redox properties of the particles and can enhance their effectiveness as catalysts [20]. There is also interest in whether hydrogen could be produced at much less severe conditions, possibly even using photochemically driven reaction pathways [21]. Under such a scenario, it may be possible to realize much more widespread production of hydrogen from an energy source that is genuinely abundant—sunshine.

Particle size plays a very unusual role regarding the catalytic properties of gold. In its bulk form, gold exhibits little or no catalytic properties—that lack of reactivity and its price have kept gold at a distance from catalytic processes. On the nano-scale, it behaves differently. Researchers at Northwestern have shown that gold nanoparticles can accelerate the oxidation of carbon monoxide at room temperature [22]. In most industrial applications, gas temperatures are typically maintained above 500°C to ensure complete oxidation of CO to CO_2. Controlling the morphology of the surface to affect a specific outcome is a frontier in catalysis that seems to be continually renewed. One may find applications in energy storage, energy conversion, and pollution control. Because the volumes needed to coat a surface with nano-particles would be very small, the cost impact of using even a rate mineral as a catalyst is greatly reduced.

10.9.4   Hydrogen and Hydrogen Storage

Hydrogen offers perhaps an obvious choice energy choice if the principal concern is the release of CO_2. It would be the ideal commodity for energy, except that it is not widely available, it is difficult to store, and there is very little capacity to distribute hydrogen (unless it is chemically bound to another element, and usually that element is carbon). Most of the world's hydrogen is obtained from natural gas using thermal steam reforming of gas, and much of that is consumed not far from the point of production. About 90% of the hydrogen is generated and consumed within the refining sector to provide a product suite tailored to consumer and regulatory requirements. In a sense, refiners are producing a "hydrogen carrier" by adding H_2 to the product stream. While it's the most abundant element in the universe, it is not abundantly available for such routine applications as vehicle refueling or power production.

Making hydrogen from inexpensive solid fuels would be ideal, but solid fuels are difficult to process, and conversion methods can be costly. Gasification of solid fuels usually requires large volumes of high purity oxygen, which in turn requires an expensive air separation technology. While this links hydrogen directly to the continued use of fossil fuels, there is an equally daunting challenge of the energy density of hydrogen—its volumetric energy density is so low as to limit its widespread commercial use. In a fuel such as methane, which is 25% hydrogen by weight, approximately ½ of the total energy content of the fuel is obtained from oxidizing of the hydrogen (the balance from the carbon). For fuels such as gasoline,

Table 10.2 Energy storage capacity (weight and volumetric)

Storage mechanism	Energy density, mass (kW h/kg)	Energy density, volumetric (kW h/l)
5,000 psig tank	2.1	0.8
10,000 psig tank	1.9	1.3
Liquid hydrogen	2.0	1.6
Metal hydride	0.8	0.6
Chemical hydride	1.6	1.5
Ammonia-anhydrous	4.79	2.95
LNG(−259°F)	12.0	6.3
Kerosene/JP-8	12	10

kerosene, and methane, the carbon acts as the hydrogen carrier, providing a modestly dense pack of liquid energy. One way to carry a lot of hydrogen is to compress the gas to high pressure (Table 10.2). This fills in the voids between molecules, but it still does not yield the kind of energy density so readily found in liquid hydro-carbon fuels. Even at 5,000 psig, hydrogen only begins to approach 20% of the energy density found in kerosene (and with a very large penalty for compression and storage requirements).

The US Department of Energy has established targets for fuel systems that can store 6% by weight of hydrogen and 45 kg of hydrogen per cubic meter. Goals for 2015 are even higher: 9-wt % of hydrogen and 81 kg of hydrogen per cubic meter. Chemical hydrides, which store the hydrogen with less demanding material require-ments, are expected to be one of the possible solutions to a method of storing hydrogen in a high-density form

Working with $LiNH_2$, $LiBH_4$, and MgH_2, Pacific Northwest National Laboratories (PNNL) researchers have been able to form new compounds with desirable energy density features. Compressed hydrogen, peaks at a volumetric energy density of less than 60 kg of hydrogen per cubic meter, a limitation reached because of non-ideal gas effects that begin to dominate at the ultra-high pressures considered. Hydrogen stored chemically as a hydride (or a hydrocarbon) can achieve signifi-cantly higher densities than compressed gases without the requirement of a very bulky container to maintain the high-pressure conditions.

Recent breakthroughs came with boron compounds coupled with ammonia. By itself, the hydrogen content of ammonia (NH_3) is 18 wt%. Ammonia borane can improve on this figure significantly, releasing between 12-wt% and 25-wt% equiva-lent, depending upon the reaction pathway. This compares quite favorably with conventional middle distillates such as kerosene, which are roughly 15–16% hydro-gen by weight. The trick is to successfully cycle the hydrogen from gas phase where it can be used for combustion or in a fuel cell, to solid-state, and back again. One way of making the transition from solid-state hydrogen to gas phase is through thermal decomposition. Simply by heating the hydrogen is released. A research team including Ping Chen of the National University of Singapore; Thomas Autrey at Pacific Northwest National Laboratory in Richland, Wash.; William I. F. David at the Rutherford Appleton Laboratory in the U.K.; and their coworkers converted ammonia borane to lithium amidoborane ($LiNH_2BH_3$) finding that the amido compound releases nearly 11 wt % of hydrogen at just 90°C.[23] In addition, the amidoborane releases hydrogen that's free from borazine, an impurity and fuel-cell poison that evolves from ammonia borane. This process can be further enhanced using catalytic materials to accelerate the rate of release [24]. The prize here is to be able to successfully store, at high energy density, a fuel that is carbon free.

10.9.5   *Methanol*

Methanol itself is not a direct mitigation option, but it is a stepping stone from one high carbon resource (coal or petroleum coke) to a chemical commodity that serves as an excellent hydrogen carrier, containing 12.5 wt% hydrogen. Reaching this gateway is likely to require two additional developments: one in gasification (where much has already been achieved), the other in gas separation.

Methanol can be produced from coal-derived synthesis gas (CO and H_2), or reformed from natural gas (which is also reformed to make CO and H_2, but with a larger component of the hydrogen coming from the fuel). Given that it can be manufactured from lower quality fuels, it can serve as a hydrogen carrier in applications where hydrogen is the desired feed. Because it has an energy density about one-half that of a hydrocarbon fuel, it isn't likely to substitute as a direct replacement for liquid fuels. While it has been considered many times as a fuel in power generation, that sector has continued to be dominated by gas, coal, and oil. In the past 10 years there have been several proposals to build and operate a unit on methanol, but in the final analysis, these never matured. In cases where mobility is a factor, perhaps requiring a battery, using methanol could potentially be a better competitor, especially if the energy conversion is achieved using a high efficiency device, such as a fuel cell. Fuel cell's are reported to reach efficiencies well over 60%, making them good candidates in the challenge to provide energy as inexpensively as possible, while limiting the impact to the environment.

10.10   Gasification

If one approach to reducing carbon emissions is to utilize a fuel with less carbon and more hydrogen, gasification offers one pathway to reaching that goal. Gasification evolved as a solution to a developing energy challenge reached at the turn of the nineteenth century. Early gasifier designs—in the late 1800s—capitalized on the abundance of low cost fuels (coal) to provide a combustible gas for illumination. District gas distribution systems were common in major urban centers, in an era prior to wide scale availability of electricity. By the beginning of the twentieth century, thousands of gasifiers dotted the landscape in Europe and North America. Electrification quickly assumed the role of providing illumination and the availability of natural gas would mark the end of this phase of gasification for residential and commercial applications.

A century later, gasification would find new applications based on a completely different set of requirements—decreasing quality of the feedstock and increasing demand of more refined products. Today's gasifiers produce syn-gas (a mixture of CO and H_2) primarily for the production of chemicals (the Great Plains gasification complex produces synthetic natural gas from low rank lignite coal). Beginning in 1980s, there were several attempts to match gasification technology (a process that could use cheap but low quality fuels) to gas turbine technology (a process that was not dependent on fuel properties like octane and Cetane ratings) and operating at higher efficiency. The combination could be a win-win since carbon could be extracted just after the gasification step supplemented with a water-gas-shift reaction. The techno-economic challenges of gas separation (carbon dioxide from fuel gases, and oxygen from the atmosphere) saddled the technology with additional complexity and high costs. Today, there are a handful of gasifiers operating as Integrated Gasification Combined Cycle (IGCC) and many of these were heavily subsidized in an attempt to build momentum toward commercialization.

The largest application of gasification technology is in the production of syngas to manufacture chemicals. For power generation, engineers still struggle with designs that will minimize the cost and complexity without sacrificing reliability. A lignite burning plant described earlier can reach over 41% thermal efficiency; the IGCC reaches 40% without including the parasitic loads to capture CO_2. However, if gasification is one-step in the path forward, what might that highway look like? Economically, gasification has a viable future for the production of commodities (like ammonia and liquid fuels), or Polygeneration (commodities like fuel in addition to electricity or steam). The idea of a "CO_2 capture ready" facility might look more like a chemical plant making SNG or Fischer-Tropsch liquids (where in both cases the CO_2 must be chemically separated in such facilities prior to the chemical production step).

10.11   Chemical Looping

A variant of gasification is the concept of Chemical Looping. Chemical looping, a series of continuous oxidation-reduction steps that converts fuel to useful energy at a very high efficiency, is a promising design approaches in fossil energy combustion (an idealized concept is shown in Fig. 10.7). That is, if it could ever be achieved at commercial scale. Most of the research has been limited to paper studies, or half-reactions in the laboratory using milligram quantities of metal/metal-oxide reactants. In a simple form, chemical looping takes fuel in one reactor vessel (on the right in figure), and air in another, while exchanging the reactive medium (typically a metal/metal oxide) between the two. The oxidation step of the reactor extracts the oxygen from the air supply, leaving a stream of enriched nitrogen. The metal (nickel for example) is reduced by the fuel, producing CO_2 and H_2O in the other reactor. The metal in both reduced and oxidized states are cycled (or looped) between each half-reactor. Energy/work is extracted between the two. The oxidation/reduction mechanism has a built in air-separation component,

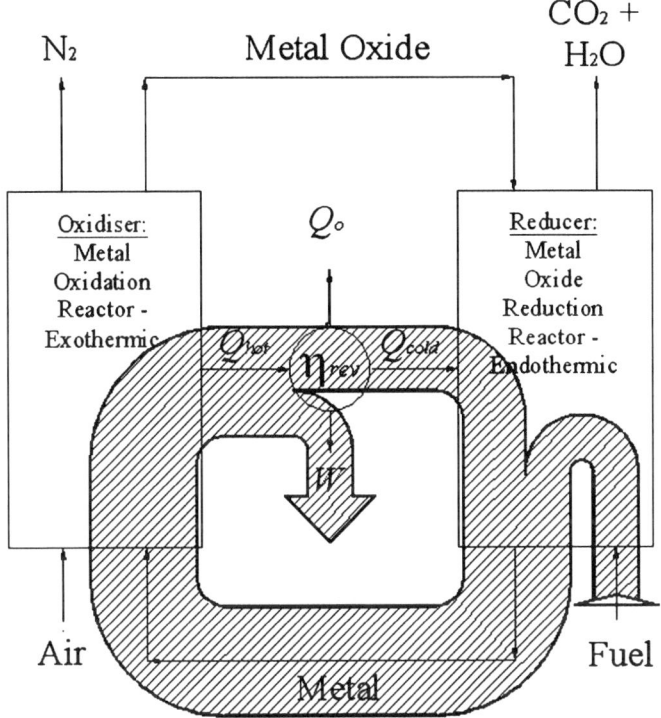

Fig. 10.7 Chemical looping conceptual design

isolating the nitrogen and argon and eliminating the need for cryogenic air separation. Conceptually, chemical looping could, in several steps, substantially reduce the largest parasitic losses found in the current approaches targeting carbon capture/reduction. It might do so at efficiencies comparable to those found in the most efficient combined cycle plants available today [25–30].

As noted by the McGlashan, et. al. a current weakness of chemical looping is its inability to handle fuels with a significant ash component [31]. For this reason, most researchers have focused their efforts on oxidation of relatively clean fuel gases, such as natural gas. The ultimate prize with chemical looping is efficiency levels that are beyond the reach of combustion driven vapor power cycles.

Work by Xian Wenguo and Chen Yingying (in Nanjing, China) suggests that it might be possible to achieve three objectives simultaneously with chemical looping [32]. They simulated the performance of a system using ASPEN software to understand the key boundary conditions limiting the process. Using a mixture of iron oxides, their work suggests that they could produce high purity hydrogen in one stream, CO_2 in another, with some excess power, and an overall net efficiency of 57%. Since hydrogen is so often cited as one of the best solutions to a reduced carbon footprint, this approach offers some very promising insight, if in fact it can be achieved. In Fig. 10.5, it was noted that a modest increase of a single percentage point would yield some 60 million tonnes reduction in CO_2 (with a vapor power cycle system). Reaching performance as high as 50% would cut most CO_2 emissions from fossil coal plants in half.

10.12   Policy Issues and Technology Choices

One cannot readily have an open discussion on environmental policy without opening the subject of energy policy. Energy policy emphasizes carbon in its reduced chemical state, environmental policy deals with oxidized states of carbon—but in each case, we are dealing with the same element—carbon. Since most of the worlds energy resources are hydrocarbon based, these two carbon forms are tightly interwoven and will be for some time. Our current approach to energy policy usually focuses on energy independence, sometimes independence from specific sources of oil, sometimes independence from any source that is not functionally renewable. Navigating the landmarks for a path forward can be almost mind numbing. Consider the debate over the energy balance of ethanol from corn. Depending upon the method of data analysis, energy balances may range from as low as 0.61 (it takes more energy to produce the ethanol than is found in the ethanol) to 1.65 (suggesting that this is a sure bet for developing a sustainable energy strategy). Both extremes can't be correct. Nor is it likely that the environmental benefits (or harm) are as easily quantified as both the promoters and detractors claim. On the environmental side of the equation, we don't have convincing evidence that the increase in N_2O emissions from crop production and changes in hydrological cycle are balanced by the uptake of CO_2 by growing specific crops in the first place.

One part of energy policy that is often left unanswered is who would put all of this new technology together? At stake are billions of dollars in research that would be funneled into hundreds of billions of dollars in demonstrations, and finally trillions of dollars of capital investments. This at a time when the most experienced demographic component of the population is within 5–10 years of retirement. Whether it be boiler makers, welders, pipefitters, machine tool operators—the population is in rapid decline for the most important trade skills needed to make all this happen. The capacity to construct and fabricate high technology mega-scale projects is quickly eroding, and little is being done to capture the loss of experience and skill sets. Special attention needs to be given to how such a great undertaking would be achieved, and who will be doing the real heavy lifting.

10.12.1   Public Utility Commission Role

Unique to the United States, Public Utility Commissioners (PUC's) play a pivotal role in how energy, and to some extent environmental, policy is diffused to the ultimate consumer, the ratepayer. They fulfill the role of gatekeeper to ensure that the most affordable energy rates are passed on to consumers, and that changes to the infrastructure do not cause deterioration in service. Part of their role is to help prevent blackouts, brownouts, and price spikes without unduly burdening either the provider or consumer. That's an unusual role, one that is seldom replicated in other parts of the world. Interestingly, this unique position has also been almost solely responsible for challenging some of the technical innovations being proposed as a solution to reducing CO_2 emissions from new power projects. Take for example, Integrated Gasification Combined Cycle, or IGCC. There are pro's and con's about the different technologies for burning coal. Supercritical plants are well known technology with reasonably well understood pricing structures; IGCC's have the capacity to reduce emissions from all pollutants to a greater extent, including CO_2, but there are only a handful of such plants operating globally that generate electricity, and nearly all received substantial subsidies in one form or another to construct the facility. Whatever the longer-term benefits may be for one technical choice over the other, in the end, the PUC will usually side with a technology that has the lowest delivered cost to the consumer—all other factors taken into consideration.

The cost issues, and complexity of the IGCC, have taken a toll on projects being developed. At one point, there were proposals to build perhaps as many as a dozen plants in North America. Many were suspended, and by 2008, only one plant was under construction and expected to be commercial within a few years. Uncertainty over CO_2 regulations for coal-based generation turned out to be a new concern. Even deep pockets are sometimes not enough to push a project forward. Several federally subsidized projects were found to have such burdensome costs that had great difficulty moving beyond the initial engineering evaluation stage.

Slicing it a bit deeper, while the IGCC may have had a difficult developmental schedule, the gasification component of the IGCC has been much more favorably received. First, it doesn't take a review by state commissioners to construct a gasification plant, and second, they are almost exclusively being used to make something other than electricity. Examples include gasification of pet-coke, production of hydrogen to make ammonia, or production of syn-gas to manufacture methanol. Because of that, gasification may still offer a route to utilization of coal (by converting coal to SNG) and using the most efficient power generation available (combined cycle or combined heat and power) to produce power. Since the CO_2 must be extracted at the point of SNG production, this is effectively a *"carbon-capture ready"* plant that could be put in place today. As shown earlier, the combined cycle gas-fired technology produces less than one-third the CO_2 footprint of a comparable coal plant. In summary, while IGCC may be an answer to how we use coal for power, the tools are in place to rearrange these components while still having the option of capturing and sequestering CO_2.

10.12.2   Operability and Reliability Constraints

Many of the technical proposals for carbon capture target treatment of 100% of the exhaust gases at each source (although limiting carbon capture to something less than 90%). Because the CO_2 extraction system typically incurs high-pressure drop, a large number of fans and pumps must be employed to move the gas and liquid flow streams (typically in counter-current flow). Pressure imbalances, flow disruptions, system controls, even small leaks, have the potential to cause a unit trip or malfunction of the emission control system. If the environmental control is tightly integrated to the overall plant control, there is increased risk of tripping the entire generating unit off-line. These risks are compounded by the size of the unit. Losing 50 MWe of electrical generation in a small community could easily be corrected by transferring additional power to the district from a much larger grid (assuming it is fully interconnected to this grid). However, if the plant is large, say 500 MWe, the impact has the potential to quickly propagate through the grid, possibly beyond region to other reliability council regions. System integration, connecting the carbon capture to the power generation and carbon removal has the potential to reduce plant availability and overall system reliability. Larger, more complex plants will likely be predisposed to experience faults owing to the significant increase in plant complexity. The degree of integration will affect overall system reliability. Doubling the overall system complexity has the distinct potential of degrading overall system reliability.

From a power generation perspective, the United States is sub-divided into a series of smaller regulatory council regions, each overseeing requirements related to transmission, distribution, interconnection, and system reliability (see Fig. 10.8). However, the country is also segmented electrically into three asynchronous

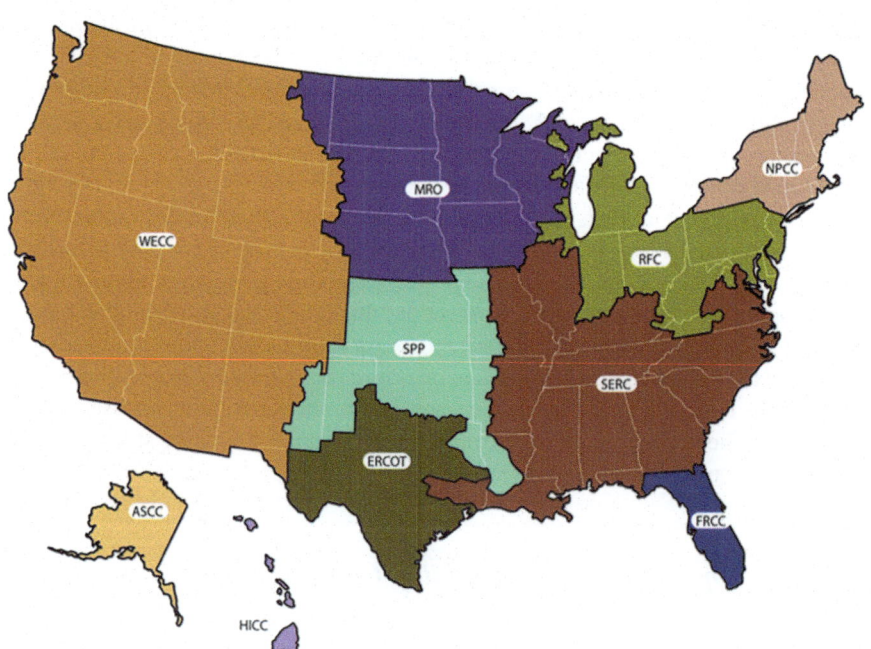

Fig. 10.8 North American reliability council regions

regions: the far west, WECC; the East, and effectively the state of Texas (ERCOT). These regions are effectively out-of-phase with one another, a barrier that makes it difficult to "ship" electrons from East-to-West, or North-to-South. Some of the regulatory issues in these regions are state controlled, through their separate regulatory commission, and some regulated at the federal level.

In a vast network designed to extract carbon dioxide from exhaust gases, the problem of handling mechanical upsets that affect power system reliability must be addressed early on. During process upsets, and they will happen, a number of issues would need to be addressed quickly. At a minimum these should include;

- What to do with the exhaust gases that were supposed to be processed by the carbon capture system,
- How to handle the enormous quantities of solvent cycling through the system,
- Does this malfunction propagate back towards the primary generator, the power plant,
- Will an environmental control malfunction transition into a larger grid disturbance? If the fans and dampers are not able to quickly redirect the gas flow and the process safely corrected, the potential for a complete unit trip increases, as well as upsets to the larger grid

A large carbon capture and sequestration system added to a block of major power generating sources is suggestive of placing another network layered on top of the existing electrical grid. As with electricity, there is likely to be little capacity to store captured CO_2 at the point of generation. The volumes would be enormous, and the most convenient form of storage (as a supercritical liquid) would demand significant quantities of high-pressure storage capacity. Because of this, when CO_2 is extracted, it will be necessary to move it off-site as fast as the recovery process extracts it, or on-site accumulation would quickly overwhelm any local above-ground storage. Similar to the electricity, the CO_2 and must be moved onto a separate grid (in this case, a CO_2 pipeline, or possibly a gas pipeline converted to CO_2 transmission). If there is a failure in the CO_2 transport network, this event will backup very quickly to the operating plant. Since the CO_2 is likely to be supercritical, it will act more like a liquid and not a gas, with substantially reduced compressibility. Thus, the viability of the carbon capture process is now tied to the capacity to withdraw the product stream as quickly as it is generated. A significant failure in any one component may require operator action that entails: (1) venting of CO_2 where it is generated, (2) shutdown of the capture system, or (3) shutdown of the generator system. These complex system issues have not been factored into the grid reliability assessments and must be part of any long-range strategy to deploy the technology.

10.12.3   Storage and Sequestration

Capturing carbon as CO_2 is one thing, but putting it in permanent storage adds an entirely new dimension. Nature has a demonstrated capacity to permanently store carbon.

CO_2 storage domes that are millions of years old have been tapped as gas sources for Enhanced Oil Recovery (EOR). We have minimal experience creating, and possibly administrating, new large scale storage domains. Addressing these issues will require development of complex frameworks to administer the regulation and monitoring injection and long-term storage of CO_2 in various reservoirs. Some critical areas that need to be addressed (and for brevity this has been limited to CO_2 specific):

1. *Monitoring of CO_2 reservoirs.* Accounting for CO_2 stored in subterranean and possibly even above ground reservoirs (soils, agriculture, etc.). How are losses accounted for, and how to quantify changes that might occur due to man-made actions (drilling into a reservoir) or released from geological storage through seismic events?
2. *Custody transfer.* Where CO_2 is recovered and shipped for long-term storage, what rules will be in place for handling the large volumes that would be transferred? Does the original generator retain any residual risk if a problem develops later?
3. *Rights of ownership above ground and below ground.* Is CO_2 stored underground a resource, or a waste? Does it represent a risk, or benefit? CO_2 is used in West Texas to recover oil from depleted oil fields. If stored in an aquifer, does it have any comparable value? If it diffuses into other reservoirs, is it producing value somewhere else, or undefined risk?
4. *Insurance.* Will private insurers fill the gap, or will it require specialized insurance instruments or funding? Is some sort of specialized legal standard required? Does it share the usual definition as "property," or is it really something else. If it really is property, with defined qualities and risks, it should be insurable.
5. *Long-term storage risks.* What happens in the event of a failure? Who should pay, and how are the effects accounted for? Should credits that were once transferred to a source in exchange for CO_2 captured be revoked if that CO_2 is eventually lost from storage? In light of recent financial upheavals, developing and accounting for the value of carbon (as CO_2) must be much more carefully thought out.
6. *Valuation.* Finding a way to value carbon dioxide has not been particularly easy. Developing a regulated trading system may provide part of the answer, but successfully benchmarking a price for CO_2 will require extensive monitoring. And protocols to make this happen are still being developed. As a starting point, one could link the value of carbon as CO_2 directly to the value of the carbon in the fuel. Successfully functioning fuel markets already exist.

Property rights, insurance, and valuation are probably the most vexing. CO_2 may be property in one sense, and at the same time, a liability. A failure to properly and clearly define its legal status could make it next to impossible for insurers to enter the markets. The financial dislocations that erupted in 2008 provide some hint of the speed and scale that markets react when exposed to uncertain and undefined risks. CO_2 sequestration will become a herculean undertaking and the stakes in sequestration's success or failure are huge, making quantification a major challenge in its own right. The desired length of sequestration is probably longer than the lifespan of any institutions that we know of that could either monitor or regulate its storage. Our strategy should be one that allows for alternative solutions in parallel, such as renewables and energy efficiency, technologies that function today and can produce measureable results much more quickly.

10.12.4   Resource Limitations: Water

Another regulatory concern is access to a key resource that substantially influences the efficiency of the underlying power process as well as any added post-combustion emission controls. This other is with water consumption and its use as a thermal reservoir for heat rejection. Water is not only used as the primary working fluid in the vapor power cycle, but it is also used as the heat sink that helps fix the lower end of the thermodynamic boundary for efficiency. In essence, it sets the value of T_{low} in Carnot expression:

$$\eta_c = 1 - T_{low} / T_{high}$$

For power plants located near a cold ocean, efficiencies are usually better than those located inland. In desert climates where water is unavailable or inaccessible, the minimum temperature for condenser discharge could be 30°C or higher during the day (a time when power demands are likely to be greatest). To compensate for the lack of cooling water, air cooled condenser technology may be the only likely alternative. This approach erodes plant efficiency since the lowest air is a poor quality thermal sink in comparison to water. The power demands to operate fans for cooling can add tens of megawatts of additional parasitic load to a utility scale generator. In turn, this will demand even greater fuel consumption to meet specific load requirements. Finally, the large heat exchanger requirement adds millions to the overall plant costs, which is recovered in the cost of the electricity.

In a post-combustion carbon capture plant, water is also required to cool the chemical solutions after regeneration (or in some cases, perhaps the coolant for the exhaust gases). And extraction of steam from the turbine cycle that is not returned will require an increase in the size of the water treatment facility for the unit for the make-up water. High purity water is also critical for any power plant, placing additional water requirements on a project. Even with a separate steam system for the CO_2 process, water treatment for the make up water will be required. Facilities that have difficulties gaining access to water now would find the problems compounded by these additional water requirements.

Another challenge associated with water is not the lack of water, but rather the elimination of water from where it could cause serious operational problems— namely in a CO_2 pipeline. Water, combined with CO_2, forms a corrosive acid, which can attack the pipeline from the inside. High corrosion rates of up to 1–2 mm per week are possible. The source of the water is likely to be the loss of dehydration at the carbon capture plant prior to pipeline injection, possibly a more likely event during transients such as startup, shutdown, or trips (that might occur for various reasons). Preventing and forecasting such process failures represents enormous uncertainty and increased risk.

10.13   Alternative Approaches

Because the challenge for reducing CO_2 is daunting, alternative plans may provide a quicker route to reducing the emission burden. The current power generation inventory provides some insight; the United States obtains almost 20% of its electricity from nuclear power, and renewables such as wind are gaining rapidly. These technologies can provide a stepping-stone to reaching a goal of reducing the carbon burden to the atmosphere until the technical challenges reviewed can be overcome.

10.13.1   Nuclear Power

Nuclear power's obvious attraction is its lack of CO_2 emissions, at least during the operational phase, but it is also expensive and problems of fuel utilization and wastes have not been adequately resolved. Costs for a new nuclear plant range from $6,000/kWe [33] to as high as $10,000/kWe [34]. Despite very complex issues, the world is currently experiencing a nuclear renaissance. However, in open markets generation options nominally compete based on the metric of Levelized Cost of Electricity (LCOE). On this basis gas fired generation yields costs of <$80/MWh compared to new nuclear costs of >$100/MWh.

Nuclear power was once thought to be so inexpensive that it would be "*too cheap to meter*" [35]. Despite the drawback of spent nuclear waste, the lack of CO_2 emissions from nuclear power continues to act as a powerful incentive to revive the industry—that and the increasing cost of some fossil fuels. A significant wave of nuclear power plants was constructed between 1960 and 1980, using two basic plant designs (Boiling Water Reactor, BWR, and Pressurized Water Reactor, PWR.) Even with a developmental head start of nuclear power in the United States Navy beginning in the 1950s, the overall fleet reliability did not reach its peak performance until the late 1990s and early 2000, decades after the peak of the construction boom. In retrospect, it took 20–30 years for owners to optimize the performance of the complex equipment with knowledgeable and trained operators, engineers, and maintenance personnel (see Fig. 10.9). As the costs of carbon capture escalate rapidly, a fair comparison could be made to which technical choice

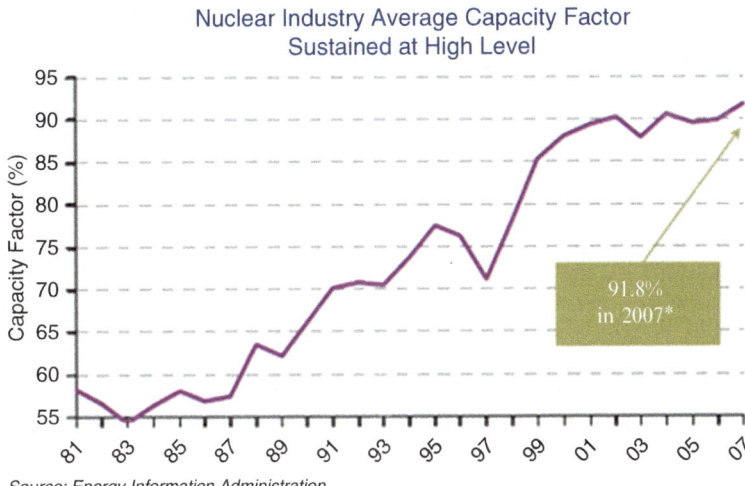

Source: Energy Information Administration
*Preliminary data for 2007

Fig. 10.9 Improvements in nuclear plant capacity factor with time. It has taken almost 25 years to go from 50% CF to 90% capacity factor (CF) (Source NEI)

is preferable: A relatively simple fossil plant design incorporating a complex system of carbon capture, or a nuclear plant with virtually no carbon emissions. While the financial question is not properly situated in the subject of R & D for emission controls, it will be a deciding factor in rating technology choices. In fact, this is a fundamental feature built into the EPA's BACT (Best Available Control Technology) selection process. However, we have not built a new reactor in nearly 20 years, making it difficult to ascertain what that final cost might be. Both the upfront cost of new nuclear generation and its end-of-life costs continue to be major impediments to expanded nuclear in the US (where market forces play a dominant role).

Since the first wave of nuclear units came on line, capacity factor (the ratio of the actual output from the fleet compared to the maximum amount) has steadily improved. But again, it was an evolving process of learning how to optimize refueling, and improving operating, maintenance, and management practices. Because of their low fuel costs, nuclear plants are operated continuously at their base load. It can be inferred from Fig. 10.9 that in just the last 17 years, there have been significant improvements in what one could describe as technology that was considered fully developed prior to the 1990.

While scores of nuclear plants are being proposed, the issue of what to do with the nuclear waste remains in hiatus. Over 90% of the energy in the original fuel is unused in conventional fission reactors and the by-products can be hazardous for thousands of years. One alternative might be to continue to "burn" the material that is now classified as waste. That is one objective of the Next Generation Nuclear Plant (NGNP). For now, much of the world's designs are essentially limited to efficiencies between 30% and 35%, far lower than a commercial natural gas combined cycle unit.

The United States Department of Energy (DOE) is proposing a new, very high temperature gas-cooled reactor (VHTR) that would yield gas temperatures as high as 900°C. The energy from the reactor could produce electricity or some degree of polygeneration, where both hydrogen and electricity are the products. The fuel will be the transuranic materials that are typically classified as high-level radioactive waste. The process, referred to as 'deep-burn' could utilize 65% of the energy content of the fuel, compared to only 3–5% of the energy in enriched uranium obtained with current reactor designs. The heat transfer medium is expected to look radically different from today, possibly relying upon liquid sodium as a coolant (in contrast to water used in LWR's) [36].

Reaching high cycle temperatures could solve one major problem with current nuclear systems designs. Today's steam turbines are very large because they operate with saturated steam instead of supercritical. Reaching supercritical temperatures could result in significant size and material reductions which would have a substantial positive benefit on overall system economics. There could be 40% reduction in turbine size by being able to move to supercritical steam temperatures. Given the enormous capital cost in a nuclear plant, such savings would make nuclear generation on this scale a much more competitive proposition.

10.13.2   Renewables: Wind

Wind represents both an opportunity and a challenge. As a renewable resource, it is growing rapidly; nearly 10,000 MW installed new capacity in the United States in 2009. Many of the desirable areas for wind are not located near the major load centers; and the fragmentation of the US transmission infrastructure does not ease the burden of moving power out of those regions with the best resources (the divisions in Fig. 10.8 are indicative of some of the limitations on moving power from one region to another). The ability to forecast wind (or weather) creates some additional uncertainty as to the reliability of the energy (or capacity) available from the wind resources.

We have made significant improvements in both reliability and scale for extraction of energy that could be considered essentially carbon free. Today's wind turbines are capable of peak outputs in the range of 2.3–3.0 MW, in contrast to the 100–500 kW scale wind turbines deployed some 20 years ago (Fig. 10.10 depicts a modern wind turbine design with a rotor diameter in excess of 100 m). While not as inexpensive as power from new gas-fired generation, it is one of the least expensive renewable sources (with levelized costs in the range of $100/MWh).

Fig. 10.10   2.3 MWe wind turbine renewable energy with no emission CO_2 emissions

10.14   Summary

Based on past experience, it is expected take a significant amount of time to deploy technology to counter anthropogenic emissions if the designs are a substantial departure from the current technology. Using established air pollution control technology as a benchmark for comparison, it has taken years to fully develop a much simpler SO_2 technology mitigation with Flue Gas Desulfurization. Pilot studies underway today have a demonstration schedule on the order of 10 years before commercial availability [37]. Further, it has taken decades to build the supplemental nuclear infrastructure, one that is essentially described as "CO_2 free."

Some have compared the task at hand to the Manhattan Project or the Apollo Project, but this may only be an appropriate comparison for the development phase of the technology itself. Deployment of this still-evolving technology might demand a comparison more like the construction of the United States interstate highway system. Construction of this project lasted from 1956 to 1991. Many of the technical innovations needed to build the system evolved from earlier designs, highways built in Pennsylvania in the 1940s and Connecticut in the 1930s. A cadre of young engineers and workers familiar with unique construction trades and equipment became widely available after the Second World War. In the end, the final cost of the system was nearly four times the original estimate, and it is in state of continuous repair. Where funding (or oversight) is limited, parts of the network are in various stages of decay and disrepair, occasionally making headline news when an entire segment, like a bridge, catastrophically fails.

Many of the newest technologies that are carbon-capture focused are only in very early stages of development. Advanced hydrogen storage, new catalysis methods, and solvents for carbon capture will require significant investments if they are to successfully reach commercialization on a broad scale. There is near unanimous expectation that such a wide reaching, yet fundamental change to our energy conversion infrastructure will require significant investment. Just upgrading and improving the United States power generation and T&D system could require over $1 trillion over the next 20 years [38]. Adding carbon controls adds an entirely new cost (and complexity) factor to the financial forecasts. To minimize the costs and the cost uncertainties, an intense, well-planned research program supplemented with full-scale demonstrations, is warranted. A cautionary note: It could take 10–15 years to move from developmental status to full commercial status, and many more years to convert (or rebuild) the entire network. Yet implementing some of the actions reviewed here (e.g., fuel switching, energy efficiency upgrades, plug in motor vehicles, renewables, nuclear power, etc.) has significant benefits to the economy as a whole, and they are technologies we are familiar with, a feature that makes them easier to utilize. Some power technologies that are essentially carbon-free (see Fig. 10.2) are available now. In a recent study on the challenges of reducing CO_2 emission, EPRI estimated that the economic benefits could be as high as $1 trillion [39]. Some of the key technical challenges to be overcome are summarized in Table 10.3.

Since the radiative warming effect of CO_2 impacts the global climate on the scale of a century (or longer), then it becomes more important to address the impact of what is released to the atmosphere today. Similar to a discounted cash flow, mitigating 50 tonnes of CO_2 released to the atmosphere in 2010 could be equivalent to mitigating perhaps 100 tonnes of CO_2 released in 2025. There is a sense of urgency, and a rewarding benefit, to early reductions. Yet we are limited in that there are no readily available technologies to inexpensively capture CO_2 emissions from point or mobile sources. Globally, there are no functioning examples where a nation, region, or city has achieved a reduction in CO_2 emissions by employing any technology that is specific for capture and sequestration. Certainly many nations are exploring control and mitigation measures at various levels. Some attempts to force the technology have focused on creating a market first, expecting that the instruments will evolve, with no appreciation that CO_2 is radically different from any other gas. Despite our best efforts, even per capita emissions of CO_2 are continuing to increase globally, making the challenge for a technical fix even more difficult.

As already noted, we have commercial products available today that could quickly achieve near term objectives of reducing CO_2 emissions from major sources. This would be done by improving efficiencies at all levels, increased utilization of natural gas (or even conversion of low quality feedstocks to higher quality SNG),

Table 10.3 Summary of critical research and development requirements

Focus area	Challenge	Objective
Energy conversion and power transfer	Fossil power generation (steam turbines, gas turbines, fuel cells), nuclear power generation, propulsion (transport), electricity distribution.	Improve overall system efficiency, alloys to reach 700°C+ on vapor power cycles. Enhanced design features to minimize the costs of super-alloys.
		Alloys resistant to fireside corrosion at ultra-supercritical temperatures.
		Safe operation of nuclear systems to allow higher operating steam temperatures for improved efficiency.
		Materials with reduced resistive losses (reduce losses to 5% of generation, maximum)
		Reduce water consumption for process cooling.
Gas separation	Air separation plants (separation of oxygen from air), CO_2 from product gases, and hydrogen from gasified fuels	Reduce energy required to produce high purity oxygen (<150 kW h/tonne), separation of CO_2 (<50 kW h/tonne), and concentrated H_2
		Improved solvents/membranes for gas separation.
		Enhanced solvent stability in the presence of oxygen; solvents not dependent upon steam or water consumption for regeneration.
		Low toxicity solvent/sorbents
		Demonstration of complex carbon capture systems in power generation applications.
Energy storage	Increase the energy density of rechargeable systems that can provide grid stability as well as adaptation to the transport sector	Increase energy density to reach a level of 10–25% of current hydrocarbon liquids.

end-use efficiency improvements (insulation, weatherization), and expanding renewables like wind. For the United States, the low-carbon energy situation has significantly improved with the exploitation of massive reserves of natural gas in tight gas formations. These gas reserves would support a significant expansion to the gas generation fleet in North America—both capacity (MWe) and energy (MWh). Technology improvements like enhanced energy storage would allow enhanced energy transfers between the power generation sector and other significant sources of CO_2, such as transportation. Right now energy storage in hydrocarbons (10 kW h/l) is far greater than any battery alternative we have today.

Beyond that, we could develop more nuclear power, although there are severe limitations on how much nuclear power could actually be constructed in the interval desired. Longer term, beyond 10–20 years, we will need to be able to deploy the new concepts that will have to be developed in the interim. These might include new cycle designs, or advanced renewable technologies (e.g. novel coal-to-energy conversion systems, direct production of hydrogen from sunlight, etc.), or geothermal engineering on a global scale to meet the world's energy demands. It will take a very long time to change the system from its current status, to something radically different, while still providing significant benefit to the world's economies.

References

1. McFarland J (2000) et. al. The Economics of CO_2 Separation and Capture, Technology 7 (1): 13–23. http://sequestration.mid.edu/pdf/net_lMcFarland.pdf
2. Biondo SJ, Marten JC (1977) A history of flue gas desulfurization systems since 1850. J Air Pollut Control Assoc 27(10):948–961
3. Johnson TL, Keith DW (2004) Fossil electricity and CO_2 sequestration: how natural gas prices, initial conditions and retrofits determine the cost of controlling CO_2 emissions. Energ Policy 32:367–382
4. Rubin ES, Chen C, Rao AB (2007) Cost and performance of fossil fuel power plants with CO_2 capture and storage. Energ Policy 34:4444–4454
5. Feeley et al (2007) DOE/NETL's carbon capture R&D program for existing coal-fired power plants, A.P. Presented at PowerGen 2007, Orlando, Dec 2007
6. Chapel DG, Mariz CL, Ernest J (1999) Recovery of CO_2 from flue gases: commercial trends. Presented at the Canadian Society of Chemical Engineers annual meeting, Saskatoon, 4–6 Oct 1999
7. Jared Ciferno, CO_2 capture ready coal plants DOE/NETL-2007/1301 final report, Apr 2008
8. 2008–2017 NERC capacity margins: retrofit of once-through cooling systems at existing generating facilities. Downloaded http://www.nerc.com/files/NERC_SRA-Retrofit_of_Once-Through_Generation_090908.pdf
9. Nsakala N et al (2001) Engineering feasibility of CO_2 capture on an existing US coal-fired power plant. Presentation at the 1st National Conference on carbon sequestration, Washington, 15–17 May 2001. http://www.netl.doe.gov/publications/proceedings/01/carbon_seq/7c1.pdf
10. James Katzer, The future of coal, An interdiscipilinary MIT Study, Massachusetts Institute of Technology MIT Coal. options for a carbon-constrained world. MIT, p 19, ISBN 978-0-615-14092-6
11. Beér JM (2007) High efficiency electric power generation: the environmental role. Prog Energ Combust Sci 33:107–134
12. Tim Heavy, California's coastal power plants: alternative cooling systems analysis, prepared by Tetra Tech, Inc. Golden, Co Feb 2008. http://www.swrcb.ca.gov/water_issues/programs/npdes/docs/cooling/fullreport.pdf
13. Gottlicher G (2004) The energetics of carbon dioxide capture in power plants. US Department of Energy, Office of Fossil Energy, NETL, Feb 2004
14. Dr. Niall McGlasahan Imperial college
15. Yu YS et al (2009) An innovative process for simultaneous removal of CO_2 and SO_2 from flue gas of a power plant by energy integration. Energ Convers Manag 50:2885–2892
16. Millward AR, Yaghi OM (2005) Metal-organic frameworks with exceptionally high capacity for storage of carbon dioxide at room temperature. J Am Chem Soc 127(51):17998–17999. doi:10.1021/ja0570032
17. Wong-Foy AG, Matzger AJ, Yaghi OM (2006) Exceptional H_2 saturation uptake in microporous metal-organic frameworks. J Am Chem Soc 128(11):3494–3495. doi:10.1021/ja058213h

18. Dai P et al (1995) Synthesis and neutron powder diffraction study of the superconductor $HgBa_2Ca_2Cu_3O_8+\delta$ by Tl substitution. Phys C Supercond 243(3–4):201–206. doi:10.1016/0921-4534(94)02461-8

19. Grant PM (2004) The supercable: dual delivery of hydrogen and electric power, power systems conference and exposition. IEEE PES 3(10–13):1745–1749. doi:10.1109/PSCE.2004.1397675

20. Chen W et al (2008) Effect of confinement in carbon nanotubes on the activity of Fischer-Tropsch iron catalyst. J Am Chem Soc 130(29):9414–9419. doi:10.1021/ja8008192

21. Esswein AJ, Nocera DG (2007) Hydrogen production by molecular photocatalysis. Chem Rev 107(10):4022–4047. doi:10.1021/cr050193e

22. Kung M, Davis RJ, Kung H (2007) Understanding au-catalyzed low-temperature CO oxidation. J Phys Chem C 111(32):11767–11775

23. Zhitao Xiong Z, Keong Yong C, Wu G, Ping Chen, Shaw W, Abhi Karkamkar, Autrey T, Jones MO, Johnson SR, Edwards Peter P, David W (2008) High-capacity hydrogen storage in lithium and sodium amidoboranes. Nat Mater 7:138–141. doi:10.1038/nmat2081

24. Keaton RJ, Blacquiere JM, Baker RT (2007) Base metal catalyzed dehydrogenation of ammonia-borane for chemical hydrogen storage. J Am Chem Soc 129(7):1844–1845. doi:10.1021/ja066860i

25. Richter HJ, Knoche KF (1983) Reversibility of combustion processes, in efficiency and costing – second law analysis of processes. ACS Symp Ser 235:71–85

26. Ishida M, Zheng D, Akehata T (1987) Evaluation of a chemical-looping-combustion power-generation system by graphic exergy analysis. Int J Energ 12:147–154

27. Ishida M, Jin N (1997) CO_2 recovery in a power plant with chemical looping combustion. Energ Convers Manag 38:S187–S192

28. Brandvoll Ø, Bolland O (2004) Inherent CO_2 capture using chemical looping combustion in a natural gas fired cycle. Trans ASME 126:316–321

29. Mattisson T, Zafar Q, Johansson M, Lyngfelt A (2006) Thermal analysis of chemical-looping combustion. Trans I Chem E A Chem Eng Res Des 84:795–806

30. Mattisson T et al (2007) Chemical-looping combustion using syngas as fuel. Int J Greenhouse Gas Control 1:158–169

31. McGlashan N, Heyes AL, Marquis AJ (2007) Carbon capture and reduced irreversibility combustion using chemical looping. In: Proceedings of GT2007 ASME Turbo Expo, Montreal, 14–17 May 2007

32. Xiang W, Chen Y (2007) Hydrogen and electricity from coal with carbon dioxide separation using chemical looping reactors. Energ Fuels 21(4):2272–2277. doi:10.1021/ef060517h

33. IHS CERA (2009) Power market fundamentals, client briefing, New York

34. http://www.powergenworldwide.com/index/videogallery.html

35. Strauss LL (1954) Speech to the National Association of Science Writers, New York, 16 Sep 1954

36. http://www.ne.doe.gov/newsroom/2008PRs/nePR042808.html

37. Zero Emission Project (ZEP) project list, Niederaussem. http://www.zeroemissionsplatform.eu/projects.html/fossil-fuel-power-plants-announced-pilot-demonstration-programmes/9-niederaussem

38. Chupka MW, Earle R, Fox-Penner P, Hledik R (2008) Transforming America's power industry: the investment challenge 2010, prepared by the Brattle Group. http://www.edisonfoundation.net/Transforming_Americas_Power_Industry.pdf

39. Revis James, Richard Reichels, Geoff Blanford, Steve Gehl. The power to reduce CO_2 emissions: the full portfolio, EPRI discussion paper, prepared for the EPRI 2007 summer seminar

40. http://mydocs.epri.com/docs/public/DiscussionPaper2007.pdf

Chapter 11
The Role of Technology in Mitigating Greenhouse Gas Emissions from Power Sector in Developing Countries: The Case of China, India, and Mexico*

Samudra Vijay and Ananth Chikkatur

Abstract China, India, and Mexico are among the top developing country emitters of CO_2. The electric power sectors in China and India is dominated by coal-fired power plants, whereas fuel oil and natural gas are the key fossil fuels in Mexico. Spurred by economic development and population growth, demand for electricity in these countries is expected to continue to rise. Meeting this increased demand will have a significant impact on emissions of greenhouse gases (GHG). While available portfolio of generation and mitigation technologies may not suffice to arrest the growth of emissions, it can help reduce the rate of emissions growth. To achieve significant reductions, multiple approaches are required, such as reducing demand by adopting end-use efficiency improvement measures, accelerating the deployment of renewable and nuclear power, and adopting cleaner more efficient generation technologies. Retrofitting the existing fleet to meet strengthened environmental standards, and accelerated fleet-turnover, coupled with adoption of state-of-the-art high efficiency generation technologies, such as supercritical and ultra-supercritical boilers and advanced combined-cycle gas turbines, should play

---

*The findings included in this chapter do not necessarily reflect the view or policies of the Environmental Protection Agency. Mention of trade names or commercial products does not constitute Agency endorsement or recommendation for use. Views expressed in this paper do not necessarily represent those of ICF International.

S. Vijay (✉)
ORISE Research Fellow, U.S. Environmental Protection Agency, Office or Research and Development, National Risk Management Research Laboratory, Air Pollution Prevention and Control Division, Research Triangle Park, NC 27711, USA
and
Sam Analytic Solutions, LLC, 614 Willingham Rd, Morrisville, NC 27560, USA
e-mail: sam@samanalyticsolutions.com

A. Chikkatur
Belfer Center for Science and International Affairs, John F. Kennedy School of Government, Harvard University, Cambridge, MA 02138, USA
and
ICF International, Fairfax, VA 22031, USA
e-mail: ap_chikkatur@yahoo.com

F.T. Princiotta (ed.), *Global Climate Change - The Technology Challenge*,
Advances in Global Change Research 38, DOI 10.1007/978-90-481-3153-2_11,
© Springer Science+Business Media B.V. 2011

an important role in meeting the increasing demand with the least amount of GHG emissions. In parallel, significant R&D efforts will have to be undertaken to adapt off-the-shelf generation technologies to suit local needs. In the medium to long term, developed countries will need to provide financial and technical support for these countries and partner with them to develop, design, demonstrate, and deploy technologies for capturing and sequestering carbon dioxide.

11.1   Introduction

The Executive Director of the International Energy Agency (IEA), Nobuo Tanaka, underlined the significance of the "huge energy challenges" facing the rapidly growing economies, particularly those of India and China. Given the strengthening science behind the human influence on climate change [1], he also pointed the need to develop a global response to meet this challenge and find ways to mitigate greenhouse gas (GHG) emissions from these growing economies. Among developing nations, China, India and Mexico were large emitters of carbon dioxide (CO_2) in 2005 [2]. While China and India's power sectors are dominated by coal-fired power generation, the Mexican electricity sector, in contrast, relies heavily on fuel oil and natural gas.

In this context, it is instructive to view total GHG emissions within the IPAT[1] framework, where Impact is GHG emissions, **P** is total population, and **Affluence** is gross domestic product per capita. **Technology** signifies carbon emission intensity (emissions per dollar of gross domestic product). Economic growth, which broadly increases per-capita consumption, is essential to improve the quality of life. Hence, the key factors by which emissions can be reduced are population growth and the emission intensity of the economy. In this chapter, we focus on the role of power generation technologies in three developing countries – China, India, and Mexico – for managing their GHG emissions. Emissions from power generation play a large role in GHG emissions from these economies.

As a modern energy carrier, electricity is a critical component of energy supply in any country. Thus, the availability of, and access to electricity, is an important element in the effort to increase standard of living. Hence, we present the current status of the power generation sector, the prospect for future growth, and its implications for GHG emissions in China, India, and Mexico. These three countries pose a challenge for the whole world, for the way in which they meet their energy challenges will have long-term ramifications for the health of the entire planet. The economies of China and India are particularly challenging, as both rely heavily on coal-fired generation. The evolving portfolio of generation technologies in these two countries will play a significant role in their contribution to the global emissions of GHGs. Mexican power sector has undergone significant structural shift in the

---

[1] The terms in IPAT identity or equation, originally coined by Ehrlich and Holdren [3], stand for Impact = Population × Affluence × Technology.

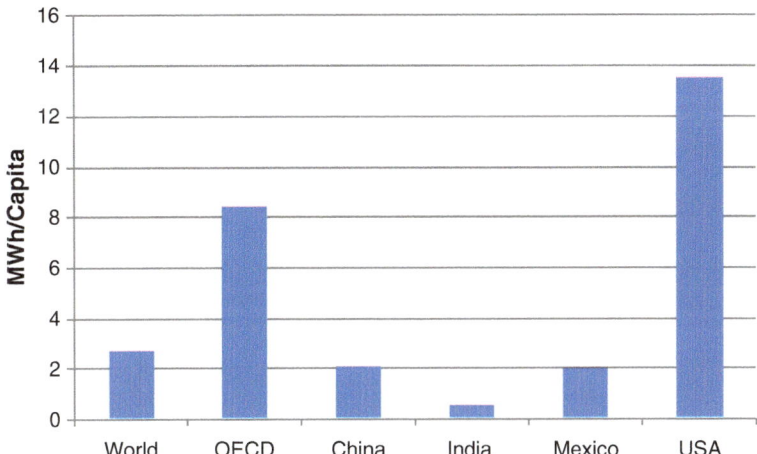

Fig. 11.1 Per capita electricity consumption (MWh/person) (2006) (Source: IEA [4])

recent past, which has reduced its GHG emission intensity from the power sector, but total emissions have continued to grow due to increased energy consumption. Further, energy security concerns have dictated that coal, a carbon-intensive fuel, be considered an important part of the future fuel-mix in Mexico.

In spite of the high growth in their generation capacity, the per capita electricity consumption of these developing countries is still low, when compared to that of developed economies and the world average electricity consumption (Fig. 11.1). Per capita electricity consumption in India is less than one-fifth of the global average, and about one-sixteenth of that of members of organization of economic cooperation and development (OECD) countries. Per capita electricity consumption in China and Mexico are similar, and are less than one-fourth of that of OECD countries [4]. To sustain economic growth and to continue to improve the quality of life of its people, the developing countries would have to undertake massive efforts to add significant amount of generation capacity in the medium to long term.

While there are some common elements among the three countries, the structure of the power sector and socio-economic drivers of demand for electricity are very different in these economies. For example, electrification rate in China (~99%) and Mexico (~95%) is high, whereas it is low in India (~62%).

11.2   Greenhouse Gas Emissions and the Role of Power Sector

In 2006, the total global emissions of GHGs from the combustion of fossil fuel were 28 billion tonnes. U.S. was the single largest emitter of GHG emissions, and it contributed slightly over one-fifth of the total global emissions, closely followed by China (20%). India and Mexico contributed 4.5% and 1.5% respectively [4].

Emissions of CO_2 from fossil fuel combustion, cement manufacturing and gas flaring from US appeared to be growing at a slower rate than that of China in 2005. GHG emissions show a sharp rise for China, a steady increase for India, and slow rate of growth for Mexico (Fig. 11.2). Recent studies suggest that GHG emissions from China surpassed that of the US in 2006, and China is now the largest emitting nation [5]. Largest share of GHG emissions in China comes from combustion of coal (Fig. 11.3).

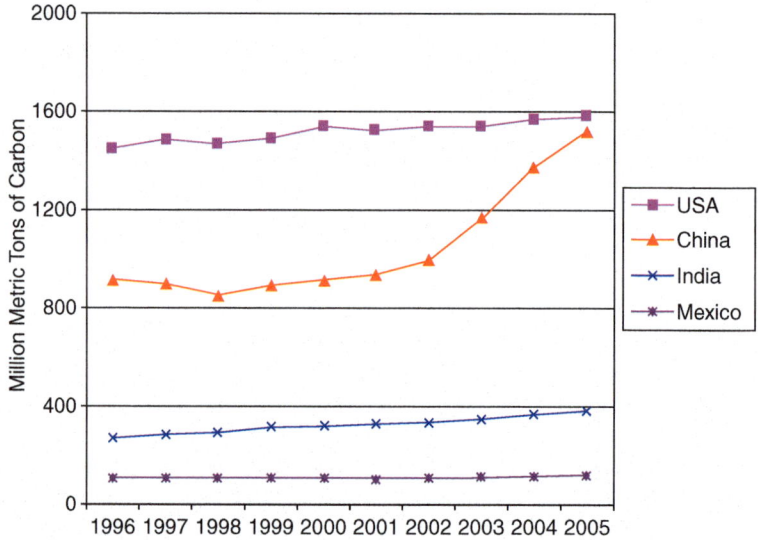

Fig. 11.2 National CO_2 emissions from fossil fuel burning, cement manufacture, and gas flaring (1996–2005) (Marland et al. [2])

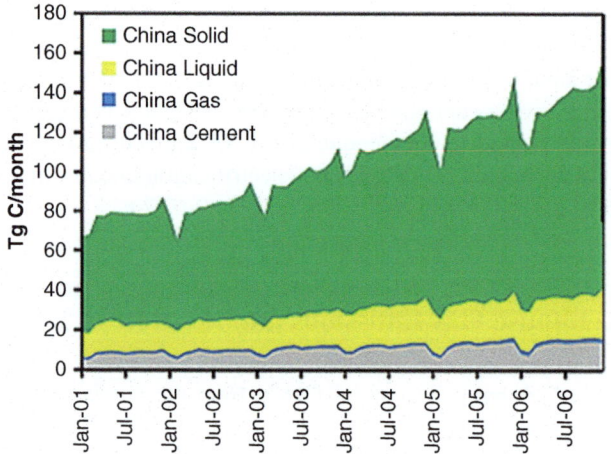

Fig. 11.3 Anthropogenic sources of CO_2 emissions in China (Source: Gregg et al. [5])

According to IEA's reference case scenario – which provides a baseline picture of how global energy markets would evolve if governments make no changes to their existing policies and measures – the global primary energy needs are projected to grow by 55% between 2005 and 2030, with China and India expected to account for 45% of this increase [6]. Majority of this increase in primary energy demand comes from economic and population growth. A large share of the global increase in primary energy demand will likely be met by increasing the production and consumption of coal for power generation – particularly in China and India. The power sectors are the main consumers of coal in China and India, and are the largest sectors responsible for GHG emissions in both countries. Electric power generation in Mexico contributed to 31% of the total GHG emissions in 2004 [7]. In Mexico, the current share of coal in the electricity generation is relatively small. However, in order to diversify its energy supply portfolio and reduce dependence on oil and natural gas, Mexico is striving to make additions to its coal-fired generation capacity. This could further increase its GHG emissions from power generation sector.

In addition to their contribution to global climate change, coal-based power plants significantly impact the local environment. Direct impacts resulting from construction and ongoing operations include [8]:

- flue gas emissions – particulates, sulfur oxides, nitrous oxides, and other hazardous chemicals;
- pollution of local streams, rivers and groundwater from effluent discharges and percolation of hazardous materials from the stored flyash;
- degradation of land used for storing flyash; and
- noise pollution during operation.

Indirect impacts of these plants result mainly from coal mining and include: degradation and destruction of land, water, forests and habitats; and displacement, rehabilitation and resettlement of people affected by mining operations.

Hence, while reducing GHG emissions from these countries is important, it is also critical that the countries reduce the local environmental impacts of their power sectors.

11.3   Power Sector in China

The power sector in China has witnessed a phenomenal growth in recent years, led by increased demand in the industrial and household sectors. The demand growth in the power sector is fueled primarily by the very high rate of sustained economic growth – the gross domestic product of China has increased at an average of 9.8% per annum since 1980. A significant share of its GDP comes from the industrial sector (~49% in 2006), which relies heavily on availability of electricity for manufacturing and other activities. Furthermore, increases in per capita income and standard of living have resulted in high electricity demand from the residential sector, especially as electricity-based consumer goods and services have rapidly penetrated the domestic Chinese market [9].

In order to meet the steep growth in the demand for electricity, installed generation capacity in China has recently risen at an unprecedented pace. The electricity sector in China has also gone through reforms that have made significant structural changes that have enabled China to meet the challenge posed by rapidly growing demand.

11.4 Structure of the Power Sector in China

With 713 GWe of installed generation capacity in 2007, China has the second largest electricity market in the world – a market that is now open to participation by private, local and foreign entities [10, 11]. The reform process in the Chinese electricity market began in the mid-1980s when non-central governmental entities were allowed to invest in generation. The process of reform was significantly advanced in 2002, when the State Power Corporation (SPC) was split into two transmission and nine-generation companies. Further, private investments by local and foreign players have been increasing under the watchful eye of the State Electricity Regulatory Commission. Central government has been actively seeking the involvement of private and state-owned entities to introduce the elements of market-based incentives to improve system efficiency and obtain needed capital investment to meet the projected growth in power generation sector [12].

Generation capacity in China is dominated by coal-fired plants, followed by hydropower. The Three Gorges project is the world's largest hydropower project. As such, China is the largest producer of renewable energy in the world, despite the fact that solar and wind power have a relatively small share of the total installed generation capacity. Coal is expected to remain the key source of primary energy for electricity sector in the near future.

11.5 Demand and Supply of Electricity in China

The demand for electricity has been growing rapidly in this decade. Electricity generation in China was 2544 TWh in 2005, with an installed capacity of 517 GWe, which grew at an unprecedented rate to 713 GW in 2007. In its reference case, the IEA expects total generation to increase at a rate of 4.9% per year, resulting in three times the electricity generation in 2005 by 2030 [9]. The Chinese projections are a bit lower, as China's state energy council expects generation to double from 2005, and installed capacity to increase to 1,120 GWe in 2020 [13]. Regional demand and supply of electricity in China is shown in Fig. 11.4. Generally, power demand in the coastal regions is higher than supply, whereas inland regions have excess supply, with the exception of the Northeast region. Eastern coastal and central inland regions have the highest demand for electricity.

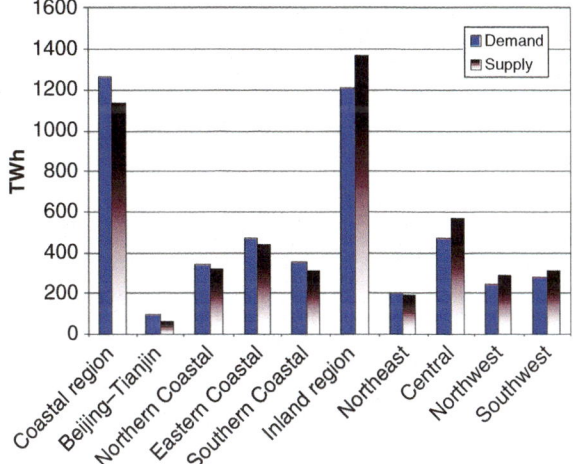

Fig. 11.4 Regional imbalances in the Chinese electricity markets (Source: Wang and Nakata [14])

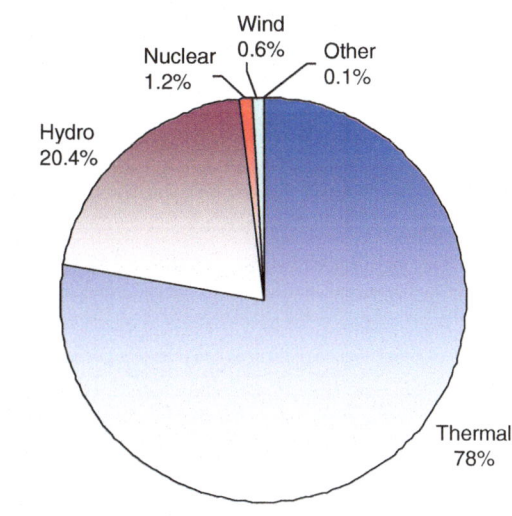

Total: 713GWe; Coal: 530 GWe(~74%)

Fig. 11.5 Installed generation capacity in China (2007) (Source: Zhao et al. [10])

China's installed generation capacity (Fig. 11.5) has doubled in a short span of 5 years, from 356.6 GWe in 2002 [15] to 713 GWe in 2007 [14]. About three-fourth of the total installed generation capacity is based on pulverized coal-fired plants. According to IEA, China added over 105 GWe (almost double the total installed generation capacity of Mexico) to its capacity in the year 2006 alone, of which about 100 GWe is from coal-fired plants [9]. While diversification of generation capacity remains a key goal of policy initiatives in China's power sector, coal is expected to remain a dominant source of generation in the near to mid-term.

11.6   Status of Generation Technology

A large number of small (100–300 MWe) subcritical coal-fired power plants provide the dominant share of China's current coal generation capacity. These plants are generally older, and have low thermal efficiency [9]. Fleet turnover of the coal fired capacity, and addition of large more efficient units have improved overall generation efficiency of coal-fired thermal power capacity in China. In the recent past, net plant efficiency has significantly improved, from under 30% in 1997 to over 34% in 2007 (Fig. 11.6). The gain in plant efficiency is a result of closing down of small inefficient units (about 14 GWe of capacity), and addition of large more efficient units, and the use of advanced technologies, such as supercritical (SC) and ultra-supercritical (USC) boilers [10].

Current Chinese fleet of coal-fired generation plants is dominated by subcritical power plants with an efficiency range of 30–36% [9]. By 2007, China had added about 8.8 GWe of ultra-supercritical generation capacity (7 × 1,000 MW and 3 × 600 MW),

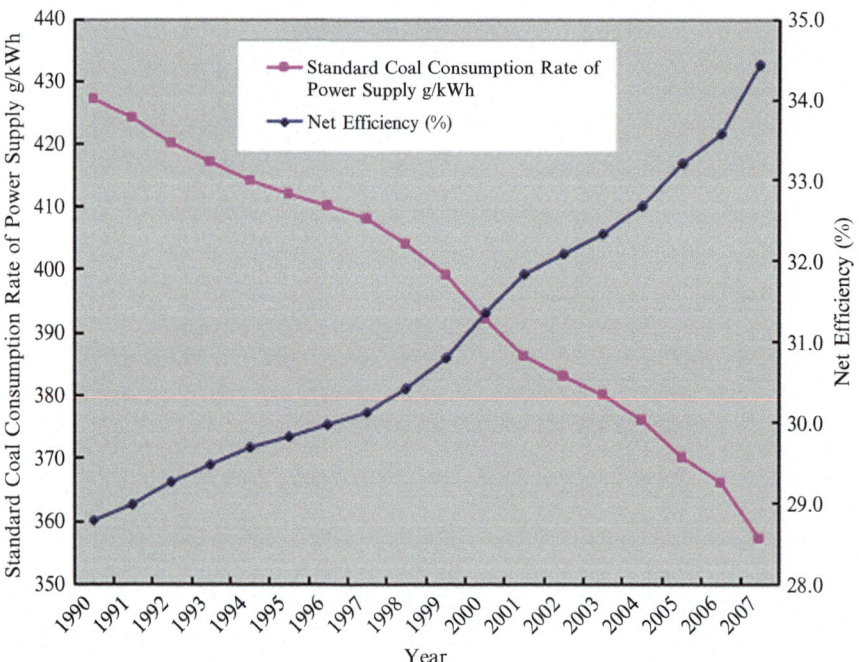

Fig. 11.6 Improvement of standard coal consumption rate and efficiency of coal-fired generation capacity in China (1990–2007) (Source: Zhao et al. [10])

with an efficiency of about 43%, and is expected to add significant amount of new capacity including supercritical and ultra-supercritical power plants [10].

11.7   Air Emissions from Power Generation in China

China's power sector's heavy reliance on coal results in significant emissions of local air pollutants, SO_2, NO_X, and PM [16]. Sulfur dioxide emissions from power plants are a result of naturally occurring sulfur content in the Chinese coal. In 2000, total SO_2 emissions in China were reported to be about 20 million tonnes, which increased to 26 million tonnes in 2005 [9]. Emissions of SO_2 from coal-fired plants are primarily responsible for acid rain problems in the southwestern cities of China [13]. China has undertaken an ambitious plan to reduce emissions of SO_2 by installing flue gas desulfurization (FGD) systems. In 2003, only about 15 GW (~5% of total installed capacity in 2003) had FGD installations. However, by 2007, over half the generation capacity had FGD installations in place [13, 17]. Nonetheless, the emission limits on new coal-fired power plants are significantly higher than its European and US counterparts [10]. To improve local air quality, China would have to tighten limits on air emissions from new coal-fired power plants, and deploy end-of-pipe controls and combustion modifications on existing capacity.

NOx emissions in China were reported to be about 12 million tonnes in 2005, with coal combustion having the largest share, followed by industry and transport sector. Most of the power plants have electrostatic precipitators installed, therefore PM emissions from power plants is not a major concern, although emissions of smaller particulates (with diameter less than 2.5 μm) would still be an issue. However, coal combustion in small and village enterprises (without adequate controls) results in large amount of particulate emissions. Mercury emissions from coal combustion, are also an important environmental concern in China. China is actively pursuing collaborative efforts with the US to control mercury emissions from coal combustion.

China's energy related CO_2 emissions in 2005 were estimated to be 5,100 million tonnes. Largest share of these emissions came from the power generation sector (Fig. 11.7). Almost all of these emissions from power sector can be attributed to the combustion of coal. In the IEA reference case, power sector emissions are expected to grow at an annual rate of 3.7% in 2030. While total energy related emissions are expected to more than double in period 2005–2030, the share of the power sector GHG emissions is expected to increase to 52% [9].

On the supply side, key options to reduce GHG emissions from the power sector include shifting the fuel-mix, accelerating generation fleet turnover, and rapidly deploying cleaner coal technologies. In the national plan for reduction of GHG emissions, China pledged to reduce its dependence on fossil fuel and adopt renewable energy technologies for generation and reduce energy intensity of its economy by 20% in 2010.

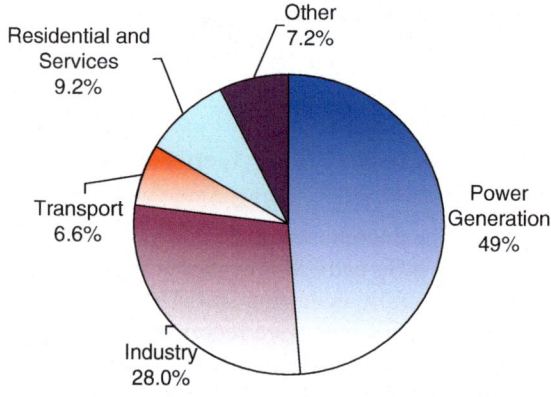

Total: 5100 Million Tonnes (2005)

Fig. 11.7 Share of CO_2 emissions from energy related sources in China (2005) (Source: IEA [9])

11.8   Power Sector in India

Per capita electricity consumption in India is relatively low (Fig. 11.1) due to a large rural population that is not connected to the grid and significant gap between demand and supply of electricity. With increased per capita income and economic growth, the demand for electricity has outpaced supply. Although, India has a significant installed base of hydroelectric power generation, fossil-fuel generated power, primarily from coal based thermal power plants, remains the key contributor to the total electricity generation. As of July 2008, total installed generation capacity of the Indian power sector was 145.6 GW. Fossil fuel based power generation had the largest share (~ 65%) followed by hydro (~25%), other renewable (~8%) and nuclear power (less than 3%). Of the fossil fueled generation capacity, 93 GW, the share of coal fired generation capacity is 83% (77 GW), about 16% natural gas, and remaining oil [18].

Thermal power plants are a key contributor to the local and global air pollutant emissions in India. In the wake of recent economic growth in the country and reforms in the power sector, coal-fired power generation is expected to experience a significant growth in the near term. Current Indian coal-fired power plant population is dominated by sub-critical pulverized coal-fired plants, using indigenous high-ash coal. The choice of technology to meet future growth in demand, air pollution control regulation, and adoption of technology will have a significant impact on emissions of criteria and GHG emissions from the power generation sector in India.

11.8.1 Institutional Structure of Indian Power Generation Sector

The Indian power sector remains dominated by government ministries and public sector corporations. The Ministry of Power is primarily responsible for the development of all aspects of electricity generation, transmission, and distribution in India. It is involved in planning, policy formulation, processing of project and investment decisions, project monitoring, human resource development, and implementation of electricity legislation [19].

In the generation sector, although currently about 60% of installed capacity is vested in the State sector, National Thermal Power Corporation (NTPC), a Central government-owned utility, is considered a de facto leader in the power sector at the national level. NTPC is the single largest thermal power utility in the country, accounting for about 20% of total capacity (27 GW) and about 28% of the total power generated in the country. NTPC is often the first utility in India to experiment with, and deploy, new technologies. For example, the first deployment of supercritical pulverized coal technology is taking place in NTPC-owned plants. It also is actively involved in developing gasification technologies for Indian coal.

Bharat Heavy Electrical Limited (BHEL), also a public sector corporation, is the key player in the electric power technology-manufacturing sector. BHEL manufactured more than 60% of the units installed in the 1970s and nearly all of the power plants constructed in the following decade. BHEL units now account for more than 60% of all units installed in India [20].

11.8.2 Recent Power Sector Reforms in India

The Indian power sector has seen dramatic institutional changes in the past two decades. First, in the early 90s, the government promoted the private sector by providing lucrative incentives for Independent Power Producers (IPPs). However, this attempt to bring in the private sector failed, and by 2003 only 5.3 GW of IPP projects were fully commissioned [21], and the overall capacity addition in the country slowed down in the mid-to-late 1990s. In the mid-1990s, the World Bank, which had previously engaged with the central utilities and had responded lukewarmly to the IPP policy [22], decided to focus on bringing about changes to the Indian power sector by offering financial support to states that would implement its policies for restructuring their electricity sectors. The main changes were to institute a regulatory commission and split the electricity boards into generation, transmission, and distribution units. In the late 90s, the Central government consolidated these state reforms through the Electricity Regulatory Commission Act in 1998 and the Electricity Act in 2003.

The Electricity Act 2003 required all state electricity boards to unbundle and privatize, while introducing at the same time wholesale competition, trading and bilateral contracts with regulation. By forcing the unbundling of vertically integrated companies, the Act intended to separate generation from transmission and distribution, with the hope that generation would be subject to market competition. The Act envisioned a new, market-driven framework where electricity would be just another commodity that can be generated, sold, and traded in the market as determined by supply and demand.

11.8.3   The Demand and Supply in India

While installed capacity and electricity generation has steadily increased over the years, the peak demand shortage of electricity has been rising. Demand-supply gap was reported to be about 9%, the peak demand shortage was reported to be over 15% in the year 2007–2008 [23]. A significant portion of the population living in rural area still do not have access to electricity. An ambitious plan to increase rural electrification and eliminate the shortage "Power for All by 2012" envisions installed capacity to increase to 200 GW.

Much of the expected growth in electricity generation in India over the next few decades will likely be based on coal, particularly domestic coal. The demand for utility-generated electricity is projected to more than double from about 520 TWh in 2001–2002 to about 1,300 TWh by 2016–2017, with an annual growth rate of about 6–7% [24]. Longer-term scenarios indicate demand to be around 3,600–4,500 TWh by 2031–2032, with the installed capacity (including captive power) to be about 800–1,000 GW by 2031–2032 [25]. Hence, it is clear that India's demand for electricity is projected to rise rapidly over the next 20–30 years.

The projected rapid growth in electricity generation over the next couple of decades is expected to be met by using coal as the primary fuel for electricity generation (see Table 11.1). Table 11.1 assumes that Indian GDP will grow at an average rate of 9%. Other resources are uneconomic (as in the case of naphtha or LNG), have insecure supplies (diesel and imported natural gas), or simply too complex and expensive to build (nuclear and hydroelectricity) to make a dominant contribution to the near-to-mid term growth [20]. Liquid fuels such as heavy oils have limited use in the power sector for economic and environmental reasons. Prospects for gas-

Table 11.1 A "middle of the road" scenario for sources of electricity generation in India [20]

Year	Electricity generation (TWh)	Hydro (TWh)	Nuclear (TWh)	Renewables (TWh)	Thermal energy (TWh)	Thermal fuel demand Coal (Mt)	NG (BCM)	Oil (Mt)
2003–2004	592	74	17	3	498	318	11	6
2006–2007	724	87	39	8	590	379	14	6
2011–2012	1,091	139	64	11	877	521	21	8
2016–2017	1,577	204	118	14	1,241	678	37	10
2021–2022	2,280	270	172	18	1,820	936	59	12
2026–2027	3,201	335	274	21	2,571	1,248	87	15
2031–2032	4,493	401	375	24	3,693	1,659	134	20

based power are limited by supply constraints, as many of the recent natural gas based power plants in the private sector have been facing fuel supply shortages. India has significant hydroelectricity resources, but there are a number of problems, including shortage of funds, inter-state water use conflicts, lack of suitable transmission infrastructure, long gestation periods, geological uncertainty in the Himalayan regions, high environmental impacts, and problems of resettlement and rehabilitation of displaced people [26]. The potential for nuclear power development is not high in the short to medium term, because of limited domestic natural uranium resources and various international restrictions that have held back the Indian nuclear power industry [27]. Electricity from renewable sources are relatively small and used mainly in niche applications; even wind power, which has grown significantly in the last decade, is mainly concentrated in a few states in India.

Thus, coal will continue to energize the Indian power sector and its role cannot be understated. Use of India's significant domestic coal resources for power generation would enhance energy security – which is an emerging priority in the country. India's domestic oil and natural gas reserves are minimal (about 0.5% of world reserves), and over three-quarters of India's petroleum consumption was met through imports in 2004–2005. Based on the Planning Commission [25] scenarios, coal-based capacity of utility power plants is likely to be in the range of 200–400 GW in 2030, up from about 68 GW in 2005.

The projected high growth of coal power has significant implications for India's GHG emissions. Coal contributed to about 62% of India's total CO_2 emissions of 817 Tg in 1994, with energy transformation (electricity generation and petroleum refining) contributing 43% [28]. Contribution of solid fuels (coal) to total fossil fuel-based emissions is now about 70%. Given coal power's rate of growth, it will continue to be the major contributor to carbon emissions from the country [2].

11.8.4 Technology, Size, Vintage, and Efficiency of Coal-Fired Power Plants in India

Nearly all of the currently installed coal-fired capacity is based on sub-critical conventional steam cycle. Size distribution of the coal-fired installed capacity in 2005 indicates that 77% of the installed capacity is smaller than 250 MW in size. Only 23% capacity is equal to or larger than 500 MW units [20]. Further, of the total 386 units, in 2005, about 10% (representing only 3% of the installed coal capacity) were more than 40 year old, and about 20% coal-fired capacity is 25 year or older. Average net efficiency of coal-fired power plants in India is reported to be 29%. Smaller, old units (less than 200 MW) have very low efficiency (<25%) and plant load factor (PLF) (<70%) [29]. The best Indian power plants – 500 MW units – operate with a net efficiency of about 33%. In comparison, the average net efficiency for the 50 most efficient U.S. coal-based power plants is 36%, with the fleet average being 32% [20].

The current "standard" for coal-power technologies in India is the BHEL 500 MW sub-critical PC unit. These units are based on assisted circulation boilers with main steam pressure of 170 kg/cm² [30]. Currently, more than 25 of these units are in operation. Many power utilities are now entering the global markets for power plants through their tender process, which has the potential for bringing in new technologies to India. For example, the two NTPC super-critical power plants are based on Russian and Korean technologies – obtained through a global tendering process.

11.8.5 *Air Emissions from Power Generation in India*

The ash content in the Indian coal is very high, resulting in high particulate emissions. Run-of-mine domestic Indian coals typically have ash content ranging from 40–50%, moisture content between 4% and 20%, sulfur content between 0.2% and 0.7%, gross calorific value between 2,500 and 5,000 kcal/kg, with non-coking steam coal being in the range of 2,450–3,000 kcal/kg [31–33].

Most of the particulate emissions come from the flue gas, although fugitive dust from coal handling plants and dried-up ash ponds also are significant sources of particulate pollution. Particulate emissions are better regulated than other pollutants, in part because of the use of electrostatic precipitators (ESPs) in all of the plants. Stack emissions of sulfur oxide and nitrogen oxide emissions are not regulated, and only ambient air concentrations are monitored and regulated for these pollutants. Although about 30% of NOx emissions in India derive from power generation, NOx emissions from coal-based plants are not regulated. Finally, the release of trace elements such as mercury (Hg), arsenic (As), lead (Pb), cadmium (Cd), etc., from power plants through the disposal and dispersal of coal ash is also a growing concern in India. The concentrations of many trace elements are high in comparison to coals from other countries [34].

In terms of CO_2 emissions, India's emissions from fossil fuels have been increasing at a compounded annual growth rate of 5% from 1990–2004 [2], although it has decreased more recently (2000–2004) to 3.8% – see Fig. 11.8. Nonetheless, India's total emissions in 2004 were still about 4.5 and 3.7 times smaller than U.S. and China emissions, respectively. In per-capita terms, India's carbon emissions in 2004 were almost 1/16 of the emissions of the United States and one-third of those of China.

11.9 Mexican Power Sector

Mexico is the sixth largest global producer of oil, and has a significant reserve of natural gas, whereas its proven coal reserves are estimated at only 1.3 billion tonnes, and the share of coal in the total energy mix is relatively small [35].

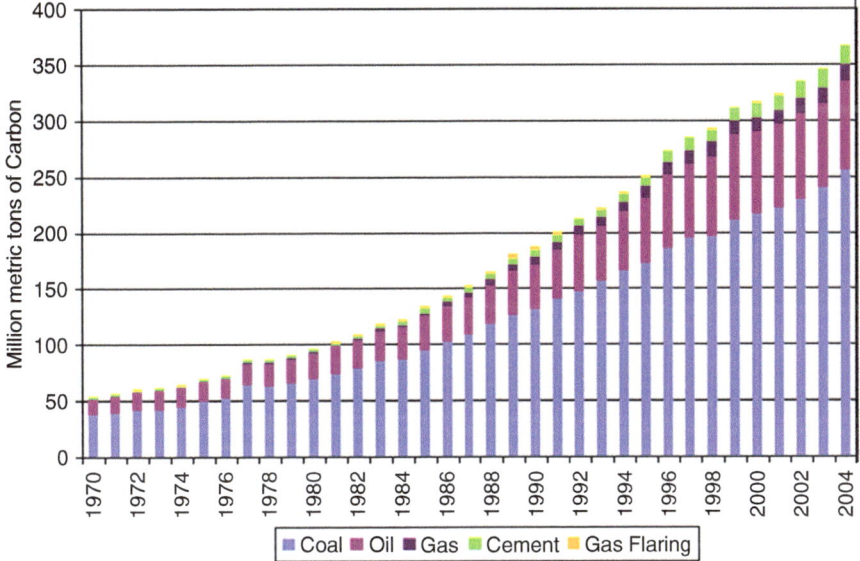

Fig. 11.8 Indian carbon emissions from fossil fuel use (1970–2004) (Source: Marland et al. [2])

Therefore, historically, the Mexican power sector has been dominated by oil – in contrast to coal playing a similar role in China and India. As a result of the reform process that began in 1990s, the Mexican power sector has undergone significant structural changes in terms of its ownership, fuel-share, and generation technology. The changes in Mexico's generation portfolio have had significant impacts on its emissions trajectory from power sector.

11.9.1 Structure of the Mexican Power Sector

Although the Mexican federal constitution limits the participation of non-governmental entities in energy-related activities, regulatory changes starting in the 1990s, have made it possible for the private sector to build, own and operate (BOO) power generation facilities. The regulatory changes leading to this shift can be found elsewhere in the literature (see, for example, [36]; [37]). Recent growth in the installed capacity in the Mexican power sector has largely come from privately owned facilities. The structural shift in the Mexican power sector had three components. First, a policy initiative, which allowed independent power producers (IPPs) to make necessary investment in the infrastructure; second, the availability of resource, i.e., natural gas for use in the power sector; and third, the availability and competitiveness of the natural gas combined cycle technology, which enabled the IPPs to quickly install and operate the newer, more efficient generation capacity.

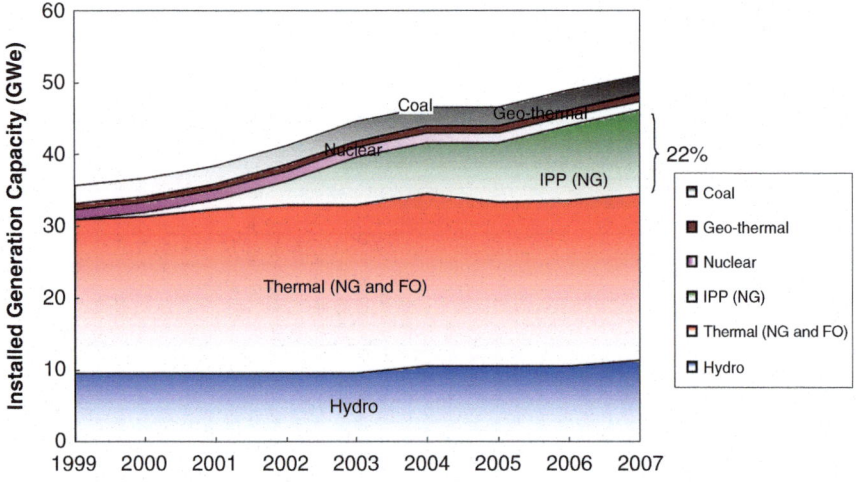

Fig. 11.9 Installed power generation capacity in Mexico (Source: SENER [39])

Historically, the Mexican power sector has been dominated by the state-owned public utilities, *Comisión Federal de Electricidad* (CFE), which is the largest state owned utility in Mexico, and *Luz y Fuerza del Centro*[2] (LFC), which is primarily a transmission company operating mainly in the Mexico City metropolitan area. LFC owns some generation assets as well. As a result of the deregulation process, independent power producers (IPPs) have started to play a key role in the power sector, beginning with the 484 MWe plant in Merida in the state of Yucatán, which came online in 2000. Since then, the first decade of this century has seen a phenomenal growth of the role of IPPs in meeting the electricity demand in Mexico (Fig. 11.9). By the end of 2007, IPPs had a share of over one-fifth of the total installed capacity and contributed slightly less than one-third of total electricity generation (Fig. 11.10).

11.9.2   Demand and Supply of Electricity in Mexico

Increase in population and growth in economic development are the two key drivers for increasing electricity demand in Mexico. Mexico's electric sector planning body estimates that the demand for electricity for domestic consumption will increase at an annual rate of 4.8% from 2007 through 2016. Total consumption of

---

[2] In October 2009, in a major move, CFE, the Mexican state owned utility announced take-over of smaller state-owned utility LFC. The key reason cited for this move is inefficiency of LFC's operations [38].

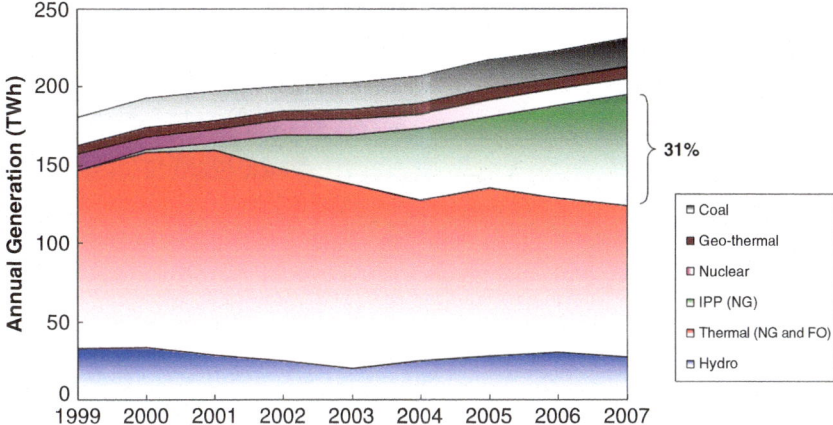

Fig. 11.10 Electricity generation in Mexico by fuel type (Source: SENER [39])

electricity is expected to increase from 197 TWh in 2006 to 318 TWh by 2016. This will require an increase of over 60% in total generation capacity in a decade. It is anticipated that a total of 22.7 GWe new capacities will need to be installed to meet the increase in demand [40].

Historically, residual fuel oil has been the principle source of primary energy for Mexican power sector. Shares of nuclear and coal have been relatively small. The share of natural gas has been increasing steadily since the turn of the century. While natural gas based combined cycle plants are expected to dominate the portfolio of new generation capacity, diversity of fuel-mix and energy security concerns have prompted the inclusion of coal as a potential generation source in the future. Mexico has successfully exploited its renewable energy resources in the past: by April 2008, its geothermal capacity was reported to be 965 MWe, and wind power generation capacity was at 85 MWe. According to the recent planning documents and public releases, there is a renewed emphasis on the role of renewables in meeting future demand, especially wind power. Mexican power sector is expected to add 507 MWe wind power generation capacity in 2011 from the state of Oaxaca [54].

11.9.3   Fuel Shares and Portfolio of Generation Technologies

Until 2000, the Mexican power sector relied heavily on fossil fuels, particularly oil-fired conventional sub-critical steam power plants, with generation efficiency in high twenties to low thirties. However, the IPPs have shifted the structure and fuel-share of the Mexican power sector by rapidly building natural gas fired combined cycle plants.

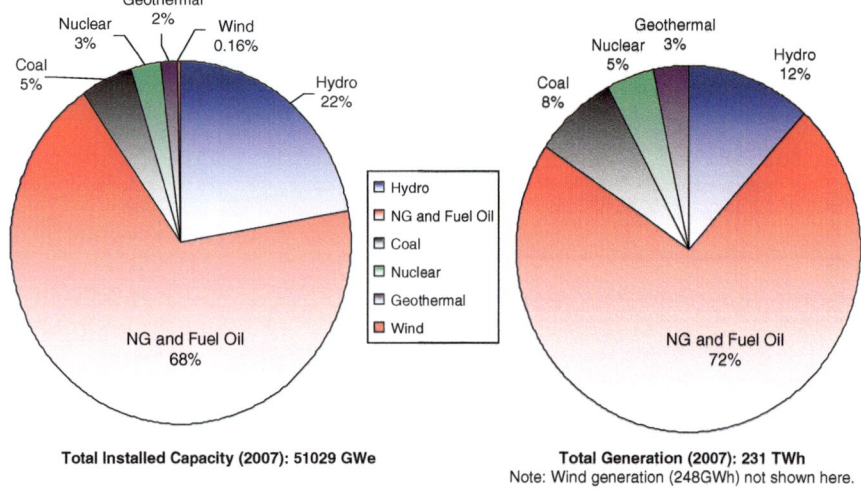

Fig. 11.11 Distributions of installed public sector capacity and generation in 2007

In 1999, before the first IPP came online, conventional oil-fired thermal plants played a key role in supplying electricity to the country. Fuel oil contributed 61% to the total fuel consumed for power generation, followed by natural gas (19%) and coal (12%), and rest nuclear and diesel [40]. Since then the share of fuel oil has continued to decline in the total fuel-mix (Fig. 11.11), and as a result there has been a significant reduction in sulfur dioxide emissions from power generation (see below). For example, units of Jorge Luque and Valle de Mexico located in the Mexico City Metropolitan Area reduced air emissions of criteria pollutants significantly [41]. In addition to being much cleaner and more efficient than fuel oil plants, the gas power plants provide reserve capacity to meet peak demand.

11.9.4 Structural Shift and Its Implications on Air Emissions

As mentioned earlier, the public sector installed capacity in Mexico is dominated by fossil fuels, with coal's share being relatively small. Hydropower generation capacity is the next largest capacity, although its contribution to total generation is relatively small (Fig. 11.11). Recent trends in Mexican power sector indicate a sharp decline of fuel oil in the generation mix, primarily driven by local air pollution concerns, and substitution of capacity with IPP-owned natural gas combined cycle plants (Fig. 11.12). This shift in the fuel share has had clear impact on emissions of criteria pollutants and GHG emissions from the Mexican power sector.

Electricity generation from the public sector peaked in 2001 and has continued to decline since then, despite increase in the overall generation. Much of this is

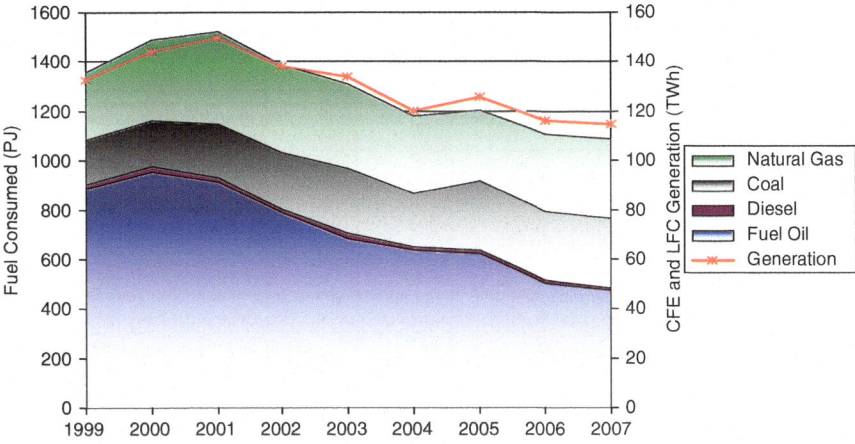

Fig. 11.12 Fossil fuel consumption and power generation trends in Mexican public sector (Source: SENER [39, 40])

because of increasing environmental pressure, wherein the combustion of high-sulfur containing residual fuel oil has been frowned upon, especially for plants located near heavily populated metropolitan areas.

The share of IPPs in the installed generation capacity has been steadily increasing in the Mexican power sector. They contributed over one-fifth of the total generation capacity, but were responsible for about one-third of total electricity generation in Mexico in 2007. These IPP plants, however, have significantly lower share of criteria pollutants and GHG emissions from the power sector (Fig. 11.13). Overall this structural shift has resulted in decreasing emission intensity from the power sector as a whole. Emission intensity of NO_x, CO_2 and PM has declined between 20% and 25%, whereas SO_2 emission intensity has seen most dramatic impact, and has decreased by about 40% from 2002 through 2007 (Fig. 11.14).

11.10   Comparing China, India, and Mexico

While China, India, and Mexico are all developing countries, there are stark differences in their socio-economic characteristics, and the institutional and technological aspects of their power sectors. In this section, we highlight some of the key differences and discuss the implications for technology development and options for GHG mitigation. Mexico's population is less than one-tenth of the population of both China and India, and these two countries have more than one billion people each, accounting for nearly 40% of global population [4]. Despite its high population, however, China has done relatively well in terms of its economy, and its per capita electricity consumption is slightly higher than Mexico (Table 11.2). Per

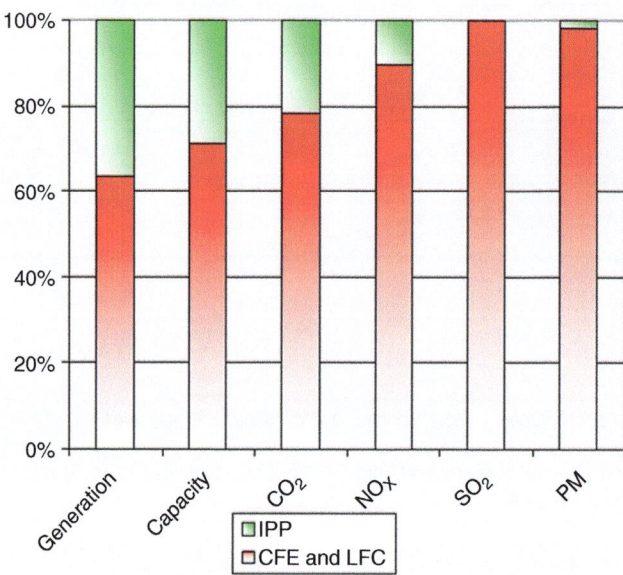

Fig. 11.13 Share of installed capacity, power generation, and emissions from independent power producers (IPPs) in Mexico (Source: Vijay et al. [42])

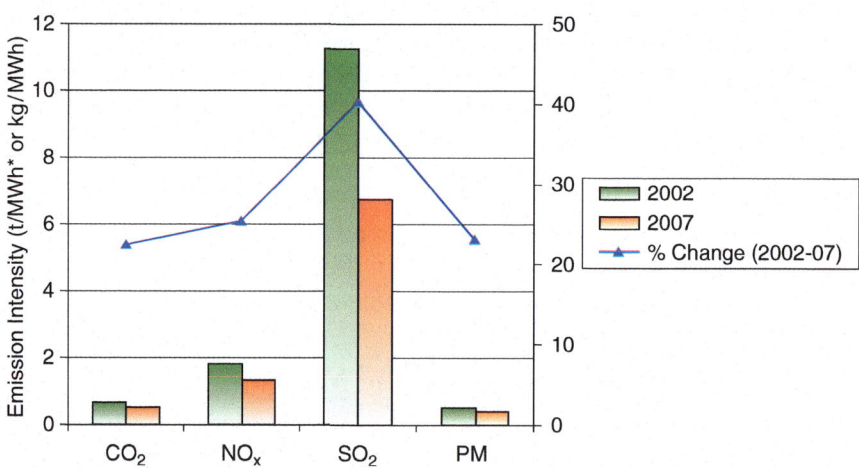

Fig. 11.14 Changes in emission intensity from Mexican power sector (2002–2007) (Source: Vijay et al. [42])

capita income (purchasing power parity adjusted) in Mexico is the highest at about USD 10,000, whereas for China, it is about two-third of that of Mexico (~USD 6,624) and India's per capita GDP is the lowest, at about a one-third of that for Mexico [4]. China has a large industrial manufacturing capacity that has driven its

Table 11.2 Key energy statistics of China, India, and Mexico Source: IEA [4, 6]

	Population (million)	GDP (PPP[a] 2000$)	Total primary energy supply (MTOE)	Par capita electricity consumption (kWh/year)	CO_2 emissions (Mtonnes)[b]	Electrification rate
China	1,312	8,685	1,879	2,040	5,607	99%
India	1,110	3,671	566	503	1,250	62%
Mexico	105	1,030	177	1,993	416	95%

[a]Purchasing power parity
[b]Fossil fuel combustion only

export-based economy, and Mexico benefitted from its participation in North American Free Trade Agreement (NAFTA) which resulted in setting up of manufacturing base across the US border. Indian economy, on the other hand is dominated by the service sector. Contribution of the agricultural production to the total GDP is relatively small, but it still remains the largest provider of employment to the rural population. Manufacturing sector in India contributes about one-fifth to the total GDP. Its service sector, driven by the information technology industry, has seen a significant growth in the recent years. In terms of energy, per capita electricity consumption in China is highest at slightly above 2,000 kWh; Mexico is only slightly lower than China; and Indian per capita electricity consumption is about one-fourth of those of China and Mexico. Electrification rate is highest in China at 99%, closely followed by Mexico at 95%. India also lags behind in this important indicator with only 62% of its population with access to electricity [4].

With over 120 billion tonnes of proven coal reserves [4], China is endowed with the most coal resources among the three countries. However, China's coal production is less than its demand, making it a net importer of coal. Similarly, India imports steam-coal and coking coal, despite having relatively large coal resources [20]. Moreover, the quality of coal in India is very poor – with high ash content, which can sometimes be as high as 50%. Mexico also has some coal reserves, but its main energy resource lies in oil and natural gas: it has 12.4 billion barrels of oil, and 14.6 trillion cubic feet of proven natural gas reserves. Mexico imports steam coal for one of its coal-fired power plants, and primarily relies on residual fuel oil from its refineries. Natural gas production has not been able to keep up with demand (driven by industrial and power generation sector), and hence Mexico is now a net importer of natural gas.

The power sector in these countries (as is true in most other countries) is driven largely by the availability of domestic resources. Chinese and Indian power sectors are primarily fueled by coal, whereas Mexican power sector is dominated by residual fuel oil from its refinery capacity and, since 2000, natural gas based combined cycle generation capacity. In the near to mid-term, it is unlikely that there will be major changes in the fuel-mix of the generation portfolio of the three countries. Despite their efforts to add more nuclear and renewable capacity, China and India are likely to continue to rely on coal as primary resource for power generation.

Mexico, on the other hand is likely to adopt a more diversified approach: while relying mostly on natural gas combined-cycle (NGCC) plants, Mexico will add more renewables and some oil and coal-fired capacity to its generation portfolio. The overall efficiency of Mexican power sector will also improve, as the NGCC plants are much more efficient than conventional oil-fired steam cycle power plants. Mexico's key challenge in the future will be to increase the operating efficiency of its NGCC plants from high 40% to high 50%. The Chinese power sector has already gained significant experience and expertise in developing and deploying the more efficient coal based USC power plants and the key technological challenge for China is likely to quickly use the experience gained in installing and operating the USC plants to accelerate its fleet turnover to make itself more energy efficient. The relatively high sulfur content in the Chinese coal would require the installation of FGDs on majority of its new capacity. The technologies in the Indian power sector lags significantly behind China and the main challenge for India is to learn from the experience of China, Japan and other countries to adapt and deploy SC and USC power plants and to accelerate its fleet turnover.

11.11 Technology Options for Mitigating GHG Emissions from Power Sector in Key Developing Countries

Economic development and the fulfillment of basic human needs such as education, sanitation, health, and communication are dependent on the availability of modern energy services. Indeed, improved standards of living in developing countries are closely associated with an increase in energy demand, particularly electricity. However, the urgent need for a continued and rapid enhancement of the availability, reliability, and affordability of modern energy services, especially electricity, needs to be consistent with sustainable development goals. As such, there are several technology and policy issues associated with reducing GHG emissions from the power sectors in the key developing economies. These issues can be categorized into following four groups: demand management, improving efficiency of existing generation portfolio, new fossil-fired power plants, and non-GHG emitting generation options such as nuclear and renewable. We explore these options and their penetration potential in short (<5 year), medium (5–15 year) and long-term (>15 year).

11.11.1 Efficiency Improvements

In the short term, demand side management (DSM) and improving end-use efficiency by encouraging and enforcing energy efficiency standards and influencing consumer choice through labeling programs are a critical first step in reducing energy demand and GHG emissions. Demand management is especially important, as the penetration of electricity-based consumer goods and electricity demand is likely to increase substantially with increased economic growth. China has taken several steps in this direction, and energy efficiency has gained prominent role in policy making in past years. The Chinese energy plan aims to improve energy intensity to reach at "international levels" [9]. The Indian Bureau of Energy Efficiency (BEE) has taken key steps in providing information to consumers through labeling programs and the agency might also introduce market mechanisms for providing incentives to improve industrial efficiency. Mexico has set up a national commission to conserve energy, known as Comisión Nacional para el Ahorro de Energía (CONAE). CONAE plays the key role in setting up appliance standards for energy efficiency, and promoting technologies such as combined heat and power to improve industrial energy efficiency, in Mexico. While these steps are a good start, consistent and greater efforts in DSM is necessary.

In addition, reduction in transmission and distribution (T&D) losses is another key challenge, especially in India where T&D losses account for a significant loss of revenue and energy. Reducing India's losses to a more manageable (though still high) 10% will release power equivalent to about 10,000–12,000 MW of capacity [43]. Such short-to-medium term measures can be very effective in improving efficiency.

11.11.2   *Improving Existing Plants*

Second, the key developing countries already have a large existing base of power generation technologies, and it is important to consider options to reduce GHG emissions from this existing fleet of power plants. These efforts require relatively smaller capital investment and shorter payback periods, as they are often based on proven and tested technical know-how. As noted in the previous sections, the generation efficiency of existing power plants in all three countries can be significantly improved, especially since there is a large portfolio of smaller, older, and inefficient plants. The efficiency of existing plants can be increased by repowering or upgrading existing facilities and improving energy auditing and management practices. The efficiency gains could yield substantial reduction in primary energy consumption and GHG emissions. In the medium-to-long term, many of the smaller plants need to be retired and replaced with new efficient technologies wherever possible. Such retirements should be particularly targeted at older and more inefficient coal-based power plants. Several power plants in Mexico utilize natural-gas fired steam cycle for power generation, which has much lower efficiency than natural gas fired combined cycle configuration. Conversion of these steam cycle units to combined-cycle plants offer considerable scope to increase fleet-efficiency and reduce natural gas consumption and GHG emissions. Even NGCC fleets introduced in this decade in Mexico and India have the potential to improve efficiency to about 60%, which is the efficiency of the best available combined-cycle plant.

In the short-to-medium term, it is also very important for all three countries to strengthen and strictly enforce the standards for air emissions (particulates, SO_x, NO_x, and mercury) – both for existing fleet as well as for new capacity. Although the control of these local pollutants would slightly reduce net efficiency and capacity, it is critical to prevent socio-environmental problems in all three countries. Furthermore, a clean flue gas is critical for any economic retrofitting of post-combustion carbon capture technologies [20]. In addition, conservation and efficient use of water is another important aspect for fossil-fuel based power plants, as these plants require large volumes of water for cooling.

11.11.3   Technologies for New Fossil Fuel Plants

In the three countries discussed in this chapter, domestic fossil fuel will continue to remain a key element of power sector. Many developing countries with ready access to such cheap domestic fuel will use them for future development, and therefore it is important to consider potential GHG mitigation options for these new fossil fuel-based plants. A key first option, especially for China and India, is to focus on advanced combustion technologies, such as supercritical (SC) and ultra-supercritical (USC) PC technologies. Some of these plants can help replace retire older inefficient plants from the existing fleet. China has already made substantial progress in this direction, and installed 8.8 GWe indigenous USC PC generation capacity. Further, China has embarked on significant future capacity addition based on USC and SC PC combustion technology. According to the IEA, as a result of the introduction of advanced steam cycle plants and the closure of smaller inefficient plants, carbon emissions intensity of coal-fired generation in China is expected to drop by about 25% by 2030. India, however, lags behind China in improving its coal-fired fleet efficiency. Currently there are only two SC PC plants under construction with imported technology. Given that high ash content in the Indian coal poses specific problems in using off-the-shelf advanced generation technologies, indigenous technology development and adaptation will be the key in the short-term to achieve self-sufficiency in this area. Advanced technology based on imported coal might be an option for India [44]. Given that India has significant lignite resources, RD&D efforts needs to be focused in the short-to-medium term on developing a supercritical cycle based on circulating fluidized bed combustion (CFBC) of lignite. In order to diversify its energy portfolio, Mexico is embarking on increasing its coal-fired generation capacity. Planning for coal-fired plants based on advanced steam cycle with efficiency as high as 45% can result in significant reduction in resource use and low GHG emissions for Mexico.

According to the reference scenario of IEA's 2008 World Energy Outlook, 75% of the projected global increase in energy-related CO_2 emissions to 2030 comes from China, India, and the Middle East, and electricity-related emissions for non-OECD emissions in 2030 are expected to double from 6.5 Gt in 2006 [6]. The business-as-usual projected increase in energy-related CO_2 emissions to 2030, assuming no new global or regional climate policies, is consistent with atmospheric CO_2 concentrations of 660–790 ppm CO_2 by 2100, which can lead to an *equilibrium* temperature rise about 6°C above pre-industrial levels [6]. Clearly, such high temperature rises would be catastrophic, and hence mitigation of energy-related CO_2 emissions (especially emissions from coal-power plants) is inevitable over the next few decades.

Hence, the introduction of carbon capture and sequestration (CCS) to new power plants will likely be take place sooner than later in China and India, and perhaps even in the NGCC plants in Mexico. A recent estimate has shown that CCS would be required globally for coal, gas, and oil plants by 2050, with rapid expansion of CCS technologies by 2100. It is estimated that about 70 million tons of CO_2 would be stored by 2020, rising to 600 million tons by 2050 and 6,000 million tons by 2100 [45].

China and India have both taken some initial steps in this direction, particularly in developing and demonstrating new gasification technologies for power generation. China has started construction of three demonstration plants that will use integrated gasification combined cycle (IGCC) technology. It has developed its own gasifiers, with focus on using it for chemicals production. The characteristics of Indian coal prevent the use of standard gasification technologies [20] and hence have been developing fluidized bed gasifier that is more amenable to Indian coals. India has plans for a pilot scale facility using these gasifiers in an IGCC plant. Scientists and engineers in both countries are also now beginning to do research on economically viable carbon capture technologies. However, it is unlikely that aggressive efforts will be directed at research, development, and deployment of carbon capture and sequestration (CCS) technology before the demonstration and deploying of CCS in industrialized countries. Even in industrialized countries, full-scale deployment of CCS requires a major effort in demonstration of economic viability of CCS, initiating the development of infrastructure for transport and storage of CO_2, and creating legal and regulatory frameworks [6]. Moreover, the timing and nature of a post-Kyoto international climate treaty will determine the pace of CCS deployment both in industrialized and developing countries [46] Some of the other key CCS issues for China and India include support for financing and reducing financial cost of CCS, joint research and development of new capture technologies as well as for adapting these technologies to the local context, and detailed assessment of storage sites. Once a viable CCS technology is demonstrated, the manufacturing prowess of China can help bring down the cost of this technology significantly.

The need for detailed storage assessment for CCS is an important issue and needs to be emphasized, as early action is critical for future deployment of CCS. The amount of storage in oil and gas reservoirs is limited and geological underground storage in saline reservoirs is currently the most promising option for storing large quantities of CO_2. However, storage in geological media requires detailed assessments of specific storage locations and capacity within these locations. Only broad first-of-a-kind estimates of storage capacity are currently available in both India and China, and there is a strong need for detailed site-specific assessment of storage mechanisms and capacity in potential on-shore and offshore locations. Furthermore, first-of-a-kind CCS need to be more conservative in their choice of reservoirs as a successful and safe first-of-a-kind CCS plants are critical for larger scale deployment in the future. Hence, it is important to embark on such detailed assessments, as well as relevant demonstration projects, as early as possible in order to inform any siting decisions of new coal power plants [46].

11.11.4 Non-Fossil Fuel Power Plants

Last, but not least, zero-CO_2 emission technologies such as nuclear and renewables, need to account for a larger fraction of new capacity in order to reduce GHG emissions. The emphasis on these technologies needs to be paramount, and wherever possible substitute for fossil fuel based plants. Nuclear power can play a key role in meeting the electricity demand without CO_2 emissions. While operating and safety performance of nuclear plants have improved, and new designs offer safer and competitive generation options, public perception of risk and safety of nuclear power, and ultimate disposal of nuclear waste still remain key challenges facing the nuclear industry. Developing countries, China and India in particular, have continued to make additions to their nuclear generation capacity. While China and India will continue to add to their existing nuclear generation capacity, share of nuclear power in Mexico is likely to continue to decline. Mexico's Laguna Verde plant has two units totaling 1,300 MWe generation capacity and no new unit is planned at this time.

According to the IAEA's power reactor database, by the end of 2008, China had 11 operational reactors (8,438 MWe) and 9 were under construction (8,220 MWe). India on the other hand has 17 reactors in operation (3,782 MWe) and 6 under construction (2,910 MWe) [47]. One key difference in the two countries is the unit size and the choice of technology: while Chinese plants are mostly 1,000 MWe pressurized water reactors (PWR), Indian plants are smaller in size, and have a mix of PWR, and indigenous pressurized heavy water reactor (PHWR) and a fast breeder reactor (FBR). India does not have abundant natural uranium resources, and has limited technological capability to enrich uranium. Hence, the new nuclear deal with the United States and the International Atomic Energy Agency (IAEA) will clearly help India in importing uranium. India also has vast amounts of thorium reserves, which can be used in combination with fissile plutonium or uranium, to produce nuclear material for power generation. However the Indian nuclear program is far from being able to exploit thorium due to technological and capital limitations. While nuclear power offers significant potential to reduce GHG emissions, public perception of risk associated with nuclear power, lack of standardization of design and capital intensity of nuclear power remain key obstacles in the short and medium term. Nuclear industry still has to find a safe and secure long-term storage of the high-level radioactive waste generated by the nuclear plants.

Contribution of renewable energy in meeting the power demand in China, India, and Mexico has continued to grow. Over 15% of China's total primary energy consumption and about 30% of India's primary energy consumption in 2005 was met by renewable energy. Biomass was the dominant energy source for meeting cooking and heating demand in rural households [6]. Electricity generation was dominated by hydropower as the renewable source; it contributed 16% of the total generation. Installed wind capacity is 1.3 GW, and expected to reach 5 GW in 2010 and 30 GW in 2030. In the short to medium term, wind and solar power will remain a marginal source of electricity generation in China. However, in the remote areas where distribution network is unavailable, solar power is likely to provide electricity. By 2030, China's expect solar generation capacity is expected to increase to about 9 GW from about 70 MW in 2005. India plans to add significantly to his hydropower capacity, and it remains one of the fastest growing markets for wind power. Biomass based power generation is also another important source of low emissions power in India. Although the growth of renewables in the future is expected to be large, they still only provide meet a small fraction of the power demand in the country. Mexico has been exploiting its geo-thermal resources for power generation. Recent renewed efforts to enhance its renewable portfolio are heavily dependent on installation of new wind-farms. However, as in the case of China and India, renewable energy is not likely to amount for large fraction of power generation in the short-to-medium term.

11.11.5   Role of Power Sector Reforms

The power sectors in China, India, and Mexico have undergone a process of deregulation and increased privatization, to varying degrees. The impact of these reforms has been mostly positive with increased competition and independent regulation helping to improve the overall institutional and financial health of the power sector. In India, the liberalization and restructuring of the power sector has to be seen as mixed at best due to poor design of the "reformed" power sector (i.e., not fully suitable for the Indian context), inept management of the reform process, and deficient governance in practice [48–51]. However, there are signs of hope, including greater scrutiny of the performance of the reforms; better understanding that successful reforms necessarily will require a tailoring to the Indian context; and institutional learning and capacity building. Regulatory institutions in India will have to be strengthened by giving them greater credibility and enabling the development of their capacity. In addition, regulators themselves must work in a cooperative manner to improve and strengthen regulatory practices and improve stakeholder participation [48]. While private investors have entered other sectors in India, they have been more hesitant to enter the energy market because of the preferential treatment given to public-sector energy companies [9]. It is, therefore, important that a transparent, predictable, and consistent investment framework be put in place. Another priority should be to reduce start-up hurdles, such as delays in acquiring land, environmental clearances, and construction permits [9].

Power sector reforms in the Chinese electricity market began in the mid-1980s when non-central governmental entities were allowed to invest in generation, and the process of reform received a boost in 2002, when the State Power Corporation (SPC) was split into two transmission and nine-generation companies, resulting in a dramatic change in the structure and ownership of the power sector. Further, private investments by local and foreign players have been increasing, overseen by the State Electricity Regulatory Commission. Central government has been actively seeking the involvement of private and state-owned entities to introduce the elements of market-based incentives to improve system efficiency. However, to meet the goals of economic growth, power sector reforms will have to be accelerated in China, to meet the needed investment in the power sector, without compromising environmental quality and health of its land and people [52].

While political efforts to introduce reforms in the energy and power sector have often been stalled by the Mexican congress, *de facto* liberalization has already taken place. The liberalization in the Mexican power sector is spurred by changes in the regulations initiated in 1993 to open power sector to private investment [53]. In 2009, power generated by IPPs contributed about one-third of the total electricity generated in the country [38]. Given that private sectors contribution was nil only 10-years ago, this indicates a significant shift towards privatization in Mexican power sector, compared to that in China and India. However, the uncertainty in the policy environment and slow pace of deregulation has been a key deterrent to increased participation by the private sector in all aspects of power sector. Specifically, the reform has been slow in other parts of power sector. To summarize, structural reforms are a key component in meeting long-term energy demand of the three economies, but the electricity sectors are far from being competitive. Weak institutional framework and lack of clear policy direction are key impediments in achieving regulatory reforms to make the power sector more competitive in these economies.

11.12 Conclusion

Technology will play a major role in mitigation and abatement of GHG emissions from power sector in China, India, and Mexico. Reducing the rate of growth of electricity demand by aggressively pursuing end-use energy efficiency measures and improving the operational efficiency and management of the existing generation fleet are the best options to change GHG emissions trajectory in the short-term. In the medium term, deployment of renewables and nuclear power needs to be accelerated. Retrofitting the existing fleet to meet strengthened environmental standards, and accelerated fleet-turnover, coupled with adoption of state-of-the-art high efficiency generation technologies, such as SC and USC boilers and advanced gas turbines should play an important role in meeting the increasing demand while emitting the least amount of GHG emissions. In parallel, significant R&D efforts will have to be undertaken to adapt the off-the-shelf generation technologies to suit local needs. Moreover, in the longer term, developed countries will need to provide financial and technical support for these countries and partner with them to develop, design, demonstrate, and deploy end-of-pipe controls for capturing carbon dioxide, and its sequestration. In general, effective GHG mitigation in these three countries requires a common understanding and equitable sharing of costs and benefits among both developing and industrialized countries.

References

1. IPCC (2007) Summary for policymakers of the synthesis report of the IPCC fourth assessment report. Intergovernmental Panel on Climate Change, Geneva
2. Marland G, Boden TA et al (2007) Global, regional, and national CO_2 emissions. In Trends: a compendium of data on global change. http://cdiac.ornl.gov/trends/emis/em_cont.htm. Accessed Sept 2007
3. Ehrlich PR, Holdren JP (1971) Impact of population growth. Science 171:1212–1217
4. IEA (2008) Key world energy statistics 2008. International Energy Agency, Paris
5. Gregg JS, Andres RJ et al (2008) China: emissions pattern of the world leader in CO_2 emissions from fossil fuel consumption and cement production. Geophys Res Lett 35:L08806
6. IEA (2008) World Energy Outlook 2008. International Energy Agency, Paris
7. SEMARNAT (2007) Estrategia Nacional de Cambio. Comisión Intersecretarial de Cambio Climático Climático. Secretaría De Medio Ambiente Y Recursos Naturales (SEMARNAT). Distrito Federal, Mexico
8. Chikkatur AP (2008) A resource and technology asssessment of coal utilization in India. Prepared for the Pew Center on Global Climate Change, Arlington
9. IEA (2007) World energy outlook 2007: China and India insights. International Energy Agency, Paris
10. Zhao L, Xiao Y, Gallagher KS, Wang B, Xu X (2008) Technical, environmental, and economic assessment of deploying advanced coal power technologies in the Chinese context. Energy Policy 36(7):2709–2718
11. Zhenhua X (2005) China's power sector on fast track. China Daily Special Supplement, 16 May 2005
12. Xu S, Chen W (2006) The reform of electricity power sector in the PR of China. Energy Policy 34:2455–2465
13. Qiu J (2008) The environmental challenge of coal combustion in China. Presentation at US. EPA, Washington, DC, 15 Apr 2008
14. Wang H, Nakata T (2009) Analysis of the market penetration of clean coal technologies and its impacts in China's electricity sector. Energy Policy 37:338–351
15. IEA (2008) Energy technology perspectives 2008: scenarios & strategies to 2050. International Energy Agency, Paris
16. Yi H, Hao J et al (2007) Atmospheric environmental protection in China: current status, developmental trend and research emphasis. Energy Policy 35:907–915
17. Zhao L (2008) Economic assessment of deploying advanced coal power technologies in the Chinese context. Presentation at ETIP seminar series, Cambridge: Energy technology innovation policy research group, Belfer Center for Science and International Affairs, 19 Feb 2008
18. MoP (2008) Power sector at a glance. Ministry of Power (MoP), Govt of India, New Delhi. http://powermin.nic.in/JSP_SERVLETS/internal.jsp. Accessed 12 Sept 2008
19. MoP (2006) Annual report 2005–06. Ministry of Power, Government of India, New Delhi
20. Chikkatur AP, Sagar AD (2007) Cleaner power in India: towards a clean-coal-technology roadmap. Discussion paper 2007–06. Belfer Center for Science and International Affairs, Harvard University, Cambridge

21. TERI (2004) TERI Energy Data Directory and Yearbook (TEDDY) 2003–04. The Energy and Resources Institute, New Delhi
22. Dubash NK, Rajan SC (2001) Power politics: process of power sector reforms in India. Econ Polit Wkly 36:3367–3390
23. MoP (2008) Annual report 2007–08: power – the building block of economy. Ministry of Power, New Delhi
24. CEA (2000) Sixteenth electric power survey of India. Central Electricity Authority, Government of India, New Delhi
25. Planning Commission (2006) Integrated energy policy: report of the expert committee. Planning Commission, Government of India, New Delhi
26. CEA (1997) Fourth national power plan 1997–2012. Central Electricity Authority, Government of India, New Delhi
27. Gopalakrishnan A (2005) Indo-US nuclear cooperation: a non-starter? Econ Polit Wkly 40(27):2935–2940
28. MoEF (2004) India's initial national communication to the United Nations framework convention on climate change. Ministry of Environment and Forests, Government of India, New Delhi
29. Chikkatur A (2005) Making the best use of India's coal resources. Econ Polit Wkly 40:5457–5461
30. CEA (2003) Report of the committee to recommend next higher size of coal fired thermal power stations. Central Electricity Authority, Ministry of Power, New Delhi
31. IEA (2002) Coal in the energy supply of India. International Energy Agency – Coal Industry Advisory Board, Paris
32. Sachdev RK (1998) Beneficiation of coal – an economic option for the Indian power industry. In: 17th World Energy Congress, Houston, 13–18 Sept 1998
33. Visuvasam D, Selvaraj P et al (2005) Influence of coal properties on particulate emission control in thermal power plants in India. In: Second international conference on clean coal technologies for our future (CCT 2005), Sardinia
34. Masto RE, Ram LC et al. (2007) Soil contamination and human health risks in coal mining environs. In: 1st international conference on managing the social and environmental consequences of coal mining in India, New Delhi, Indian School of Mines University, Dhanbad
35. EIA (2007) Country analysis briefs: Mexico. Energy Information Administration, U.S. Department of Energy, Washington
36. Flores-Montalvo A (2005) Private vs. public ownership of power generation in Mexico: should environmental policymakers care? Doctoral thesis, Massachusetts Institute of Technology
37. Breceda M (2000) Debate on reform of the electricity sector in Mexico. Report on its background, current status and outlook, North American commission on environmental cooperation, Montreal
38. SENER (2010) Sector Eléctrico Nacional – Generación Bruta. Secretaría de Energía. Distrito Federal, Mexico
39. SENER (2008) Sector Eléctrico Nacional – Generación Bruta. Secretaría de Energía, Distrito Federal, Mexico
40. SENER (2008) Consumo de Combustibles Secretaría de Energía. Distrito Federal, Mexico
41. Molina LT, Molina MJ (2002) Air quality in the Mexico megacity: an integrated assessment. Kluwer, Dordrecht/Boston/London
42. Vijay S, Sanchez S et al. (2008) Structural shift in the Mexican power sector – implications for air emissions. In: International Energy Workshop 2008, Paris, 30 June 2008
43. CEA (2007) Report of the working group on power for 11th plan. Central Electricity Authority, Government of India, New Delhi. http://cea.nic.in/planning/WG%2021.3.07%20 pdf/03%20Contents.pdf
44. Chikkatur AP, Sagar AD (2010) Rethinking India's coal-power technology trajectory. Econ Polit Wkly 44(46):53–58

45. Edmonds J, Wise M et al (2007) Global energy technology strategy – addressing climate change: phase 2 findings from an International Public-Private Sponsored Research Program. Pacific Northwest National Laboratory, Richland
46. Chikkatur AP, Sagar AD (2009) Positioning the Indian coal-power sector for carbon mitigation: key policy options. Pew Center for Global Climate Change, Arlington
47. IAEA (2008) Power reactor information system. http://www.iaea.or.at/programmes/a2/. Accessed 12 Dec 2008
48. Dubash NK, Rao DN (2007) The practice and politics of regulation: regulatory governance in Indian electricity. Macmillan, New Delhi
49. Dubash NK, Singh D (2005) Of rocks and hard places: a critical overview of recent global experience with electricity restructuring. Econ Polit Wkly 40:5249–5259
50. Sharma DP, Chandramohanan Nair PS, Balasubramanian R (2005) Performance of Indian power sector during a decade under restructuring: a critique. Energy Policy 33:563–576
51. Singh A (2006) Power sector reform in India: current issues and prospects. Energy Policy 34:2480–2490
52. IEA (2006) China's power sector reforms. International Energy Agency, Paris
53. Ibarra-Yunez A (2008) Electricity reform in Mexico: challenges for an integrated North American market. The Centre for International Governance Innovation. http://www.portalfornorthamerica.org/teaching-resources/electricity-reform-mexico-challenges-integrated-north-american-market#bibliography
54. SENAR (2011) SECTOR ELÉCTRICO NACIONAL - PROYECTOS DE GENERACIÓN EN PROCESO DE CONSTRUCCIÓN. Secretaría de Energía, Distrito Federal, Mexico. http://www.energia.gob.mx/res/PE_y_DT/ee/Proyectos_de_Generacion_en_proseso_de_Construccion.xls

Chapter 12
Potential Adverse Environmental Impacts of Greenhouse Gas Mitigation Strategies*

C. Andrew Miller[†] and Cynthia L. Gage

Abstract The Fourth Assessment Report released by the Intergovernmental Panel on Climate Change (IPCC) in 2007 was unequivocal in its message that warming of the global climate system is now occurring, and found, with "very high confidence" that it was "very likely" that the observed warming was due to anthropogenic emissions of greenhouse gases (GHGs). To address the problem, the IPCC developed an outline of approaches to reduce GHG emissions to desired levels. The expected changes in technologies and practices needed to mitigate emissions of GHGs will lead to changes in the impacts to the environment associated with energy production and use. Some of these changes will be beneficial, but others will not. This chapter identifies some of the potential environmental impacts (other than the intended mitigation of climate change) of implementing GHG mitigation strategies, but will not attempt to quantify those impacts or their costs. Included are discussions of the impacts of implementing energy efficiency and conservation measures, fuel switching in the power generation sector, nuclear and renewable energy, carbon capture and storage, use of biofuels and natural gas for transportation fuels, and hydrogen and electricity for transportation energy. Environmental impacts addressed include changes in air emissions of nitrogen oxides, sulfur dioxide, and particulate matter; impacts to water quality and quantity; increased mining of coal to meet the power demands of carbon capture systems and of metals to meet demands for vehicle batteries; and impacts to ecosystems associated with biofuel production and siting of other renewable energy systems.

---

*The findings included in this chapter do not necessarily reflect the views or policies of the Environmental Protection Agency. Mention of trade names or commercial products does not constitute Agency endorsement or recommendation for use.

[†]© US Government 2011

C.A. Miller (✉) and C.L. Gage
Air Pollution Prevention and Control Division, U.S. Environmental Protection Agency,
Office of Research and Development, National Risk Management Research Laboratory,
Research Triangle Park, NC, USA
e-mail: miller.andy@epa.gov

12.1   Background

The Fourth Assessment Report (AR4) released by the Intergovernmental Panel on Climate Change (IPCC) in 2007 was unequivocal in its message that warming of the global climate system is now occurring, and found, with "very high confidence" that it was "very likely" [1] that the observed warming was due to anthropogenic emissions of GHGs [1].[2] To address the problem of climate change, the IPCC also reported an outline of approaches to reduce GHG emissions to desired levels [2]. These approaches focused on energy production and use, and included improved energy end use and production efficiency, use of low-carbon fuels and renewable energy sources, development of policies such as land use planning that encourage system-wide energy efficiency, control of non-CO_2 GHG emissions, and development and application of carbon capture and storage (CCS) technologies.

These expected changes in technologies and practices will lead to changes in the impacts to the environment, some of which are beneficial and others of which are not. Energy conservation, for instance, will generally have beneficial environmental impacts. However, most of the available means of GHG mitigation result in some degree of adverse environmental impact. This is most clearly seen in the recent scientific (and increasingly public) debate about whether biofuels have a positive or negative influence on the environment [146, 147, 148, 149].

This chapter attempts to identify some of the adverse environmental consequences associated with GHG mitigation for a wide range of mitigation approaches, but will not attempt to quantify those impacts or their costs. This discussion will not evaluate specific GHG mitigation strategies, but is intended to identify the potential environmental impact (other than the intended mitigation of climate change) of implementing those strategies. The purpose is to provide a starting point for further and more in-depth evaluations of the environmental impacts with the aim of avoiding as many adverse impacts as possible as GHG mitigation approaches are implemented. A comprehensive discussion of the adverse environmental impacts associated with all possible, or even likely, mitigation options cannot be covered in a single chapter. Rather, our intent is to discuss selected options across key sectors, as a means to highlight that there can be adverse impacts associated with potential strategies to mitigate climate change.

We will also note that we do not address here the range of mitigation strategies and their associated impacts for the highly diverse industrial sector. Some of the

---

[1] The IPCC defined terms of confidence and probability for use in their reports. "Very high confidence" was defined as having at least a 90% chance of being correct. "Very likely" was defined as having a likelihood of the stated outcome of greater than 90%.

[2] The widely accepted list of greenhouse gases (and that used by the IPCC) are carbon dioxide (CO_2), methane (CH_4), nitrous oxide (N_2O), hydrofluorocarbons (HFCs), perfluorocarbons (PFCs), and sulfur hexafluoride (SF_6).

material below has relevance to the industrial sector. For instance, the discussions of carbon capture and sequestration and use of alternative fuels in the power generation sector are largely relevant to industrial sector fuel combustion. The environmental impacts associated with process changes, use of more efficient motors and other equipment, and tradeoffs between changing components versus changing systems are not discussed here. We have also omitted discussion of the agricultural sector outside the section on biomass as an energy source. Although we recognize the importance of the industrial sector relative to GHG mitigation, we have chosen to focus on power generation, transportation, and residential/commercial energy end use in the available space.

It has been well recognized for some time that many of these approaches are likely to have multiple environmental benefits, and that accounting for improvements in air quality and similar environmental indicators will further increase the benefits associated with many of the GHG mitigation approaches [3]. Energy efficiency, for instance, will reduce the need for fossil fuel combustion in general, which will reduce emissions of the full suite of pollutants generated by fossil-fuel electric generating stations, motor vehicles, and industries. Similarly, use of natural gas rather than coal will result in lower emissions of sulfur dioxide (SO_2), nitrogen oxides (NO_x), and particulate matter (PM).

The recent Energy Technology Perspectives report published by the International Energy Agency (IEA) stated that achieving the desired objectives of reducing GHG emissions and minimizing the impacts of increased GHG concentrations in the atmosphere "will require a transformation in how we generate power; how we build and use homes and communities, offices and factories; and how technologies are developed and deployed in the transport sector" [4]. The report amplified the basic conclusions of numerous other studies that have pointed out the need for significant changes in how energy is produced and used. The range and extent of technological change needed to reduce GHG emissions are tremendous, if the reductions that are anticipated to be necessary are to be achieved [5].

Some caveats are in order before we begin to discuss the consequences themselves. First and most importantly, identification of the consequences in this discussion should be considered as a preparation for avoiding or mitigating them, rather than as an argument to delay or avoid adoption of serious GHG mitigation approaches themselves. Although actions to mitigate climate change are being developed and proposed to avoid long term and potentially catastrophic consequences of global warming, the specific adverse environmental impacts of the proposed changes in consumer behavior, technology, and energy use must also be considered in the development of GHG mitigation strategies. The adverse impacts of the mitigation approaches are likely to be experienced more quickly and more immediately than the impacts on warming due to a major climate change mitigation program, and even though we may be working to minimize the impacts of climate change, we must also continue to protect human and environmental health.

Second, we must emphasize that the purpose of this discussion is to identify some of the potential adverse environmental impacts that have largely been overlooked during the broader discussions of how to mitigate GHG emissions.

Although many of the GHG mitigation approaches can provide significant environmental benefits beyond GHG reductions, it is well beyond the scope of this effort to attempt to quantify or even identify the full range of possible beneficial or adverse consequences.

The types of mitigation approaches and the extent to which they are deployed, and thus the extent of any adverse environmental impacts, will depend upon the total level of required GHG reduction. Because the level of GHG emission reduction will depend upon economic growth, population growth, and the carbon intensity of energy production over several decades, the approach typically taken to evaluate possible emission reduction requirements has focused on different scenarios of future changes.[3] The IPCC, for instance, has developed a series of emissions scenarios [6] that describe possible future conditions and these have provided a widely used set of consistent scenarios that researchers and policy makers can use for evaluating different mitigation strategies. This evaluation will focus on the key technologies identified in the IPCC scenarios.

Finally, it must be emphasized that adoption of mitigation technologies does not take place in a vacuum. Changes in the mix of technologies for energy production and use are made in the context of the existing energy system, and the impacts of such changes are properly evaluated in comparison to the range of options that are available at the time those changes are made. Most commonly, the comparison is to a business-as-usual (BAU) option. Although the BAU option is often considered to be a "no action" approach, it is more accurately characterized as a "continuing action" approach, and the environmental impacts of mitigation approaches are most appropriately compared against the impacts of continuing current practices. As noted above, however, the current discussion is not meant to be such a comprehensive evaluation of the broader advantages and disadvantages of mitigation technologies, but rather to point out that most, if not all, mitigation approaches have their own environmental impacts that cannot be ignored, even when the net environmental benefit is positive (and often strongly so).

12.2   Scale and System Influences

One of the more challenging aspects of the need to respond to the environmental impacts of climate mitigation strategies is scale. Most of the consequences that have been evaluated so far have been largely based on only a very limited range of experience and at a handful of locations. Although the U.S. Environmental Protection Agency is evaluating some of these impacts, such as how air quality and climate are linked, such evaluations are scarce [7]. Like mitigating GHG emissions

---

[3] It must be noted that the amount of CO_2 reductions required must also consider estimates of the desired levels of CO_2 in the atmosphere in combination with current concentrations and the projected levels if no mitigation actions are taken.

themselves, a critical factor is the scale of adoption. It is unclear how widespread adoption of climate change mitigation approaches will impact the environment, based on measurements at a few plants or on estimates from modeling studies. Again, biofuels offers a good example of the issues of scale. When U.S. bio-ethanol production was on the order of just 1–2 billion gallons per year, it was not an issue of scientific or policy debate. That level of production, even with corn as a feedstock, was largely sustainable (although it did not provide a significant GHG mitigation impact). As policies changed and set goals of 20 times that level of production, it became evident that such an increase could result in significant environmental problems globally.

To be effective in minimizing the rate of global temperature increase, the various approaches must be implemented nationally and globally at an accelerated rate. The IEA BLUE Map scenario (which would cut projected GHG emissions in 2050 by 50% relative to 2005 emissions) evaluated in the recent IEA Energy Technology Perspectives report estimates that, *each year*, 17.5 GW of new coal-fired power plants with CCS, 32 GW of new nuclear plants, 5 GW of biomass-burning power plants, 14,000 new wind turbines, *and* 215 million m^2 of solar PV panels will need to be installed to achieve the targeted GHG reduction goals [4]. Although these estimates are global, it is clear that the level of plant and infrastructure construction and introduction of new technologies is significantly higher than our existing regulatory systems have had to address in the past. Indeed, it is arguably a larger technical change than has happened in human history, particularly in terms of the fraction of the global population that will potentially be affected. The rate and magnitude of these changes will result in numerous environmental challenges, both foreseen and unpredicted.

In addition, to a much greater extent than conventional pollution mitigation strategies, mitigation of GHG emissions and the impacts of those mitigation strategies, will cut across economic sectors and geographical regions. It is imperative, then, to recognize and understand how specific technological, behavioral, and economic changes may have impacts beyond the point at which mitigation approaches are applied. For instance, if natural gas is used in substantial amounts to displace coal for electricity generation, the resulting increase in prices of both natural gas (due to increased demand) and electricity (due to higher generating costs) could result in industries turning to coal as a less expensive primary energy source. The net reduction of GHG emissions may therefore be less than originally estimated if such interactions are not considered, and local and regional air quality could be adversely impacted if appropriate steps are not taken to prevent increased emissions.

Additional environmental impacts are likely to be associated with changes to the energy infrastructure that will be needed to effectively utilize renewable and intermittent energy sources such as wind and solar. More transmission lines will be needed to transport energy from high-wind areas (which can shift over the course of a year or even a day) to locations where electricity demand is high. Additional pipelines may be needed to transport biofuels from relatively dispersed conversion plants to urban areas where fuel demand is greatest. Intermittent energy sources are also likely to need increased energy storage and recovery systems to address the potential differences in electricity production and consumption peaks; these systems may also have their own environmental impacts.

As we utilize material that we now consider wastes as feedstocks for other production processes, material transport patterns are likely to change, resulting in changes in emission patterns. In general, it is unlikely that 100% of waste material will be economically viable as process feedstocks, at least in the near term. Thus, the remaining wastes, although reduced in volume, may well be more difficult to process or dispose of in safe and environmentally sound ways.

The global nature of the problem and the interconnections of modern energy and economic systems will result in environmental impacts beyond the boundaries of any single locality, state, or nation. It is imperative that the international impacts on land use, energy use, and material production, all of which have implications for the environment, be evaluated and recognized during development of GHG mitigation strategies.

12.3   Electric Power Generation

12.3.1   Carbon Capture and Sequestration: Carbon Capture

Current carbon capture technologies require significant energy to remove CO_2 from the exhaust gas, which in turn requires significant increases in fuel consumption to generate adequate power to meet both end use demand and the needs of carbon capture technologies [8]. The IPCC's Special Report on CO_2 capture and storage [9] estimates that use of CCS would result in increased emissions of NO_x and NH_3, as well as increased solid wastes requiring disposal. Table 12.1, based on Table 3.5 of the Special Report, summarizes the changes in effluents.

Table 12.1 Material influents and effluents estimated for several carbon capture and sequestration approaches (Adopted from IPCC special report on carbon dioxide capture and storage [9])

Influent/effluent	Pulverized coal with CCS		IGCC with CCS		NGCC with CCS	
	Baseline (kg/MWh)	% Increase with CCS (kg/MWh)	Baseline (kg/MWh)	% Increase with CCS (kg/MWh)	Baseline (kg/MWh)	% Increase with CCS (kg/MWh)
Resource inputs						
Fuel	297	31%	312	16%	133	17%
Lime-stone	20.7	33%	–		–	
Ammonia	0.61	31%	–		–	
CCS reagents	0	(2.76)	0	(0.005)	0	(0.80)
Emissions						
Solid residues	58.8	39%	36.0	16%	0	0
SO_x	0.29	−99.7%	0.28	18%	0	0
NO_x	0.59	31%	0.09	11%	0.09	22%
NH_3	0.01	2,200%	–		0	0.002

The increased fuel use, particularly for coal plants, will result in increased mine production with its inherent environmental impacts. Included in these impacts will be increased CH_4 emissions [10], mine overburden, coal wastes, potential for water contamination, and surface subsidence. Many of these issues were addressed in detail in a 1999 life cycle assessment of coal-fired power generation by Spath et al. [11], which included air, water, and solid waste effluents for three types of power plants, and included emissions from power production, mining, and transportation [11].

For installation of a CCS system at a conventional power plant, the Spath et al. results can be used to estimate impacts by assuming a 30% increase in coal consumption as estimated in the IPCC CCS report. The major air emissions of concern from mining would be methane and other hydrocarbons, nitrous oxide (N_2O), and the criteria pollutants (particulate matter, SO_2, NO_x, and CO). Releases to water include chlorides, fluorides, sulfates, and iron and other metals, usually at low levels per kWh of electricity generation. The pollutants with the largest mass emission per unit electricity output are dissolved matter, oils, suspended matter, and chemical oxygen demand (COD). Spath et al. also address solid wastes, with mining accounting for a measurable fraction of hazardous, municipal, and industrial, and unspecified solid wastes. Table 12.2 shows the major effluents from mining operations

Table 12.2 Average life cycle emissions per kWh of electricity produced from an average coal-fired power plant, assuming river transport of coal from mine to plant [11]

	Total effluent (g/kWh)	Effluent due to mining (g/kWh)	Effluent due to mining, adjusted for CCS (g/kWh)
Emissions to air			
Carbon dioxide (CO_2)	1.02×10^3	8.67×10^0	1.13×10^1
Methane (CH_4)	2.00×10^0	1.99×10^0	2.59×10^0
Total particulate matter (unspecified)	1.01×10^1	9.31×10^{-1}	1.21×10^0
Sulfur oxides (SO_x as SO_2)	6.70×10^0	7.50×10^{-2}	9.76×10^{-2}
Non-methane hydrocarbons (including VOCs)	1.98×10^{-1}	6.66×10^{-2}	8.66×10^{-2}
Nitrogen oxides (NO_x as NO_2)	3.34×10^0	3.67×10^{-2}	4.78×10^{-2}
Carbon monoxide (CO)	2.61×10^{-1}	2.69×10^{-3}	3.49×10^{-3}
Nitrous oxide (N_2O)	4.30×10^{-3}	8.75×10^{-4}	1.14×10^{-3}
Organic matter (unspecified)	6.95×10^{-4}	9.39×10^{-5}	1.22×10^{-4}
Aldehydes	4.64×10^{-4}	5.23×10^{-5}	6.80×10^{-5}
Ammonia (NH_3)	2.22×10^{-4}	1.16×10^{-5}	1.51×10^{-5}
Emissions to water			
Dissolved matter (unspecified)	1.57×10^{-1}	8.87×10^{-3}	1.15×10^{-2}
Oils	1.87×10^{-3}	1.05×10^{-4}	1.36×10^{-4}
Suspended matter (unspecified)	1.91×10^{-4}	1.52×10^{-5}	1.97×10^{-5}
Chemical oxygen demand (COD)	1.94×10^{-4}	9.70×10^{-6}	1.26×10^{-5}
Emissions of solid waste			
Waste (unspecified)	2.15×10^1	3.50×10^0	4.55×10^0
Waste (municipal and industrial)	1.24×10^{-1}	1.70×10^{-2}	2.20×10^{-2}
Waste (hazardous)	3.31×10^{-9}	1.23×10^{-10}	1.60×10^{-10}

associated with coal-fired power production. The values given here are for the study's average plant plus an assumed 30% factor for increased coal consumption. These figures include the impact associated with direct emissions as well as emissions from underground mining activities.

Babbitt and Lindner [12] also evaluated life cycle impacts from coal-fired electricity generation, and identified many of the same issues as Spath et al., although the Babbitt and Lindner report estimated that mining contributed a significantly higher fraction (76%) of total dissolved solids to water than did Spath et al. (less than 5%). Other life cycle studies have been more focused on air emissions and particularly GHG emissions [13, 14], largely with the purpose of enabling direct comparison between CCS and other mitigation alternatives. More recently, Koornneef et al. [15] evaluated the environmental impacts of CCS systems in the Netherlands from the perspective of institutional and procedural aspects of the screening and scoping phases of environmental impact statements and strategic environmental assessments, both of which are required in the Netherlands prior to approval of major projects. Although the study focused on the procedural aspects of CCS projects, it also identified many of the same potential impacts as noted in the earlier studies.

12.3.2 Carbon Capture and Sequestration: Geological Sequestration

Carbon capture must also include an approach for sequestering the captured CO_2. Current approaches to carbon sequestration rely on storage of CO_2 in underground geological formations, usually either deep saline formations or depleted petroleum reservoirs. Direct injection of liquid CO_2 into oceans has also been proposed, but this approach has not been proven at even large pilot scale [16]. Geological sequestration has been demonstrated to work in enhanced oil recovery applications and in a handful of large-scale pilot tests. Other approaches, such as ocean capture of CO_2 by natural processes, are often considered to be geoengineering approaches and are more fully discussed in Chap. 9.

Wilson et al. [17] identified several environmental risks associated with such geological CO_2 sequestration approaches and placed these risks in taxonomy according to the processes involved and whether the risks were local or global. A further study identified research needs and policy approaches to mitigate these risks [18].

In summary, those risks were identified as:

- Suffocation of humans or animals above ground
- Effects on plants above ground
- Biological impacts below ground on roots, insects, and burrowing animals
- Contamination of potable water directly, through increased mobilization of metals or other contaminants, or by displaced brines
- Interference with deep-subsurface ecosystems
- Ground heave
- Induced seismicity

The Department of Energy's carbon sequestration roadmap also pointed out the need to ensure protection of groundwater resources and minimize CO_2 leakage [19]. The roadmap identified other components of a sequestration program that will have environmental impacts, including construction of CO_2 pipelines and the need for drilling, perhaps in areas that are not now subject to deep well drilling for oil or gas. The DOE roadmap and the regional partnerships that are developing pilot sequestration projects are also evaluating monitoring, mitigation, and verification (MM&V) activities and approaches that will be required to minimize the environmental, legal, and financial risks associated with geological sequestration.

12.3.3 Nuclear Power

Many of the environmental issues associated with nuclear energy have been discussed at length, and the body of literature on these issues is substantial. There have been a few reports that have addressed life cycle GHG emissions [20, 21], but the majority of the recent literature has been related to nuclear waste and spent fuel storage. Examples of this large body of work are two recent reports by the National Research Council which addressed spent fuel storage [22] and transport [23] safety and security issues. Despite the large number of studies that have been conducted, concerns remain about fuel cycle safety and implementation of waste disposal. An MIT study, "The Future of Nuclear Power," concluded that, "We know little about the safety of the overall fuel cycle, beyond reactor operation" [24]. The study raised further concerns about geological waste disposal, but noted that the key issues seem to be less about technical feasibility than about actual implementation of the concept.

Beyond the issues associated with spent fuel disposal, a significant increase in nuclear power generation will lead to a number of other environmental impacts. Uranium mining and processing will require safe disposal of overburden and residues having low levels of radioactive materials. There is potential for contamination of water bodies during mining, depending upon methods and locations. The availability of cooling water is also a concern, particularly with competing water demand for residential, commercial, and other industrial uses.

Two other issues unique to nuclear power are of significant concern. The first of these is the potential for a catastrophic accident. The Chernobyl accident demonstrated the extent to which such an accident can impact the environment over large areas and for long periods of time [25–27]. Although commercial reactor designs in the U.S. are significantly different and cannot fail in the same manner as the Chernobyl reactor, the potential for radioactive contamination of large areas remains a possibility. The second issue is that of proliferation of nuclear material and its potential use in nuclear weapons, including "dirty bombs" that are used to disperse highly radioactive material using conventional explosives. The MIT report included this aspect of nuclear power in their evaluation and concluded that, "The current international safeguards regime is inadequate to meet the security challenges of the expanded nuclear deployment contemplated in the global growth scenario" [24]. These two issues will need to be addressed in detail as part of any significant increase in the use of nuclear power for GHG mitigation.

Even so, the British Department for Business, Enterprise, and Regulatory Reform (BERR) concluded in a recent white paper that, with the exception of nuclear waste storage and disposal, "the environmental impacts of new nuclear power stations would not be significantly different to those of other forms of electricity generation and that they are manageable, given the requirements in place in the UK and Europe to assess and mitigate the impacts" [28]. The extent to which the environmental impacts associated with nuclear power are comparable to those from other forms of power generation will likely depend strongly upon the degree to which these requirements are successfully adopted and followed.

12.3.4   Natural Gas

Natural gas, which is primarily CH_4, is an attractive option (relative to other fossil fuels) for reducing CO_2 emissions from electricity generation. Existing natural gas power generation technologies are now in widespread use, including natural gas combined cycle (NGCC) units, which can achieve thermal efficiencies of over 50% [29]. Natural gas is less carbon intensive than coal or petroleum, and can therefore reduce the CO_2 per unit of electric energy output, but has a higher global warming potential (GWP) of about 20 times that of CO_2. In addition, use of natural gas can reduce emissions of NO_x, PM, and other pollutants as well.

Natural gas can also be used as a transportation fuel, and has considerable potential for reducing GHG emissions. Wang and Huang evaluated life cycle emissions associated with increased natural gas use as a replacement for petroleum fuels, and estimated that, compared to gasoline-fueled vehicles, GHG emissions could be decreased by 20% for dedicated liquefied natural gas (LNG) vehicles, by nearly 50% for spark ignition LNG hybrid electric vehicles, and as much as 90% for spark ignition hybrid electric vehicles operating on gas diverted from natural gas flares [30].

In most cases studied by Wang and Huang, life cycle emissions of volatile organic compounds (VOCs), NO_x, and PM were reduced when using natural gas compared to emissions from gasoline-fueled vehicles, although NO_x emissions were estimated to increase (as much as 30%) for natural-gas-fired flex fuel vehicles and dedicated compressed natural gas vehicles. There has been concern over emissions of some hazardous air pollutants, particularly aldehydes, in comparison to petroleum-fueled engines, especially older diesel engine [31]. As newer engine and control systems are put into service, these differences are likely to decrease.

The production and use of natural gas results in some loss of CH_4 into the atmosphere, making it important to evaluate life cycle GHG emissions rather than focusing only on CO_2 emission reductions. Recent studies focused on the U.S. estimated leakage and other fugitive emissions to be between 1.1% and 1.4% [13, 32]. These and additional studies that have estimated life cycle GHG emissions indicate that total emissions across the full life cycle are substantially lower than those from the production and use of coal [33, 34]. Globally, the available data are not as detailed. Estimates for fugitive GHG emissions from petroleum and natural gas production are approximately 2.3% of total world natural gas production, which will be higher than emissions from natural gas production alone. The Energy Information Administration estimates that about 125×10^{12} cu ft (2.21×10^6 Gg at standard temperature and pressure) of gross natural gas was produced globally in 2004 [35], compared to approximately 50,000 Gg of world-wide CH_4 emissions estimated by EPA as fugitive emissions from combined petroleum and natural gas production and transport [36]. Global emissions vary considerably by country, with emissions from some countries being considerably higher than the 2.3% global average [36]. In general, however, increasing production of natural gas has the potential to result in additional fugitive CH_4 emissions, although measures can be employed to reduce those emissions.

Natural gas may also contain trace amounts of mercury, although there is considerable uncertainty with regard to the level of Hg in natural gas and the distribution of such contamination across different natural gas fields [37]. In general, it is very likely that displacement of coal by natural gas for electricity production will result in lower total Hg emissions, but local concentrations could increase in natural gas production fields and around processing facilities.

Additional potential impacts such as drilling fluid contamination of water bodies, increased vehicle traffic into production areas, and disturbance of ecosystems are likely to occur with increased exploration and drilling operations. If new or increased pipeline capacity is required, similar impacts will occur related to natural gas transport. There has also been considerable concern, including on the part of the U.S. Congress, regarding the increasing production of natural gas in "unconventional" formations such as the Marcellus shale formations [38]. Production of natural gas from these formations often relies on the use of hydraulic fracturing, which has the potential for contamination of water supplies by the materials used in the fracturing process.

There has been considerable discussion regarding the availability of natural gas reserves, leading to interest in methane hydrates as a possible additional source of natural gas. Methane hydrates are typically found in permafrost and deep-sea beds, and proposals for exploitation of these materials has led to concern regarding the potential for catastrophic releases of methane to the atmosphere and the subsequent increase in warming potential [39–42]. When evaluating the potential risks and benefits of such approaches, it may be appropriate to incorporate the potential for catastrophic change by using risk analysis approaches that specifically incorporate such possibilities [43].

12.3.5 Wind Power

Wind-powered generation systems do not directly emit pollutants into the environment during operation. Some air pollutant emissions, mostly from internal combustion engines, may occur during maintenance operations, but such emissions would be minimal. Much of the environmental impact associated with a wind turbine is likely to occur during construction, major maintenance, and decommissioning phases, all of which would be due to increased vehicular traffic and operation of on-site machinery. Because wind turbines are often located in remote areas, construction and use of roads to the turbines would result in environmental damage through soil disturbance, potential changes in water runoff and road dust emissions, and impacts to wildlife habitats.

It must be noted that these environmental impacts will necessarily be multiplied due to the large number of units that are projected to be installed. If large numbers (more than 100) of turbines are installed at a single site, these impacts may change significantly (either positively or negatively). For instance, a concentrated wind energy site may have continual maintenance operations that would result in on-going vehicular and machinery emissions, but could also result in less land area used for roads. As is the case with solar energy, wind energy will require additional electricity transmission lines, which will also result in environmental impacts associated with construction, habitat and soil disturbance, and periodic vehicle traffic.

On a unit energy output basis, wind generation has been estimated to have the lowest life cycle CO_2 emissions of any of the alternatives (see White and Kulcinski [44], cited in Meier et al. [45]).

The primary on-going environmental impacts of wind energy sites are noise, changes to visual resources, and collisions of birds and bats [46]. The Programmatic Environmental Impact Statement (PEIS) prepared by the Bureau of Land Management (BLM) was developed to evaluate the potential impacts associated with a proposed Wind Energy Development Program on BLM lands in nine western U.S. states [47]. The PEIS provides a comprehensive overview of the potential environmental impacts associated with construction, operation, and decommissioning of wind turbines from a general perspective, as opposed to evaluating the site-specific impacts.

The noise and visual resource issues have significant impacts when wind turbines are located near populated areas, although impacts to visual resources may be greater in currently undeveloped areas where people are accustomed to views without wind turbines. It is unclear how wind turbine noise may impact wildlife. The BLM PEIS notes that several studies have shown reduced densities of some species in habitats adjacent to roads, and at least one study attributed the reduction to elevated noise levels. In other studies, the BLM PEIS notes that the adverse noise effects on raptors were "in many cases" temporary, with the raptors becoming "habituated to the noise" [47].

The BLM PEIS reviewed several studies of bird and bat mortality due to collisions with turbines, towers, and transmission lines, and cited studies in locations from California to Massachusetts that estimated the number of bird fatalities to be from 0 to 4.45 fatalities per turbine per year. There was little apparent correlation between sites, with two large (2,900+ turbines each) California sites varying in mortality rates by roughly an order of magnitude (0.33–2.31 fatalities per turbine per year), and a site in West Virginia with on 44 turbines having a mortality rate of just over 4 fatalities per turbine per year [47]. It appears that there are significant site-specific factors that determine the bird impact rate, but those factors are currently not known.

Large-scale implementation of wind energy can also impact the broader climate by extraction of energy from the atmosphere [48]. The impacts on wind patterns and possibly precipitation patterns are estimated to be measurable, but are expected to have negligible effects on global surface temperature. Considerable research is needed to more accurately evaluate these impacts.

12.3.6 Solar Photovoltaic

The key environmental impacts associated with significant adoption of solar energy are the land area required, impacts due to the production and disposal of the solar collectors, and impacts associated with increased requirements for energy storage. As an indication of the scale of production that may be necessary, the IEA BLUE Map scenario projects a need to install, each year until 2050, an average of 215 million m^2 (about 86 mi^2) of solar panels world-wide to meet the 4,750 TWh/year of solar electricity generation they project will be needed to displace a portion of fossil-fuel-generated electricity [4]. This is in addition to concentrated solar thermal power plants.

In the U.S., the average amount of peak solar energy that reaches the surface is on the order of 1 kW/m^2. For photovoltaic (PV) systems, conversion efficiency is typically no higher than about 20%, so each m^2 of PV could produce no more than about 200 W of power. When combined with the changing azimuth of the sun over the course of a day, it is reasonable to assume that the net energy produced per m^2 would be roughly 1 kWh. Turner estimated 0.63 $kWh/m^2/day$ of electricity from flat plate collectors of 10% efficiency, and also estimated that the actual area would be twice that amount for infrastructure (supports, power lines, access, etc.) [49]. Higher efficiency collectors or tracking collectors would further reduce the amount of area required per kWh. Displacement of 1 GW of peak power would therefore require about 5 million m^2 (about 2 mi^2) of land area. Much of this surface area is currently located on rooftops in distributed generation systems, which reduces the amount of land surface that would need to be dedicated to PV collectors [4]. This approach may change as new PV technologies are developed that would be more effective in central solar facilities, but it is likely that both systems integrated into buildings and central facilities will have a role in future PV generation.

An additional environmental issue is the production process and the materials needed to produce PV cells. Work at the National Photovoltaics Environmental Research Center at Brookhaven National Laboratory has identified a number of environmental risks associated with life cycle production and use of PV power systems [50, 51]. A key concern is the use of cadmium-tellurium (CdTe) compounds in PV systems and the potential for increased Cd emissions in the mining and refining of Cd and the manufacture, utilization, and disposal of PV modules.

Some of the other hazardous materials present in PV manufacturing include arsenic compounds, carbon tetrachloride, hydrogen fluoride, and hydrogen sulfide, lead, and selenium compounds [52]. These compounds can be released in the event of fires, either at the manufacturing site or at PV installations, although there is some disagreement with respect to whether they would be released in a residential fire scenario [53, 54].

Disposal of used PV modules provides another route for introduction of hazardous materials into the environment. Fthenakis states that recycling is technologically and economically feasible, and estimates that there would be minimal environmental impact if such recycling and recovery is conducted properly [55].

The IEA, through its Photovoltaic Power Systems Programme, has recently started a task to evaluate the health and safety issues associated with PV production and use, including life cycle emissions estimates (http://www.iea-pvps.org/tasks/task12.htm). Much of the information in life cycle assessments is quite dated, as pointed out by de Wild-Scholten and Alsema, particularly given the advances in the semiconductor industry over the past decade [56]. Studies by Moskowitz and Lawrence et al. [57–59] are well over 15 years old, and do not account for changes made in PV production in the intervening time since. Recently, the U.S. Bureau of Land Management announced its efforts to develop a Programmatic Environmental Impact Statement to evaluate solar energy development on federal lands. This effort will involve evaluating the environmental impacts of tilting PV and solar-thermal systems in six states in the western U.S. [60].

12.3.7 Biomass Power

Although there has been considerable recent evaluation of the impacts of using biomass as a feedstock for the production of transportation fuels, biomass has long been considered as a fuel for electricity production. In general, the types of biomass feedstocks that have been considered for power production (such as waste woodchips and bagasse) are different than those most often proposed for transportation biofuels, which have focused on either sugars and starches (particularly sugar cane and corn) or biomass specifically grown as energy sources, such as switchgrass or poplar.

The key environmental issues associated with using biomass for electricity production are changes in air pollutant emissions and the impacts due to biomass production. Biomass used for power production can be used in conventional boilers, usually co-fired with coal, or alone in boilers specifically designed to burn biomass. In some cases, biomass is used in applications where the desired product is process heat rather than electricity, but even for these cases, biomass displaces fossil fuels and therefore can have an advantage relative to CO_2 emissions.

Gasification and direct combustion of biomass are not without potential environmental problems. In direct combustion systems, particularly stoker-grate furnaces, the amount of carbon remaining in ash can be quite high. Bottom ash may have an unburned carbon content as high as 50%, which may make the bottom ash more difficult to reuse. In such cases, it is not uncommon for the ash to be fed back into the boiler with the biomass to achieve higher carbon burnout. Fly ash composition can also be quite different than coal ash, with the potential for arsenic (As), cadmium (Cd), chromium (Cr), copper (Cu), lead (Pb), and Hg [61]. The high alkali metal content of many biomass feedstocks can also result in increased slagging and fouling of the boiler, which can reduce the thermal efficiency and therefore increase the total life-cycle GHG emissions associated with the production and use of biomass. Emissions from wood combustion, particularly from systems that are not well tuned to optimize the combustion process, can include organic compounds, including polycyclic aromatic hydrocarbons (PAHs), at higher levels per unit energy input than coal-fired systems. In units where carbon burnout is poor, direct biomass combustion systems can emit higher amounts of carbon monoxide (CO) and potentially even black carbon (BC), although BC emissions may not be as great a concern for industrial biomass use as for smaller, less well-controlled applications such as residential wood combustion.

Biomass feedstocks include dedicated energy crops, agricultural residues, forest management residues, and urban wood waste. Dedicated energy crops can result in displacement of other crops, eventually leading to significant changes in land use. It is likely that any significant biomass production specifically as an energy source will lead to more intensive agricultural or forest management practices. These more intensive practices will be more likely to have the types of adverse environmental impacts associated with modern intensive agricultural production, such as increased runoff of fertilizers, herbicides, and pesticides; higher irrigation demand; and potential soil degradation. Additional impacts associated with increased biomass demand can occur in situations in which existing forests or other growth is displaced to meet that demand, whether for fuel or other uses [62, 63]. A full life cycle analysis of the types of land changes and displaced growth is needed to fully evaluate the impact on net GHG mitigation, as well as impacts such as changes in wildlife habitat, local air quality, and water quality and quantity.

In addition to GHG emissions and reductions, land use can also impact climate change through changing the surface albedo, which can have significant global warming impacts [64]. The magnitude and even the direction of albedo changes associated with changes in land use for increased biomass production for energy are unclear, but should be recognized and incorporated into climate estimates as possible and appropriate.

Use of agricultural and forestry residues will not likely have significant land use impacts, but could result in other environmental impacts. Harvesting of agricultural residues could impact soil quality if those residues had previously been plowed back into the soil. Use of forest residues could also result in changes to forest soil quality through the removal of material that would otherwise decompose naturally, as well as change ecosystems in other ways by altering a component of the food chain. Disturbance of forest soils can also result in loss of CO_2 in the soil to the atmosphere, reducing the GHG benefit of this source of biomass. In both cases, additional fuel is likely to be needed for collection and transport of the residues, resulting in additional emissions of CO_2 and other air pollutants.

Urban wood waste – such as tree trimming residues, pallets, and construction materials – is typically disposed of in landfills, so collection and combustion of this feedstock could have significant benefits by reducing waste methane production, landfill size, and potential leaching of associated metals into ground and surface water. A potential drawback is the presence of treated wood or metals associated with wood waste, which could be emitted from the combustion process.

Because biomass requires a much greater land area to produce the same energy content as fossil fuels, there will be greater emissions from transportation from fuel production to plant site (or alternatively, a larger number of plants located closer to the fuel production). In addition, biomass is typically seasonal, which means that the fuel must be stored following harvest and used at a steady rate over the course of the year. Both the increased transportation and storage requirements can have adverse environmental impacts, whether through increased air emissions from trucks or potential runoff and leaching from storage facilities. Reduced plant sizes may be required to match available fuel supply, which could also have implications for the types of pollution controls (and subsequent effectiveness of those controls) installed at the plant.

12.3.8   Geothermal Power

There are a number of environmental impacts that are associated with geothermal power production [65]. Some of these, such as impacts during site preparation, well drilling, and construction are limited to initial development of the power plant. Others, such as potential emissions to the air and water and generation of solid wastes, will continue throughout the life of the plant. During operation, hydrogen sulfide (H_2S), ammonia (NH_3), and trace quantities of Hg are the compounds of greatest concern. Most geothermal plants re-inject the geothermal fluid back into the ground, but potential remains for leaks or spills into nearby water bodies. There is significant potential for water contamination from drilling fluids used during the drilling process.

Other impacts, such as ground subsidence from the long-term extraction of the geothermal fluid, contamination of water by toxic compounds, and disposal of solid wastes removed from the geothermal fluid, may also occur.

Emissions of air pollutants can occur from geothermal plants because of the presence of these compounds in the geothermal fluid. Emission rates depend upon both the concentration of these compounds in the fluid and the design of the plant (dry steam, flash steam, or binary designs). Bloomfield and Moore reported emissions of H_2S, NH_3, and CH_4 from several U.S. geothermal power plants of different designs in a report on emissions of CO_2 from geothermal power [66]. They reported average emissions of 0.82 kg (0.18 lb) of H_2S per MWh. These emissions can usually be reduced through application of control technologies, particularly for H_2S abatement.

Numerous documents are available that provide guidance for geothermal developers and identify appropriate measures to minimize adverse environmental impacts.

Among the organizations that provide guidance are the International Finance Corporation (associated with the World Bank) and the United Nations Environment Program [65, 67].

12.4 Transportation

12.4.1 Electric Vehicles

Large-scale production and use of all-electric vehicles have environmental implications in two key areas. The first is the increased consumption of electricity, leading to environmental issues noted above for electric power generation. Although a recent study indicated that the existing power generation system can accommodate significant penetration of plug-in hybrid electric vehicles (PHEVs), the study also pointed out that "Higher system loading could impact the overall system reliability when the entire infrastructure is used near its maximum capability for long periods" [68]. The study also noted that, if significant numbers of vehicles were being charged during load "valleys," the impact on the generating system could be to emphasize more base-load type generation as opposed to the smaller units installed to meet peak demands. This could result in overall greater energy and environmental efficiencies; although the report also noted that maintenance scheduling could be more difficult in the absence of current cyclical load patterns.

Overall, it is estimated that GHG and other emissions would be reduced by broad use of PHEVs, based on a national average 33-mile round trip length. The reductions would vary by region, and in some locations with very high coal generation, total emissions could increase. In the U.S. Northeast, for instance, GHG emissions could be reduced by 39% and NOx emissions by 59%, while in the Northern Plains, GHG emissions could increase by 1% and NOx emissions by 35% due to the high reliance on coal-fired generation [68]. Although these results are for PHEVs, the general results will hold for all-electric vehicles as well, because the study assumed that the PHEVs would be able to operate in all-electric mode for 33 miles, with longer trips being powered by fossil fuels.

The second implication is related to the increase in production and disposal of the electric storage system. Current all-electric vehicles use batteries as the energy storage system, although there has been considerable work to develop ultra capacitors that would allow rapid charging and higher power density. The most common type of battery now in use is the nickel-metal-hydride (NiMH) battery, with lithium (Li) ion, Li ion polymer, valve regulated lead acid (VRLA), and nickel-cadmium batteries also are under development for vehicle use. Increased use of NiMH batteries will necessarily require significant increases in nickel (Ni) production and the associated impacts associated with Ni mining and refining. Lave et al. noted that the key issues may be with the other metals needed for NiMH or Li ion batteries, cerium and cobalt, which tend to be present as minor trace elements in other metal ores and are more difficult to extract. This would lead to additional environmental consequences as large amounts of ore would need to be processed to obtain the necessary amounts of metals, even at the relatively low levels of penetration assumed by Lave et al. in their analysis (11% of the fleet for NiMH and 22% for lithium ion) [69]. Similarly, Andersson and Rade evaluated the long-term resource constraints associated with substantial penetration of electric vehicles. Such increased demand would likely result in increased metal prices and subsequent increases in metal mining and refining [70]. Vimmerstedt et al. evaluated the recycling of lead from lead-acid batteries that could be used in vehicle propulsion systems, and noted that lead recovery facilities may need to install additional infrastructure such as backup power generation to ensure adequate environmental protection [71]. Current battery recycling programs recover 90–95% of lead in batteries, but presumably this would increase as replacement of propulsion batteries moved to professional shops rather than being done by vehicle owners at home.

Lithium ion batteries have received considerable attention as a promising battery technology. If anticipated improvements are achieved using Li, it is likely that lithium extraction and processing will increase substantially, with a corresponding increase in potential releases of Li into the environment. Lithium has long been used to treat aggressive behavior [72]. The potential health effects of its long-term use have been evaluated in this context, with the finding that there are few effects even at intake levels that are much higher than would be typical of environmental exposure [73]. However, there is some indication that there are adverse impacts associated with Li concentrations in the ambient environment [74], although additional studies are needed to more fully understand the potential impacts.

12.4.2   Plug-in Hybrid Vehicles

The environmental issues associated with plug-in hybrid vehicles are the same as those associated with all-electric vehicles, plus the issues associated with internal combustion engines. It is likely, however, that the issues surrounding metals used in batteries would likely be somewhat reduced, given that the batteries for hybrids would be smaller. In addition, the air pollutant emissions would likely be reduced as well compared to a conventional internal combustion engine, given that the engines for hybrid vehicles are designed to be smaller and operate within a narrower and more efficient range. Even so, significant penetration of plug-in hybrid vehicles could result in upstream impacts associated with metal mining and refining, and downstream impacts associated with battery recovery, similar to those associated with all-electric vehicles.

12.4.3   Biofuels and Other Non-Petroleum Fuels

The environmental impacts of switching from petroleum-based fuels to those from other feedstocks must include the impacts across the complete life cycle of primary energy production, feedstock logistics, conversion to useful fuel, distribution, and storage of the fuel, and fuel end use.

For biofuels, the feedstock production and logistics impacts are the same as those discussed for biomass above: impacts on land use (and the potential for GHG emissions due to such changes), water quality and quantity issues, potential impacts to soil quality, air emissions from feedstock collection and transport, and potential ecosystem impacts due to changes in the biomass being grown and the disruption during production and harvest.

A significant increase in the ethanol content of fuel for motor vehicles will result in environmental impacts during the distribution and storage phase of the full bio-fuel life cycle. Increased transport of ethanol, either through pipelines or by truck or rail will result in increased spills of the fuel. The impacts of such spills include potential major fish kills if spilled into open water bodies, contamination of ground-water, mobilization of inorganic compounds such as iron and manganese, and the potential for generating noxious odors during decomposition [75]. When released into groundwater, the presence of ethanol and gasoline can result in increased ben-zene concentrations compared to gasoline alone due to the changes in degradation chemistry caused by the ethanol [75–77].

Increasing levels of ethanol in gasoline also changes the profile of compounds emitted to the air from engine operation. In general, higher ethanol concentrations tend to result in higher emissions of aldehydes, particularly formaldehyde and acet-aldehyde [78]. Although emissions of other pollutants, including organic com-pounds, may decrease due to the higher oxygen content of ethanol-gasoline mixtures, some studies have estimated that wide-scale use of high-ethanol content fuel such as E85 (85% ethanol, 15% gasoline) could result in higher ambient ozone and aldehyde concentrations and therefore higher mortality rates compared to those that would be projected for conventional gasoline or low-ethanol content fuels (E10) [79].

Given the potential environmental impacts associated with biofuel production and use, it is crucial to determine whether biofuels result in a net reduction in fos-sil fuel consumption and fuel-related greenhouse gas emissions. The current tech-nology used for the majority of biofuel production, ethanol from corn, appears to provide a slight net gain in energy compared to petroleum, and can provide a net reduction in GHG emissions [80–83]. Although the gain in total energy is rela-tively modest for corn-based ethanol (20–25%), corn ethanol does displace a sig-nificant (80–90%) volume of petroleum, because the majority of energy inputs to ethanol production are natural gas and electricity [81, 84]. The remaining fossil energy input is from coal and natural gas used in the feedstock production and fuel conversion stages of the life cycle, and these uses have the same impacts as noted above. In general, the extent to which life-cycle GHG emissions are reduced when using corn-based ethanol may depend upon two key factors: the amount of coal-derived energy used to power the conversion plant and the net changes in land use (noted above).

The Energy Independence and Security Act of 2007 (EISA) has provided considerable incentives for accelerated development of technologies to convert biomass to motor vehicle fuels [85]. Technologies to convert biomass to "advanced biofuels" (i.e., fuels that have 50% or greater reduction in life-cycle GHG emissions compared to petroleum fuels) are being developed at a particularly rapid rate due largely to the lack of any technology with a clear competitive advantage. For ethanol, those technologies are focusing on conversion of cellulosic biomass through either thermochemical or biochemical processes. Most new process designs appear to be working toward minimization of effluents from their particular process, but true zero-emissions systems are unlikely to be available at a commercial scale in the near term.

Effluents from cellulosic ethanol plants are likely to include conventional air pollutants as well as CO_2,[4] wastewater, and solid residues. Lignin-based residues will likely be one of the more substantial byproducts of cellulosic ethanol production, although it also has considerable value as a feedstock for non-fuel bio-based products or a as a fuel for heat and power generation. Another potential solid effluent is gypsum, which results from the use of lime to neutralize sulfuric acid used to hydrolyze the biomass feedstock to separate the lignin from the cellulose and hemicellulose. Although other hydrolyzation processes may be used that would avoid the generation of gypsum, acid-based processes will generate significant quantities of gypsum that will require disposal.

Biochemical processes have been developed to use organisms and enzymes that have, in many cases, been developed specifically to enhance ethanol production [86, 87]. There is also considerable interest in feedstocks that have been bred or modified to maximize their energy production potential [88–90]. These biological materials include genetically modified organisms (GMOs) and other modified biologicals that are not found naturally in the environment. The impacts of these materials if released into the environment are not understood, and therefore represent a key gap in our ability to evaluate the environmental risks and possible mitigation approaches to such releases. The industry is changing rapidly, and this pace of development is likely to continue for the foreseeable future. It is therefore likely that the number and types of these new biological materials will continue to increase, with unknown potential for adverse environmental impacts.

The existing regulatory and technical infrastructure is much better developed relative to thermochemical processing. Many, if not most, petrochemical processes rely on thermochemical processing, and these processes are technically mature and have been subject to regulatory oversight for many years. On the other hand, the biochemical processes now being developed for large-scale biofuel production are not as well developed and are less well characterized from a regulatory perspective. Although thermochemical cellulosic ethanol plants may well increase emissions of air and water pollutants and solid wastes, at least in locations that currently have little, if any, previous industrial process emissions, these changes are likely to be similar in kind to emissions from other thermochemical processes that have been

---

[4] In this case the CO_2 emissions would be a mixture of fossil- and biomass-based CO_2. The fermentation process generates significant levels of CO_2, but this CO_2 is biomass-based and therefore not a net addition to the atmospheric carbon cycle.

used commercially for decades. Beyond the question of emissions into water across the biofuel life cycle, biofuels (and particularly ethanol) will also impact water quantity. Recent life-cycle studies suggest significant increases in water consumption of up to 20 times that required for production of petroleum fuels [91, 92], with one study concluding that corn-based ethanol will require over 1,000 gal of water to produce 1 gal of ethanol in the US [92].

A full life cycle assessment of the environmental impacts of biofuels will also need to include issues such as water consumption by feedstock production and conversion to fuel, soil productivity, and other parameters that are impacted by intensive agricultural practices. Finally, there have been considerable advances in the area of conversion of biomass to hydrocarbon fuels [93]. These processes may have different environmental impacts than the biochemically-based fermentation processes that produce alcohols, and both process developers and regulatory agencies need to be aware of such changes and how they may need to be addressed.

12.4.4 Hydrogen as a Transportation Fuel

Hydrogen (H_2) has been long considered as a possible fuel for use in transportation and distributed generation, primarily because of its ability to generate electricity with high-energy conversion efficiencies while producing almost nothing other than water as the process byproduct. From a climate perspective, it is important to recognize that H_2, like electricity, requires significant inputs of other forms of energy to produce. Most H_2 is currently produced from natural gas, although other methods of production from fossil fuels include thermal cracking, partial oxidation, and gasification, all of which generate large amounts of CO_2 [94, 95]. One of the primary by-products of steam-methane reforming (SMR) is CO_2 which is presently vented to the atmosphere. In order to achieve mitigation from using hydrogen as a transportation fuel, this CO_2 by-product must be captured and sequestered. Depending on the efficiency of the reformer, carbon monoxide may be another emission with environmental concern. Hydrogen from the SMR process is at low pressure, and thus energy is required to compress the gas to a pressure required for refueling vehicles. Generally, there is about 25% more energy required to produce hydrogen as there is to produce gasoline [96]. Both of these energy demands will mean more fossil-fuel use and associated emissions.

Other concerns are associated with the methane feedstock. Use of natural gas as a feedstock for hydrogen production will compete with natural gas usage as a fuel in electricity production. The potential for fuel switching to cheaper coal and oil in electric generation would lead to increased pollutant emissions associated with these fuel sources [97]. Increased demands for methane can lead to increased fugitive emissions from methane operations. In addition, demand will lead to greater import quantities and thus the potential for ecosystem impacts as LNG facilities are constructed to receive the imports [98].

Use of renewable energy sources (solar, wind, and biomass) for H_2 production has been tested, but these are currently not economically viable, particularly given the need for development of a transport, distribution, and storage infrastructure for H_2. Emerging approaches include photobiological, photolysis, and enzymatic methods for H_2 production, but these processes remain in the concept stage [99]. In general, increased H_2 production requires an increased use of natural gas, with the associated environmental impacts associated with its production, transport, and processing [95]. Life cycle emissions of GHGs may be reduced if H_2 is used in high-efficiency fuel cells, but these are not likely to be available in the near term.

Production of H_2 fuel cells will require significant increases in the total amount of platinum (Pt) consumed worldwide, with a similar increase in mining and the environmental impacts associated with mining, processing, and transport. Although it has been estimated that the world reserves of Pt are sufficient to meet projected demand [100], most Pt is produced by South Africa and Russia and imported to the U.S. and other industrial nations. This has led to concern over Pt supply bottlenecks [101], and even to evaluation of the potential for production of Pt from waste generated from nuclear power reactors [102]. Although it is unlikely that Pt will be derived from spent nuclear fuel, these proposals do highlight the fact that a switch to a new technology will result in shifts in the critical infrastructure needed to support that technology, and that development and maintenance of that infrastructure may lead in unforeseen directions.

Fundamentally, it must be recognized that H_2 is not a primary energy source but is instead a means of energy storage and transport. Thus, the net environmental impacts associated with use of H_2 will necessarily include the impacts associated with the production, storage, and transport in addition to the impacts associated with H_2 use. It is crucial, then, to evaluate H_2 energy systems over the entire life cycle, including the sources of energy used to generate H_2, to fully understand the environmental impacts associated with H_2 use.

If H_2 is used in a significant way as a transportation fuel, consideration must be given to the impacts of H_2 emissions from leaks during production, fueling, and operation. Tromp et al. estimated that anthropogenic H_2 emissions could increase by a factor of 4–8, and total emissions by a factor of 2–3, if it were used to completely replace petroleum combustion [103]. They modeled the impacts on atmospheric chemistry and estimated that such increased H_2 could result in stratospheric ozone depletion by up to 20%, affecting the atmospheric lifetimes of CO and CH_4 due to changes in OH^- radical concentration, and increase noctilucent clouds (with subsequent impacts on albedo). They also point out that H_2 interacts with soils, and changes in atmospheric H_2 concentrations could have unforeseen impacts on soil microbial populations.

12.5   Increased End-Use Efficiency

Improvements in energy end-use efficiency are expected to be achieved through increased use of electronics in a myriad of applications, such as in buildings, vehicles, and improved delivery routing [104–106]. This trend will accelerate as

GHG mitigation strategies put additional emphasis on energy efficiency, but will also accelerate the already-growing problem of waste electronics disposal. Although use of electronics can have significant positive impacts on life-cycle efficiency and GHG reduction, their disposal can create local problems associated with the toxic materials that are used in electronics manufacture [107–109]. New approaches are being developed to reduce the environmental impact of electronics disposal and increase material recycling [110], but these approaches are not widely used, particularly in developing economies [107].

Increasing end-use energy efficiency will also provide opportunities for use of new materials, and possible use of those materials on a large scale. A key example of these new materials are nanomaterials, which are being evaluated for use in construction as coatings, concrete and steel additives, components of composite materials, and glass products, among others [111, 112]. Nanomaterials are developed from modification of materials at scales of 100 nm and smaller. The environmental impacts of these materials are largely unknown, with research in this area being in its early stages. Even so, there are indications that exposure to nanomaterials may cause adverse health impacts [113, 114]. While it is unclear whether such impacts will occur in real-world exposures, concerns have been raised about nanomaterials making their way into drinking water, with little information about their potential effects [115, 116]. Although there is, and will be, considerable pressure to move these new materials into practice, the lack of clear understanding of possible impacts has raised concerns about the pace at which these new technologies are being deployed [117].

End-use efficiency improvements can incorporate more than direct energy use reductions to include reductions in other resources. Reducing material and water reduces the energy and, therefore, the GHG emissions associated with the production and transport of those resources. It should be noted that end-use efficiency is likely to be less effective when broadly implemented in practice than would be indicated by technology-specific efficiency improvements. This is due in part to consumers' response to the reduction in energy prices that occurs as demand falls – the "rebound effect" [118]. It should also be noted that reductions in energy consumption do not always translate directly into the same level of GHG reductions. This is due to the tendency of energy suppliers to respond to drops in demand by reducing use of their highest-cost energy sources, which are often among the lower-carbon sources.

In the following sections, we discuss a few of the approaches for improving energy end-use efficiency in residential and commercial buildings. The particular approaches do not cover the full spectrum of available options, but are representative of some of the potential environmental impacts associated with these types of approaches.

12.5.1 Residential and Commercial: Lighting

One approach to increasing efficiency of lighting is the use of compact fluorescent lamps (CFLs). These are compact versions of fluorescent lamps that are designed to fit into fixtures developed for conventional incandescent light bulbs, and use 70–75% less energy than incandescent bulbs. EISA set efficiency standards for incandescent bulbs [85], which will require development of more efficient incandescent bulbs but which CFLs can now achieve. A potential disadvantage of CFLs is that they contain a small amount (approximately 2–6 mg) of mercury (Hg), which can be released if the bulb is broken or if it is disposed of improperly or incinerated (a concise overview of this issue is provided by the Congressional Research Service [119]). In situations where a single bulb is broken, such as in a residential setting, the amount of Hg present is estimated to be low enough to avoid any health risk. Recent research has identified the potential for using nanomaterials as a means to absorb Hg in the event a bulb breaks, which would further reduce any immediate risk of individual exposure to Hg in residential or commercial settings [120].

Even with improvements to minimize individual exposures, improper disposal of significant numbers of CFLs could lead to an avoidable increase in Hg emissions to the environment. Over 270 million CFLs were sold in 2007, accounting for approximately 20% of the U.S. light bulb market. If all the Hg in these bulbs were to be released to the environment, that would account for 0.6–1.8 tons of Hg. If 100% of the U.S. light bulb market were in the form of CFLs that would result in a potential 3–9 tons of Hg that could be released into the environment if all the CFLs were disposed of improperly (an unlikely scenario), eventually on an annual basis as a steady state of failure and replacement was achieved. This compares to over 70 tons per year of Hg emissions to the air from U.S. stationary sources, as estimated from the Toxics Release Inventory [121].

Light emitting diodes (LEDs) are being developed for and applied to lighting applications, with significant decreases in energy consumption compared to large (100 W) incandescent bulbs and without the issue of Hg associated with CFLs. Although high-brightness LEDs are currently used for only a very small fraction of general illumination, they represent the fastest-growing segment of the LED market [122]. As LEDs are increasingly used for general illumination, they will add to the manufacturing and disposal issues noted above for electronics in general. In particular, LEDs typically contain arsenic, indium, and phosphorous, which may need to be more closely monitored in production and disposal locations. There has not been any indication that these elements are sources of personal exposure during use.

12.5.2 Residential and Commercial: Space Cooling and Refrigeration

For space cooling (which includes heat pumps) and refrigeration, the dominant technologies are those based on the vapor compression cycle. These technologies are powered by electricity, making it important to optimize system components and design for energy efficiency. This dual impact of vapor compression systems on climate change – the direct impact from refrigerant losses and the indirect impact due to electricity usage to power the system – led to development of the Total Equivalent Warming Impact (TEWI) methodology to quantify the combined impact in units of CO_2 [123]. Direct emissions are converted to CO_2 equivalents using GWPs of the different refrigerants, and indirect emissions are estimated based on electricity usage and the CO_2 emissions associated with electricity generation. Whether direct or indirect effects dominate the TEWI calculation depends on the system design. For household refrigerators with small refrigerant losses, indirect effects dominate, while for supermarket systems with large refrigerant charges and high refrigerant losses, direct effects dominate. As of year 2000, the direct emissions of refrigerants are estimated to be responsible for 13% of the warming already in place [124].

Presently, the primary refrigerants sold in the marketplace are hydrochlorofluorocarbons (HCFCs), hydrofluorocarbons (HFCs) and perfluorocarbons (PFCs) which have significantly higher 100-year GWPs (GWP_{100}) than CO_2, which has, by definition, a GWP_{100} of 1 (see Table 12.3). Due to the Montreal Protocol, the market is transitioning out of HCFCs and into HFCs. In 2008, EPA estimated that emissions from refrigeration and air conditioning systems represented 86% of the combined HFC and PFC emissions from all sectors in the U.S. economy [125]. A majority of those were HFC-134a emissions from motor vehicle air conditioners (MVACs). The European Union has signed a directive which bans the use of HFC-134a in new MVAC systems beginning in 2011. Replacement refrigerant must have a GWP_{100} of less than 150, and the primary candidate is CO_2 (R-744), a "natural" refrigerant.

Natural refrigerants are those which occur in nature's biological and chemical cycles without human intervention and include CO_2, ammonia, hydrocarbons (such as propane, butane, and isobutane), air, and water. For space conditioning and refrigeration, the natural refrigerants of primary interest are CO_2, ammonia, and hydrocarbons (GWP_{100} <20). Although natural refrigerants have no ozone depletion potential and low GWPs, they are not without other environmental concerns including toxicity, flammability, or in some cases lower operating efficiencies.

Systems using CO_2 will operate at higher pressures and in some designs may have lower operating efficiencies. Thus it will be important to understand the TEWI consequences of substituting a less efficient CO_2 system for a more efficient HFC system [126]. The carbon dioxide used as a refrigerant is generally of industrial or scientific grade, and is typically recovered from the waste streams of industrial processes.

Table 12.3   100-Year global warming potential (GWP_{100}) values for various refrigerants on a per unit mass basis

Refrigerant	Compound	GWP_{100}
HFC-134a	1,1,1,2-Tetrafluoroethane	1,300
HCFC-22	$CHClF_2$	1,700
R-32	Methylene fluoride	550
R-125	CHF_2CF_3	3,400
R-407c	Mixture of HFC-134a, R32, and R125	1,800
R-407a	Mixture of R32 and R125	2,100
R-717	NH_3	<1
R-744	CO_2	1

Ammonia systems are typically equally to slightly more efficient than HFC systems. However, ammonia has the potential to impact human health due to its toxicity at concentrations above 300 ppm. Permissible exposure limits are 25–35 ppm. It is also moderately flammable, which is a concern for human safety. Ammonia releases can also impact ecosystems through the formation and deposition of ammonium hydroxide. It can also combine with sulfates to form secondary PM. However for these last two impacts to be measurable it may require a catastrophic loss of the ammonia charge. While ammonia is considered a natural chemical, it is manufactured through chemical processes. Thus increased production of ammonia to meet refrigerant demands will increase the environmental impacts associated with ammonia manufacturing.

Hydrocarbons are also slightly more efficient than HFCs, but are highly flammable. Use of hydrocarbons as refrigerants has been limited to systems with very small refrigerant charge, such as household refrigerators. In systems requiring larger charges, they are used in secondary loop systems, a design where the hydrocarbon refrigeration cycle cools a secondary fluid in the safety of an equipment room, and then the secondary fluid is piped to the occupied area to deliver the cooling. The addition of this secondary loop reduces the overall efficiency of the process and thus requires careful evaluation of the TEWI consequences. In addition, the hydrocarbon refrigerants may also play a role in the atmospheric formation of ozone or PM. It is questionable whether the levels of leakage would be significant enough to measurably impact ozone or PM concentrations. However, this potential should be considered when evaluating potential refrigerant alternatives.

As has been mentioned before, life-cycle evaluations of mitigation technologies will be an important aspect of determining their environmental acceptability. For natural refrigerants, the life-cycle energy usage is very favorable. The embedded energy required to reclaim, clean, liquefy and transport refrigerant-grade carbon dioxide is estimated to be 1 kg CO_{2eq} per kg. The ammonia production process has a carbon equivalent of 2 kg CO_{2eq} per kg of ammonia. This is in contrast to the fluorocarbon production process which is about 9 kg CO_{2eq} per kg of fluorocarbon refrigerant (Arthur D. Little [127]).

12.5.3 Residential and Commercial: Improving Building-Operating Efficiency through Building Shell Improvements

The greatest opportunities for improving residential and commercial energy efficiency may be in improvements to the building shell itself, as opposed to improvements in lighting, HVAC systems, and other internal components. Although complete replacement of building shells may seem to be impractical compared to internal system improvements, the recent IEA report noted that the costs of demolition and reconstruction are of the same order of magnitude as new construction.

Even though demolition and reconstruction may offer greater opportunities for improving building shell efficiencies, CO_2 will be generated as a result.

The IEA report notes that the CO_2 generated during demolition and reconstruction (above that generated during renovation) is not likely to be quickly recovered by reduced energy consumption by the new building, unless the new construction is at a "high energy efficiency standard" [4]. Palmer et al. [128] studied this issue in some detail, and found that there would be a net CO_2 reduction after 7–15 years. This study evaluated life cycle emissions, including embodied energy and construction and demolition (C&D) waste disposal, but did not evaluate other environmental impacts that could result from the increase in C&D waste disposal.

New materials have also been proposed for use in building shells to improve operational energy efficiency, including phase-change materials (PCMs) in wallboard [129]. While some of these materials use paraffin as the PCM, other compounds have also been proposed. These include methyl esters, methyl palmitate, methyl stearate, capric acid, lauric acid, coconut fatty acids [130], styrene maleic anhydride, hexadecane, octadecane and formaldehyde [131]. The environmental implications of these materials in PCM wallboard have not been evaluated, and could result in undesirable emissions of organic compounds into the indoor environment, or could create increased organic emissions during production or generate unwanted emissions following disposal.

The building industry has also evaluated the use of insulated concrete forms (ICFs) as an alternative to wood frame structures to improve building operating efficiency. The potential reduction in GHG emissions when using alternative approaches requires a comparative life cycle analysis to evaluate the total changes in GHG emissions during material production, use, and disposal. The Portland Cement Association conducted a relatively thorough life cycle analysis for a model home located in five U.S. cities, and estimated that the net GHG emissions (and most other environmental impacts) would be reduced over a 100-year building life when using ICFs rather than wood frame construction [132]. The analysis did not account for changes in emissions in the disposal stage, and also did not account for the potential sequestration of carbon in the wood used in the building, but did account for emissions during material production. The most significant area in which ICFs performed worse than wood was in emissions of polystyrene, which is used in the ICF insulation.

Building shell improvements combined with on-site power generation using renewable energy sources (primarily wind, solar, or biomass), when designed as a system, can result in buildings that require no energy from external sources ("net-zero buildings"), and can even lead to a net generation of electricity as on-site demand and production allows. The environmental impacts of the specific technologies have been described above. Net-zero buildings are sometimes considered to be sustainable, although additional evaluations of broader environmental impacts are needed to determine the sustainability of net-zero or energy-producing buildings [133].

A further approach to improving the environmental performance of building shells is the application of "green roofs." In this approach, grass or other vegetation is planted on available flat rooftop space, generally resulting in reduced heating load and acting as a sink for CO_2. The use of green roofs could result in an increased level of nutrients in runoff during storm events if appropriate designs are not used or if the systems are improperly operated [134].

A separate environmental issue that must be considered in conjunction with increased building operating efficiency is indoor air quality. Initial efforts to increase building energy efficiency starting in the 1970s relied heavily on reducing ventilation rates and reducing leakage of air into and out of buildings. However, this strategy sometimes resulted in "sick building syndrome" in which building occupants became ill due to decreased indoor air quality [135–138]. More holistic approaches to reducing building energy consumption have been developed, and are now being applied in new construction [139, 140]. Reduced use of materials that emit organic compounds and improved ventilation systems that minimize energy loss while maintaining indoor air quality are two key approaches to minimizing indoor air quality impacts once associated with improved energy efficiency [141, 142].

12.6   Municipal Solid Waste Management

Municipal solid waste (MSW) is often considered to be a "renewable" energy source because its composition is largely waste food, paper, and other materials formed from biogenic carbon. Although plastics are also a component of MSW and are produced primarily from petroleum feedstocks, they make up less than 12% of the 251 million tons of total MSW mass in the U.S. [143]. Energy recovery using MSW directly is achieved through combustion of the waste in waste-to-energy (WTE) plants. Emission levels from these units are generally lower on a per-kWh basis than coal plants due to more stringent emissions reduction requirements for WTE plants.[5] Even so, the energy generation potential from MSW is limited: even if all the MSW in the U.S. were to be disposed of in WTE facilities, the total impact on national electricity production would be about 2% of total electricity generation.

The environmental impacts of these plants can be significantly lower than those of land filling, the other major MSW disposal option. This is particularly true if landfills do not capture methane generated by waste decomposition. For both MSW landfills and disposal of WTE combustion residues, appropriate landfill design and maintenance is needed to ensure protection of water supplies from potential leaching of metals and other materials from the landfill into ground and

---

[5] A plant with a spray dryer and fabric filter is estimated to emit 0.46 lb CO, 0.062 lb PM, 0.55 lb SO_2, and .021 lb HCl per ton of waste combusted. Uncontrolled NO_x emissions are estimated at 3.6 lb per ton of waste, which can be reduced by 40% using non-selective catalytic reduction. Hg emissions are limited to 0.08 mg/dscm and dioxin/furan emissions to 13 ng/dscm under the 1995 rule for new municipal waste combustors.

surface water. It should also be noted that collection and use of landfill gas as an energy source can both reduce net GHG emissions by conversion of methane to CO_2 (much of it from renewable rather than fossil carbon) and by displacing other, often fossil, energy consumption. Methane can also be recovered from wastes other than MSW, such as waste water treatment solids and agricultural wastes. For all of these sources, use of the methane reduces fugitive emissions of methane and displaces CO_2 from fossil fuels, providing a multiple benefit.

12.7   Research Needs

The research priorities to address climate and climate mitigation strategies will need to shift significantly and quickly. Davies and Rejeski [144] addressed this issue in the context of EPA and nanotechnology, but the same fundamental issues of the pace and scope of technological change also hold true for climate issues and for other organizations. They noted that:

> EPA has spent its entire existence in a rearguard battle to mitigate the impacts of technologies born during the Industrial Revolution. The internal combustion engine, invented in the late "19th century", was just one. The same story could be told about the basic technologies used by the chemical industry or in manufacturing or electricity generation, to name just a few.
>
> EPA has made impressive progress in restoring clean air and clean water, but if it is to deal with new technological challenges, it will need new approaches.

As seen in the discussions above, new technologies are expected to be key components of climate mitigation technologies. Research programs will need to be flexible enough to address these emerging disciplines and responsive enough to incorporate breakthroughs and new results.

Key research needs, ordered by the stage of research from fundamental to applied rather than by priority, are:

1. Fundamental research is needed to better understand the basic physical, chemical, and biological processes involved with new technologies and their applications. Improved understanding of catalyst chemistry, biological processes for fuel production, nanomaterial behavior, and combustion processes for alternative fuels are examples of areas in which a more thorough fundamental understanding is needed to evaluate the potential for GHG mitigation technologies to adversely impact the environment.
2. Research is needed to understand where emerging technologies may be used in climate mitigation applications (and in which situations those technologies are not suitable for use), how they can benefit mitigation efforts, and how they may impact the environment.
3. To fully evaluate the environmental impacts of climate change mitigation approaches, research is needed to quantify the health and ecosystem effects associated with the effluents and other stressors caused by mitigation approaches. In some cases, these effects are relatively well understood, particularly where the

changes are in amount and location. In other cases, the stressors will also have changed, which will require developing an understanding of the effects associated with exposure to those stressors. Emissions of nanomaterials and genetically modified organisms are specific examples of such changed stressors. Many of the impacts identified above are to ecosystems, which have historically been of secondary importance to human health impacts. Given the increasing emphasis on use of biomass and the impacts of land use changes, additional emphasis on ecosystem health is warranted.

4. Support for technology development, demonstration, and deployment is crucial to the success of GHG emission mitigation [4, 5]. The need for information on the long-term environmental impacts must not be overlooked in the planning and implementation of the research efforts. These efforts must also include measurements of the types and amounts of effluents to air, water, and soil; direct changes in land use and impacts on ecosystems; and evaluation of the impacts associated with the construction and end-of-life disposal of the system being evaluated.

5. Information is needed in the near term to support regulatory decisions that must be made during the permitting stages of new technology demonstrations. Demonstration projects that are designed and operated in such a way that they cause measurable adverse environmental impacts will construct for themselves a significant barrier to further development and commercial acceptance. It is crucial, then, to have the support of the regulatory agencies for such projects, and that support will hinge upon the availability of the best possible data and information on environmental impacts associated with the construction and operation of these new technologies.

6. There is a critical need to develop consistent data and work with existing methods already developed for life cycle analyses and assessments (such as ISO 14040 and related standards [145]) of mitigation technologies and strategies. The life cycle analyses must go beyond evaluation of life cycle GHG or net energy consumption and address other environmental impacts to the extent possible. In many, if not most, cases, this will require significant collection of data on environmental emissions and impacts for use in life cycle assessments.

7. Climate change by its very nature is a consequence of the global scale of emissions, and many of the environmental impacts may also be of concern because they are so widespread. Research and mitigation of environmental impacts other than climate change have historically been focused on much smaller scales, particularly at the plant, neighborhood, and urban scales. The tools and approaches used for evaluations at these smaller studies are not necessarily appropriate for continental and global scale evaluations. In addition, when the scale of the system being studied increases, the types and extent of cross-systems interactions also tends to increase, resulting in significantly more complex problems. Research on such large scales will require different analytical and measurement tools and a greater degree of cross-disciplinary interaction.

8. Because the scale of the climate issue is global, research must provide the information necessary to address the environmental impacts on an international level. Interactions among researchers and technology developers across national

boundaries are crucial to ensuring that environmental impacts are not simply transferred to different locations, and to enable potential international agreements to mitigate GHG emissions that are protective of local environments and particularly of sensitive and special ecosystems.

9. Finally, many of the approaches for reducing GHG emissions involve direct action by, and interaction with, the general public. The use of compact fluorescent light bulbs is one example of a mitigation measure that directly involves situations in which the general public would need to act to mitigate potential environmental impacts associated with that measure. Although there are efforts to provide information on energy efficiency, climate footprint, and other parameters directly related to GHG emission mitigation, similar information about the potential adverse environmental impacts of these approaches is not always available. Guidance on disposal of advanced lighting technologies and embedded electronics components, awareness of material and product recycling, and other information that can be directly used by the general public needs to be more fully developed and communicated.

Although there is growing research in many developing areas, the level of research support is significantly below what is likely to be needed. The IEA report estimates that, globally, about $16 trillion in research, development, and deployment investments in energy technologies, including the cost of deploying new technologies at full scale, will be needed from now until 2050, or about $400 billion per year [4]. To be most effective, these investments need to be made earlier rather than later, which means that investments in the first 10–20 years should be at significantly higher levels than the $400 billion. These investments include not only fundamental research, but also support for technology development, demonstration, and deployment costs. However, it is not clear whether these investments include the research needed to understand the environmental impacts associated with emerging energy technologies, and it is certain that they do not include research associated with agriculture, afforestation, and reforestation, and other non-energy mitigation approaches.

12.8 Concluding Remarks

Mitigating GHG emissions will require significant changes in technologies and practices on a broad scale over a long term, particularly in energy production and use, from current conditions. These new technologies, practices, and behaviors will change how our activities ultimately impact the environmental. Although many of the changes will be beneficial (even beyond reductions in GHG emissions), we must be aware of the adverse impacts that will also occur and be prepared to minimize those impacts to the extent possible. A summary of the impacts discussed above are shown in Table 12.4. A level of research need (high, medium, low) for each of the topics listed in the table has been assigned by the chapter authors and book editor, and reflect our judgment of the relative importance of each topic.

Table 12.4 Summary of potential environmental impacts associated with GHG mitigation technologies

Sector	Technology	Potential impacts	Research need
Electricity generation	Carbon capture (coal)	Lower generating efficiencies lead to higher fuel use and increased generation of effluents; water consumption; increased impacts of coal mining	High
	Carbon sequestration	Risks associated with groundwater contamination; long-term or acute accidental CO_2 release; impacts to local and underground ecosystems; potential seismic impacts	High
	Nuclear power	Nuclear waste management	Medium
	Increased natural gas use	Increased release of CH_4; impacts of drilling, processing, and transport operations	Medium
	Wind power	Impacts to land cover and habitat due to wind turbine footprint and increased power lines; noise; appearance issues	Low
	Solar photovoltaic	Increased semiconductor production and potential Cd and Te releases over life cycle; impacts to land cover and habitat due to PV panel footprint and increased power lines	Medium
	Biomass power	Impacts to land cover and habitat due to increased harvesting; increased use of fertilizer and pesticides and runoff to water, water consumption; increased transportation, harvesting, and conversion emissions; emissions from storage of harvested feedstocks	High
	Geothermal power	Effluents of waste water and solids; H_2S and Hg emissions; ground subsidence; water contamination	Low
Transportation	Electric vehicles	Increased electricity consumption (see above impacts); life cycle impacts of battery production and storage (concern about Ni, Li, Pb emissions)	Medium
	Biofuels	Same issues as biomass power, use of genetically modified materials in feedstocks and conversion processes; biofuel spills and leakage into groundwater; increased organic emissions from vehicles	High
	Hydrogen vehicles	Life cycle GHG emissions due to production from primary energy sources; increased Pt mining to meet fuel cell needs; construction of H_2 production, distribution, and storage infrastructure; potential damage to stratospheric ozone layer	Low
Residential and commercial energy efficiency	Lighting	Hg from CFL disposal; increased emissions from semiconductor production for LED lighting systems (As, In, P)	Medium
	Space cooling, refrigeration	Fugitive emissions of refrigerants with high GWP_{100}; toxicity of ammonia as replacement refrigerant	Medium
	Building shell improvements	Increased construction and demolition debris disposal; potential for increased indoor exposure to organic compounds	Low
Waste management	Waste to energy	Emissions of metals and organic compounds	Low

Under each of these topics, it is possible that there are more detailed issues that should be considered more or less important relative to the overall rating.

It is equally crucial to clearly communicate these impacts to decision makers, stakeholders, and the general public. Because there will be adverse impacts of varying degrees related to nearly all GHG mitigation strategies, it is critical to recognize and communicate that there are no zero-impact answers, only approaches that have fewer or greater impacts and risks. Public and advocacy group resistance has been noted for wind power, increased power lines needed to connect renewable energy sources, nuclear energy, biofuels, and waste to energy plants. These approaches will all be needed to effectively mitigate GHG emissions. An effective communications strategy that clearly addresses concerns and identifies the relative risks associated with adopting these approaches (or not adopting them) is a critical need for long-term success of any large-scale GHG mitigation strategy.

This discussion does not endeavor to list all the adverse environmental consequences associated with climate mitigation strategies, and it does not provide a comprehensive analysis of those presented. Indeed, each of these issues would be a topic suitable for a more complete review of the state of the science. It is certain that there will be environmental consequences associated with the evolving mitigation responses that we will not, and perhaps could not, have predicted. It is even likely that a number of the technologies that will ultimately play key roles in mitigating GHG emissions have not yet been developed and are not being anticipated. Whether anticipated or not, the ancillary adverse environmental impacts are likely to act on much shorter time scales than climate change. Coupled with the fact that strategies that successfully mitigate climate change will, at best, result in changes in climate that are only slightly worse than present, it is likely that one of the most visible impacts of GHG mitigation strategies will be the adverse impacts discussed above. Thus, it is imperative to remain alert to these impacts and proactively identify and publicize both the impacts and approaches to minimize them.

References

1. IPCC (2007) Climate change 2007: the physical science basis. Contribution of working group I to the Fourth Assessment Report of the Intergovernmental Panel on Climate Change. Cambridge University Press, Cambridge/New York
2. IPCC (2007) Climate change 2007: mitigation. Contribution of working group III to the Fourth Assessment Report of the Intergovernmental Panel on Climate Change. Cambridge University Press, Cambridge/New York
3. STAPPA/ALAPCO (1999) Reducing greenhouse gases & air pollution: a menu of harmonized options. State and Territorial Air Pollution Program Administrators and Association of Local Air Pollution Control Officials, Washington, DC
4. IEA (2008) Energy technology perspectives. IEA, Paris
5. Princiotta F (2007) Mitigating global climate change through power-generation technology. Chem Eng Prog 103:24–32
6. IPCC (2000) Emissions scenarios. Cambridge University Press, Cambridge

7. USEPA (2008) Notice of public comment period: 2007 interim report of the U.S. EPA Global Change Research Program assessment of the impacts of global change on regional U.S. air quality: a preliminary synthesis of climate change impacts on O_3. Fed Regist 73:39695–39696

8. Herzog HJ, Drake EM (1996) Carbon dioxide recovery and disposal from large energy systems. Annu Rev Energ Environ 21:145–166

9. IPCC (2005) Special report on carbon dioxide capture and storage. Intergovernmental Panel on Climate Change, Cambridge, England

10. Kirchgessner DA, Piccot SD et al (2000) An improved inventory of methane emissions from coal mining in the United States. J Air Waste Manag 50:1904–1919

11. Spath PL, Mann MK et al (1999) Life cycle assessment of coal-fired power production. In: NREL/TP-570-25119. National Renewable Energy Laboratory, Golden

12. Babbitt CW, Lindner AS (2005) A life cycle inventory of coal used for electricity production in Florida. J Clean Prod 13:903–912

13. Jaramillo P, Griffin WM et al (2007) Comparative life-cycle air emissions of coal, domestic natural gas, LNG, and SNG for electricity generation. Environ Sci Technol 41:6290–6296

14. Ney RA, Schnoor JL (2002) Greenhouse gas emission impacts of substituting switchgrass for coal in electric generation: the Chariton Valley biomass project. Center for Global and Regional Environmental Research, Iowa City

15. Koornneef J, Faaij A et al (2008) The screening and scoping of environmental impact assessment and strategic environmental assessment of carbon capture and storage in the Netherlands. Environ Impact Asses 28:392–414

16. Haugan PM, Drange H (1992) Sequestration of CO_2 in the deep ocean by shallow injection. Nature 357(6376):318–320

17. Wilson EJ, Johnson TL et al (2003) Regulating the ultimate sink: managing the risks of geologic CO_2 storage. Environ Sci Technol 37:3476–3483

18. Wilson EJ, Friedmann SJ et al (2007) Research for deployment: incorporating risk, regulation, and liability for carbon capture and sequestration. Environ Sci Technol 41:5945–5952

19. USDOE (2007) Carbon sequestration technology roadmap and program plan 2007. National Energy Technology Laboratory, Pittsburgh

20. Fritsche UR (2006) Comparisons of greenhouse-gas emissions and abatement cost of nuclear and alternative energy options from a life-cycle perspective. Institute for Applied Ecology, Darmstadt

21. Fthenakis VM, Kim HC (2007) Greenhouse gas emissions from solar electric and nuclear power: a life cycle study. Energ Policy 35:2549–2557

22. NRC (2006) Safety and security of commercial spent nuclear fuel storage. National Academies Press, Washington, DC

23. NRC (2006) Going the distance? The safe transport of spent nuclear fuel and high-level radioactive waste in the United States. National Academies Press, Washington, DC

24. MIT (2003) The future of nuclear power. Massachusetts Institute of Technology, Cambridge

25. Ipatyev V, Bulavik I et al (1999) Forest and Chernobyl: forest ecosystems after the Chernobyl nuclear power plant accident: 1986–1994. J Environ Radioactiv 42:9–38

26. Moller AP, Mousseau TA (2006) Biological consequences of Chernobyl: 20 years on. Trends Ecol Evol 21:200–207

27. Rytomaa T (1996) Ten years after Chernobyl. Ann Med 28:83–87

28. BERR (2008) Meeting the energy challenge: a white paper on nuclear power. CM 7296. Department for Business, Enterprise & Regulatory Reform, London

29. Davison J, Adams D (2007) Capturing CO_2. International Energy Agency Greenhouse Gas R&D Programme, Cheltenham

30. Wang MQ, Huang HS (2000) A full fuel-cycle analysis of energy and emissions impacts of transportation fuels produced from natural gas. Report ANL/ESD-40, Argonne National Laboratory, Argonne, IL

31. Kado NY, Okamoto RA et al (2005) Emissions of toxic pollutants from compressed natural gas and low sulfur diesel-fueled heavy-duty transit buses tested over multiple driving cycles. Environ Sci Technol 39:7638–7649

32. ARI (2008) Greenhouse gas life-cycle emissions study: fuel life-cycle of U.S. Natural Gas Supplies and International LNG. Advanced Resources International, Washington, DC
33. Hondo H (2005) Life cycle GHG emission analysis of power generation systems: Japan case. Energy 30:2042–2056
34. Zhang Y, McKechnie J et al (2009) Life cycle emissions and cost of producing electricity from coal, natural gas, and wood pellets in Ontario, Canada. Environ Sci Technol 44:538–544
35. USDOE (2007) International energy outlook 2007. Energy Information Administration, Washington, DC
36. USEPA (2006a) Global anthropogenic non-CO_2 greenhouse gas emissions: 1990–2020. EPA-430-R-06-003, Office of Atmospheric Programs, Washington, DC
37. USEPA (2001) Mercury in petroleum and natural gas: estimation of emissions from production, processing, and combustion. In: EPA/600/R-01/066. Office of Research and Development, Research Triangle Park
38. Arthur J, Bohm B et al (2009) Evaluating the environmental implications of hydraulic fracturing in shale gas reservoirs. In: SPE Americas E&P environmental and safety conference, San Antonio, 23–25 March 2009
39. Kvenvolden KA (1989) Methane hydrates and global climate. Glob Biogeochem Cy 2:221–229
40. Kvenvolden KA (1993) Gas hydrates-geological perspective and global change. Rev Geophys 31:173–187
41. Kvenvolden KA (2002) Methane hydrate in the global organic carbon cycle. Terra Nova 14:302–306
42. MacDonald GJ (1990) Role of methane clathrates in past and future climates. Clim Change 16:247–281
43. Wright EL, Erickson JD (2003) Incorporating catastrophes into integrated assessment: science, impacts, and adaptation. Clim Change 57:265–286
44. White SW, Kulcinski GL (2000) Birth to death analysis of the energy payback ratio and CO_2 gas emission rates from coal, fission, wind, and DT-fusion electrical power plants. Fusion Eng Des 48:473–481
45. Meier PJ, Wilson PPH et al (2005) US electric industry response to carbon constraint: a life-cycle assessment of supply side alternatives. Energ Policy 33:1099–1108
46. NRC (2007) Environmental impacts of wind-energy projects. National Academies Press, Washington, DC
47. BLM (2005) Final programmatic environmental impact statement on wind energy development on BLM-administered lands in the Western United States. U.S. Department of the Interior, Bureau of Land Management, Washington, DC
48. Keith DW, DeCarolis JF et al (2004) The influence of large-scale wind power on global climate. Proc Natl Acad Sci USA 101:16115–16120
49. Turner JA (1999) A realizable renewable energy future. Science 285:687–689
50. Fthenakis VM (2004) Life cycle impact analysis of cadmium in CdTe PV production. Renew Sust Energ Rev 8:303–334
51. Fthenakis VM, Kim HC et al (2008) Emissions from photovoltaic life cycles. Environ Sci Technol 42:2168–2174
52. Fthenakis VM (2003) Overview of potential hazards. In: Markvart T, Gastaner L (eds) Practical handbook of photovoltaics: fundamentals and applications. Elsevier, Amsterdam
53. Alsema E (1996) Environmental aspects of solar cell modules: summary report. Netherlands Agency for Energy and the Environment, Utrecht
54. Fthenakis VM, Fuhrmann M et al (2004) Experimental investigation of emissions and redistribution of elements in CdTe PV modules during fires. In: 19th European PV solar energy conference, Paris, 7–11 June 2004
55. Fthenakis VM, Duby P et al (2006) Recycling of CdTe photovoltaic modules: recovery of glass, cadmium and tellurium. In: 21st European photovoltaic solar energy conference, Dresden, 4–8 Sept 2006

56. de Wild-Scholten M, Alsema E (2004) Towards cleaner solar PV environmental and health impacts of crystalline silicon photovoltaics. Refocus 5:46–49
57. Lawrence K, Morgan S et al (1981) Environmental, health, safety, and regulatory review of selected photovoltaic options: copper sulfide/cadmium sulfide and polycrystalline silicon. Solar Energy Research Institute, Golden
58. Moskowitz PD, Fthenakis VM et al (1987) Public health issues in photovoltaic energy systems: an overview of concerns. Solar Cells 19(3–4):287–299
59. Moskowitz PD (1992) Environmental, health and safety issues related to the production and use of CdTe photovoltaic modules. Int J Sust Energ 12:259–281
60. BLM (2008) Notice of intent to prepare a programmatic environmental impact statement to evaluate solar energy development, develop and implement agency-specific programs, conduct public scoping meetings, amend relevant agency land use plans, and provide notice of proposed planning criteria. Fed Regist 73:30908–30912
61. Demirbas A (2005) Potential applications of renewable energy sources, biomass combustion problems in boiler power systems and combustion related environmental issues. Prog Energ Combust 31:171–192
62. Fargione J, Hill J et al (2008) Land clearing and the biofuel carbon debt. Science 319: 1235–1238
63. Searchinger T, Heimlich R et al (2008) Use of U.S. croplands for biofuels increases greenhouse gases through emissions from land-use change. Science 319:1238–1240
64. Barnes CA, Roy DP (2008) Radiative forcing over the conterminous United States due to contemporary land cover land use albedo change. Geophys Res Lett 35:6
65. IFC (2007) Environmental, health, and safety guidelines for geothermal power generation. International Finance Corporation, Washington, DC
66. Bloomfield KK, Moore JN (1999) Production of greenhouse gases from geothermal power plants. In: Geothermal Resource Council 1999 Annual Meeting, Reno, 17–20 Oct 1999
67. UNEP (2005) Environmental Due Diligence (EDD) of renewable energy projects: guidelines for geothermal energy systems: release 1.0. United Nations Environment Program, Nairobi
68. Kintner-Meyer M, Schneider K et al (2006) Impacts assessment of plug-in hybrid vehicles on electric utilities and regional U.S. power grids. Part 1: technical analysis. Pacific Northwest National Laboratory, Richland
69. Lave L, MacLean H et al (2000) Life-cycle analysis of alternative automobile fuel/propulsion technologies. Environ Sci Technol 34:3598–3605
70. Andersson BA, Rade I (2001) Metal resource constraints for electric-vehicle batteries. Transp Res D Tr E 6:297–324
71. Vimmerstedt L, Hammel C et al (1996) Impact of increased electric vehicle use on battery recycling infrastructure, United States
72. Shader RI, Jackson AH et al (1974) The antiaggressive effects of lithium in man. Psychopharmacology 40:17–24
73. Leonard A, Hantson P et al (1995) Mutagenicity, carcinogenicity and teratogenicity of lithium compounds. Mutat Res Rev Genet Toxicol 339:131–137
74. Kszos LA, Stewart AJ (2003) Review of lithium in the aquatic environment: distribution in the United States, toxicity, and case example of groundwater contamination. Ecotoxicology 12:439–447
75. NEIWPCC (2001) Health, environmental, and economic impacts of adding ethanol to gasoline in the Northeast States, vol 3. In: Water resources and associated health impacts. New England Interstate Water Pollution Control Commission, Lowell
76. Buscheck TE, O'Reilly K et al (2001) Ethanol in groundwater at a Northwest terminal. In: In situ and on-site bioremediation: the Sixth International Symposium, San Diego, 4–7 June 2001
77. Mackay DM, De Sieyes NR et al (2006) Impact of ethanol on the natural attenuation of benzene, toluene, and o-xylene in a normally sulfate-reducing aquifer. Environ Sci Technol 40:6123–6130

78. Poulopoulos SG, Samaras DP et al (2001) Regulated and unregulated emissions from an internal combustion engine operating on ethanol-containing fuels. Atmos Environ 35: 4399–4406

79. Jacobson MZ (2007) Effects of ethanol (E85) versus gasoline vehicles on cancer and mortality in the United States. Environ Sci Technol 41:4150–4157

80. Farrell AE, Plevin RJ et al (2006) Ethanol can contribute to energy and environmental goals. Science 311:506–508

81. Graboski MS, McCormick RL (1998) Combustion of fat and vegetable oil derived fuels in diesel engines. Prog Energ Combust 24:125–164

82. Hammerschlag R (2006) Ethanol's energy return on investment: a survey of the literature 1990–present. Environ Sci Technol 40:1744–1750

83. Hill J, Nelson E et al (2006) Environmental, economic, and energetic costs and benefits of biodiesel and ethanol biofuels. Proc Natl Acad Sci USA 103:11206–11210

84. Shapouri H, Duffield JA et al (2003) The energy balance of corn ethanol revisited. Trans ASAE 46:959–968

85. Energy Independence and Security Act of 2007, Public Law 110–140, 19 Dec 2007

86. Himmel ME, Ding S-Y et al (2007) Biomass recalcitrance: engineering plants and enzymes for biofuels production. Science 315:804–807

87. Panesar PS, Marwaha SS et al (2006) Zymomonas mobilis: an alternative ethanol producer. J Chem Tech Biotechnol 81:623–635

88. Dinus RJ, Payne P et al (2001) Genetic modification of short rotation poplar wood: properties for ethanol fuel and fiber productions. Crit Rev Plant Sci 20:51–69

89. Sticklen M (2006) Plant genetic engineering to improve biomass characteristics for biofuels. Curr Opin Biotech 17:315–319

90. Vogel KP, Jung HJG (2001) Genetic modification of herbaceous plants for feed and fuel. Crit Rev Plant Sci 20:15–49

91. King C, Webber M (2008) The water intensity of the plugged-in automotive economy. Environ Sci Technol 42:5834 5834

92. Mubako S, Lant C (2008) Water resource requirements of corn-based ethanol. Water Resour Res 44:5

93. NSF (2008) Breaking the chemical and engineering barriers to lignocellulosic biofuels: next generation hydrocarbon refineries. National Science Foundation, Chemical, Bioengineering, Environmental, and Transportation Systems Division, Washington, DC

94. Kothari R, Buddhi D et al (2004) Sources and technology for hydrogen production: a review. Int J Glob Energ 21:154–178

95. Spath P, Mann M (2001) Life cycle assessment of hydrogen production via natural gas steam reforming. In: NREL/TP-570-27637. National Renewable Energy Laboratory, Golden

96. Cannon J (1997) Clean hydrogen transportation: a market opportunity for renewable energy. Renewable Energy Policy Project, Washington, DC

97. Yeh S, Loughlin DH et al (2006) An integrated assessment of the impacts of hydrogen economy on transportation, energy use, and air emissions. Proc IEEE 94:1838–1851

98. NAS (2004) The hydrogen economy: opportunities, costs, barriers, and R&D needs. National Academy Press, Washington, DC

99. Wilhelm E, Fowler M (2006) A technical and economic review of solar hydrogen production technologies. Bull Sci Tech Soc 26:278–287

100. Carlson EJ (2003) Precious metal availability and cost analysis for PEMFC commercialization. U.S. Department of Energy, Office of Energy Efficiency and Renewable Energy, Washington, DC

101. Osakwe R (2006) PEM fuel cells and Russia's supply of platinum: trading one bottleneck for another? Stanford Stud J Russ E Eur Eurasian Stud 2:17–36

102. Sano Y, Shinoda Y et al (2004) A strategic recovery of rare-metal fission products in spent nuclear fuel. Nucl Technol 148:348–357

103. Tromp TK, Shia R-L et al (2003) Potential environmental impact of a hydrogen economy on the stratosphere. Science 300:1740–1742

104. IEA (2007) Fuel efficiency for HDVs: standards and other policy instruments: Towards a plan of action. In: IEA/International transport forum workshop on standards and other policy instruments on fuel efficiency for HDVs, Paris, 21–22 June 2007
105. Laitner JA, Ehrhardt-Martinez K (2007) Advanced electronics and information technologies: the innovation-led climate change solution. American Council for an Energy-Efficient Economy, Washington, DC
106. Sachs HM (2005) Opportunities for elevator energy efficiency improvements. American Council for an Energy-Efficient Economy, Washington, DC
107. Cointreau S (2006) Occupational and environmental health issues of solid waste management: special emphasis on middle- and lower-income countries. UP-2. World Bank, Washington, DC
108. Hilty LM (2005) Electronic waste – an emerging risk? Environ Impact Asses 25: 431–435
109. Widmer R, Oswald-Krapf H et al (2005) Global perspectives on e-waste. Environ Impact Asses 25:436–458
110. Cui J, Forssberg E (2003) Mechanical recycling of waste electric and electronic equipment: a review. J Hazard Mater 99:243–263
111. Mann S (2006) Nanotechnology and construction. Nanoforum, Stirling
112. USDOE (2007) Nanomanufacturing for energy efficiency: workshop. Industrial Technologies Program, Washington, DC
113. Helland A, Wick P et al (2007) Reviewing the environmental and human health knowledge base of carbon nanotubes. Environ Health Perspect 115:1125–1131
114. Lam CW, James JT et al (2006) A review of carbon nanotube toxicity and assessment of potential occupational and environmental health risks. Crit Rev Toxicol 36:189–217
115. Boxall AB, Tiede K et al (2007) Engineered nanomaterials in soils and water: how do they behave and could they pose a risk to human health? Nanomedicine 2(6):919–927
116. Colvin V (2003) The potential environmental impact of engineered nanomaterials. Nat Biotechnol 21(10):1166–1170
117. Elvin G (2006) Risks in architectural applications of nanotechnology nanowerk. http://www.nanowerk.com/spotlight/spotid=1007.php. Accessed 27 June 2008
118. Greening LA, Greene DL et al (2000) Energy efficiency and consumption – the rebound effect – a survey. Energ Policy 28:389–401
119. Luther L (2008) Compact Fluorescent Light Bulbs (CFLs): issues with use and disposal. Congressional Research Service, Washington, DC
120. Johnson NC, Manchester S et al (2008) Mercury vapor release from broken compact fluorescent lamps and in situ capture by new nanomaterial sorbents. Environ Sci Technol 42:5772–5778
121. USEPA (2006b) 2004 TRI public data release e-report. U.S. Environmental Protection Agency. http://www.epa.gov/tri/tridata/tri04/ereport/2004eReport.pdf. Accessed 30 July 2008
122. Mills A (2006) Strategies in light 2006: record LED sales but price erosion. III Vs Rev 19:35–39
123. Fischer S, Hughes P et al (1992) Energy and global warming impacts of CFC alternative technologies. ORNL/M-1869. Oak Ridge National Laboratory, Oak Ridge
124. IPCC/TEAP (2005) Safeguarding the ozone layer and the global climate system: issues related to hydrofluorocarbons and perfluorocarbons. Cambridge University Press, Cambridge
125. USEPA (2008a) Inventory of U.S. greenhouse gas emissions and sinks: 1990–2006. EPA 430-R-08-005, Office of Atmospheric Programs, Washington, DC
126. Khaligi B, Sumantran V et al (1999) Environmental impact study of alternative automotive refrigerants. In: The conference on climate change and ozone protection, Washington, DC, 27–29 Sept 1999
127. Little AD (2002) Global comparative analysis of HFC and alternative technologies for refrigeration, air conditioning, solvent, foam, aerosol propellant and fire protection application. Alliance for Responsible Atmospheric Policy, Arlington

128. Palmer J, Boardman B et al (2006) Reducing the environmental impact of housing. Environmental Change Institute, Oxford
129. Feustel HE, Stetiu C (1997) Thermal performance of phase change wallboard for residential cooling application. LBL-38320. Lawrence Berkeley National Laboratory, Berkeley
130. Rudd A (1993) Phase-change material wallboard for distributed thermal storage in buildings. ASHRAE Trans 99:339–346
131. Lee SH, Yoon SJ et al (2006) Development of building materials by using micro PCM. In: Sixth Korea-China joint workshop on clean energy technology, Busan, 4–7 July 2006
132. Marceau ML, VanGeem MG (2002) Life cycle assessment of an insulating concrete form house compared to a wood frame house. Portland Cement Association, Skokie
133. Helgeson JF, Lippiatt BC (2009) Multidisciplinary life cycle metrics and tools for green buildings. Integrat Environ Assess Manag 5(3):390–398
134. Berndtsson JC, Emilsson T et al (2006) The influence of extensive vegetated roofs on runoff water quality. Sci Total Environ 355(1–3):48–63
135. Burkart W, Chakraborty S (1984) Possible health effects of energy conservation: impairment of indoor air quality due to reduction of ventilation rate. Environ Int 10:455–461
136. Fisk WJ (2000) Health and productivity gains from better indoor environments and their relationship with building energy efficiency. Annu Rev Energ Environ 25:537–566
137. Spengler JD, Sexton K (1983) Indoor air pollution: a public health perspective. Science 221:9–17
138. Turiel I, Hollowell CD et al (1983) The effects of reduced ventilation on indoor air quality in an office building. Atmos Environ 17:51–64
139. Chan AT, Yeung VCH (2005) Implementing building energy codes in Hong Kong: energy savings, environmental impacts and cost. Energ Buildings 37:631–642
140. Wang Z, Bai Z et al (2004) Regulatory standards related to building energy conservation and indoor-air-quality during rapid urbanization in China. Energ Buildings 36:1299–1308
141. Gilbert NL, Guay M et al (2008) Air change rate and concentration of formaldehyde in residential indoor air. Atmos Environ 42:2424–2428
142. Howard EM, McCrillis RC et al (1998) Indoor emissions from conversion varnishes. J Air Waste Manage 48:924–930
143. USEPA (2008c) Municipal solid waste: plastics. U.S. Environmental Protection Agency. http://www.epa.gov/garbage/plastic.htm. Accessed 8 July 2008
144. Davies T, Rejeski D (2007) Overseeing the unseeable. Environ Forum 24:36–40
145. ISO (2006) ISO 14040:2006: environmental management – life cycle assessment – principles and framework. International Organization for Standardization, Geneva
146. Koonin SE (2006) Getting serious about biofuels. Science 311:435
147. Kanter J (2008) Europe may ban imports of some biofuel crops. New York Times. New York, NY
148. Scharlemann JPW, Laurance WF (2008) How green are biofuels? Science 319:43–44
149. Time (2008) The clean energy scam. Time: 171, March 27

Index

Printed by Printforce, the Netherlands